Mastering Statistical Process Control
A Handbook for Performance Improvement Using Cases

Control Chart Reference Table

Chart	Application	Average	Warning limits	Action/Control limits	Notes
Attributes charts used for data that are counted					
np	Number of defectives (i.e. items meeting specified requirement) with constant sample size, n	$n\bar{p}$	$n\bar{p} \pm 2\sqrt{n\bar{p}\left(1 - \dfrac{n\bar{p}}{n}\right)}$	$n\bar{p} \pm 3\sqrt{n\bar{p}\left(1 - \dfrac{n\bar{p}}{n}\right)}$	Examples: Number of items failing inspection Number of trains arriving late
p	As for np, but sample size can vary	$\bar{p}$	$\bar{p} \pm 2\sqrt{\bar{p}(1-\bar{p})/n}$	$\bar{p} \pm 3\sqrt{\bar{p}(1-\bar{p})/n}$	Example: proportion of late deliveries
c	Number of defects/errors/flaws in a sample of constant size	$\bar{c}$	$\bar{c} \pm 2\sqrt{\bar{c}}$	$\bar{c} \pm 3\sqrt{\bar{c}}$	Example: number of flaws per m²
u	As for c, but the sample size can vary	$\bar{u} = \dfrac{\sum c}{\sum n}$	$\bar{u} \pm 2\sqrt{\dfrac{\bar{u}}{n}}$	$\bar{u} \pm 3\sqrt{\dfrac{\bar{u}}{n}}$	Example: Number of accidents per million hours worked
Variables charts used for data that are measured					
X (Individuals)	Used to monitor the process average when data are received one-at-a-time and there is no natural grouping	$\bar{X}$	$\bar{X} \pm 2S_{mr}$ $S_{mr} = \bar{R}/1.128$	$\bar{X} \pm 3S_{mr}$ $S_{mr} = \bar{R}/1.128$	Examples: % downtime per day/week etc contamination rates (e.g. ppm) lengths, times, weights
MR	Used with the X chart for monitoring the variability	$\bar{R}$	Not normally used	$3.27 \times \bar{R}$	$\bar{R}$ is the average of the moving ranges, often denoted MR to avoid confusion
$\bar{X}$ (Average)	Used for monitoring averages	$\bar{\bar{X}}$	$\bar{\bar{X}} \pm \frac{2}{3}A_2\bar{R}$ $\bar{\bar{X}} \pm \frac{2}{3}A_3\bar{s}$	$\bar{\bar{X}} \pm A_2\bar{R}$ $\bar{\bar{X}} \pm A_3\bar{s}$	For use with the R chart For use with the s chart
R (Range)	Used with the $\bar{X}$ chart to monitor variability with small sample sizes	$\bar{R}$	$D_7\bar{R}$ (UWL) $D_8\bar{R}$ (LWL)	$D_4\bar{R}$ (UCL) $D_5\bar{R}$ (UAL) $D_3\bar{R}$ (LCL) $D_6\bar{R}$ (LAL)	$\bar{X}$/R and $\bar{X}$/s charts have the similar applications. They are used when data are recorded in natural groups such as work shifts. Examples: Average length/weight/concentration
s (Standard Deviation)	Alternative to the R chart used with large or varying sample sizes	$\bar{s}$	$B_7\bar{R}$ (UWL) $B_8\bar{R}$ (LWL)	$B_4\bar{R}$ (UCL) $B_5\bar{R}$ (UAL) $B_3\bar{R}$ (LCL) $B_6\bar{R}$ (LAL)	
Median	A simpler alternative to the $\bar{X}$ chart	$\bar{\tilde{X}}$	$\bar{\tilde{X}} \pm \frac{2}{3}A_4\bar{R}$	$\bar{\tilde{X}} \pm A_4\bar{R}$	Usually used with the R chart

Cusum and **weighted cusum** charts can be used with most data types and are particularly powerful for identifying small process changes.

The **Multivariate** chart is an attributes chart which shows categories of causes. For example, the total number of accidents may be reported on a c chart. If there is a table on the chart showing the different types of incident occurring it is a multivariate c chart.

The **Difference** chart charts the difference between two values, usually a measured value and a target value. For example, the difference between planned and actual arrival times. The formula will be the same as for the X chart. It is usually used when the target values frequently change.

The **Z** chart is used in a similar situation to the Difference chart, and in addition assumes that the standard deviation is changing. The z values are calculated as (x-average)/s. On the chart the average will be zero and the warning and control limits set at +/− 2 and +/− 3. Constants are given in the appendix.

Mastering Statistical Process Control

A Handbook for Performance Improvement Using Cases

Tim Stapenhurst

ELSEVIER
BUTTERWORTH
HEINEMANN

AMSTERDAM • BOSTON • HEIDELBERG • LONDON • OXFORD • NEW YORK • PARIS • SAN DIEGO
SAN FRANCISCO • SINGAPORE • SYDNEY • TOKYO

Elsevier Butterworth-Heinemann
Linacre House, Jordan Hill, Oxford OX2 8DP, UK
30 Corporate Drive, Burlington, MA 01803, USA

First published 2005

British Library Cataloguing in Publication Data
A catalogue record for this book is available from the British Library

Library of Congress Cataloguing in Publication Data
A catalogue record for this book is available from the Library of Congress

ISBN 0 7506 6529 7

For information on all Elsevier Butterworth-Heinemann
publications visit our web site at www.books.elsevier.com

Transferred to digital printing in 2009.

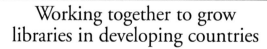

Working together to grow
libraries in developing countries

www.elsevier.com | www.bookaid.org | www.sabre.org

ELSEVIER   BOOK AID
             International   Sabre Foundation

Contents

List of Figures		xi
List of Charts		xv
List of Case Studies and Examples		xx
Reference of Charts		xxii
Preface		xxv
Acknowledgements		xxvii
Introduction		xxix
The aim of the book		xxix
The structure of the book		xxx
How to use this book		xxxiii

Part 1	**An Introduction to the Theory of SPC**		**1**
1	Statistical process control		3
	A word on processes …		3
	… And a word on variation		4
	Some statistical measures		5
	Why is understanding variation important to management?		7
	Summary of the implications of process variation		10
	Tampering (over-control) and its effect on performance		11
	Control charts: the tool for understanding process performance		14
	Dispelling some myths of SPC		16
	Are there situations where SPC is not appropriate?		19
	The relationship between SPC and Six Sigma		19
	Summary		20

Part 2	**Exploding Data Analysis Myths**		**21**
2	Problems with monthly report tables, goals and quartiles		23
	Introduction		23
	Comparing pairs of numbers: a trap for the unwary		23
	Death by numbers: the Saga of the monthly report		25
	Who wins the prize? How not to compare regional performance statistics		27
	Falsifying the data (and how to spot it): one result of setting targets		29
	Querying the top quartile: Does it mean anything?		32
	Summary		36

3 Exploring the mis-information in moving average charts   37
    How they fail to respond to process changes, out-of-control points,
    trends and seasonality   37
    Introduction   37
    Analysis   38
    Coping with seasonality and trends   43
    What moving averages actually monitor   44
    Summary   46
4 The problems with year-to-date figures   47
    Introduction   47
    Analysing YTD against plan   47
    Analysing this year's YTD against last year's YTD   50
    Why YTD charts do not work?   52
    Analysing the YTD average   53
    Comparing YTD and YTD average charts with control charts   54
    Summary   57

Part 3 Putting SPC into Practice – The Cases   **59**
    The sources of the case studies   59
    Control charts in the real worlds are not always so clear   60
    A word on chart formats   60
    Layout of and information in the case studies   61
    How to use the case studies?   62
5 Investigating variation in chemical concentration   63
    How control charts were used to identify, investigate and prove the
    cause of fluctuations in results   63
6 Improving examination results by analysing past performance
    and changing teaching methods   69
7 Demonstration that moving averages are poor indicators
    of true process performance   75
    Monitoring the frequency of incidents   75
8 Monitoring rare events   85
    How a sudden but uncertain change in safety record was shown
    to be significant   85
9 Comparing surgical complication rates between hospitals   91
10 Comparing the frequency of rare medical errors between centres   103
11 Metrics proposal for a training administration process   119
12 Reducing problems during borehole drilling   133
    An example of monitoring two metrics on one chart   133
13 Applying control charts to benchmarking in the drilling
    industry   141
14 Comparing the results of using different charts to analyse a
    set of data   153
    An application to a batch production process   153

15   Using control charts to analyse data with a trend   165
      An application to cost management   165
16   Identifying a decrease in the use of hospitality suites   175
17   Increase in reject rate at manufacture due to inspectors' fear of
      losing their jobs   187
18   Comparison of test results of production process   195
      From a batch production process to identify a key cause of variation
      and that the process is not capable of producing within specification   195
19   Categorising, de-seasonalising and analysing incident data using
      multivariate charts   213
20   Comparison of time-spent training across different facilities
      of an organisation   233

Part 4 Implementing and Using SPC   **251**
21   Understanding and interpreting a control chart   253
      Introduction   253
      The normal distribution and the standard deviation   253
      The importance of the standard deviation   255
      Definition of a process in a state of control   256
      Interpreting a control chart   257
      Possible causes of control chart signals   260
      A note on the British and American control chart limits   260
      Summary   260
22   Selecting the appropriate control chart   263
      Introduction   263
      Variables or attributes data?   263
      Defects or defectives data?   266
      Equal or variable size samples?   268
      Choosing between the p and np chart   268
      Choosing between the c and u chart   269
      Choosing between X and $\bar{X}$ charts   270
      Monitoring the mean and variability   271
      Use of X/MR charts in place of c, u, np and p charts   272
      Median and mid-range charts   272
      Median moving range charts   273
      Difference charts   273
      Z charts   274
      R or s charts?   274
      Cumulative sum (cusum) charts   274
      Selecting the most powerful chart   275
      Summary   275
23   Procedures and formula for drawing control charts   277
      Introduction   277
      Frequency of measurements   278
      Setting up charts   278

Part I Variables charts   **279**
The $\bar{X}$/R charts   279
The s (standard deviation) chart   286
The median/R chart   287
Difference charts   289
Z charts   289
X/MR charts   290
Comments   293
Moving mean/moving range charts   293
Part 2 Attributes charts   **294**
p charts   294
np charts   299
c charts   300
u charts   302
Multivariate charts   304
24   An introduction to cusum (cumulative sum) charts   307
Introduction   307
Basic cusum charts   307
Weighted cusum charts   317
Summary   320
25   Issues for the more advanced SPC users   321
Introduction   321
The number of observations required to identify a
process change (average run length)   321
Identifying and dealing with non-normally distributed data   326
Identifying and dealing with auto-correlation   328
Dealing with rare events data   329
Analysing data in groups and subgroups   330
Summary   336
26   Data analysis tools   337
Introduction   337
Histograms   337
Run charts   343
Bar charts   344
Ranked bar charts and Pareto charts   347
Check sheets   348
Scatter diagrams   349
Summary   351
27   Setting up a processing monitoring system   353
Introduction   353
Deciding what to chart   354
Creating a framework for measurement   357
Collecting, charting, analysing and deciding on appropriate action   359
A word on process improvement   364
Summary   366

28 Potential process performance metrics   367
    Introduction   367
    The power of different chart types   367
    A word on normalisation   368
    Sample size/frequency of measurement   369
    Outcome vs. process measures   369
    Generic metrics applicable to a wide variety of organisations and sectors   370
    Activity-specific metrics   371
    Industry-specific metrics   371

Part 5 Developing SPC Skills: Organisational Review Questions,
         Workshops and Exercises   **375**
29 The Rods Experiment   377
    A practical case study that can be used for training   377
    Introduction   377
    Data generation and collection   379
    Check sheet   380
    Histogram   380
    Creating a run chart   382
    Completing the X/MR control chart   383
    Introducing the process change. Monitoring and analysing in real time   384
    Drawing and interpreting an $\bar{X}$/range chart   387
    Drawing and interpreting a cusum chart   389
    The difference chart   391
    The Z chart   392
    Comments   394
30 Organisational review questions, workshops and exercises   395
    Reviewing what is happening in your organisation today   395
    Selecting performance indicators   396
    Selecting the correct control chart   397
    Data workshops and case studies   399
    Discussions   404
31 Answers to exercises in Chapter 30   405
    Selecting the right control chart   405
    Control chart interpretation   407
    Downtime workshop   407
    Repairs workshop   407
    Days taken to raise invoices workshop   410
    Blending workshop – Part 1   410
    Rejected tenders   414
    Surgical complications   415

Part 6 An Introduction to Six Sigma   **421**
Luis Miguel Giménez
32   An Introduction to Six Sigma   423
       Introduction   423
       What is Six Sigma?   423
       The basis of Six Sigma   424
       The three key roles in Six Sigma: Management, Specialists and Staff   424
       The meaning of quality in Six Sigma   424
       The two key measures in Six Sigma   425
       Selecting improvement projects   425
       The Six Sigma improvement methodology   425

       Bibliography, references and other resources   *439*
       Experimental Resources   *441*
       Appendix A   *443*
       Glossary of terms and symbols   *445*
       Index   *455*

List of Figures

1.1	A process: everything required to turn an input into an output for a customer	3
1.2	Observations vary from one another, and build into a distribution	4
1.3	The normal distribution	4
1.4	Distributions can vary in only three ways	5
1.5	In practice, control charts work with many different shapes of distribution ... including these	5
1.6	Common measures of location and variability	6
1.7	A process in control. What management likes: boring predictability. The same today, tomorrow and every day	7
1.8	A process out of control. It is interesting, exciting, unpredictable and great for fire fighting. Not so good for planning though	8
1.9	Common and special causes of variation	9
1.10	Process improvement process	10
1.11	Examples of tampering	11
1.12	(a) Golf-driving practice without tampering	12
	(b) Golf-driving practice with tampering	13
	(c) Run chart of the effect of tampering on golfing scores	14
1.13	Control chart signals	15
2.1	(a) Bar chart and (b) run chart of non-availability of equipment	24
2.2	(a) One of the most common shapes for a histogram of data in a state of control; (b) another common shapes for a histogram of data in a state of control and (c) histogram of percentage of time spent on projects	29
2.3	(a) Bar chart of the performances of 20 facilities (a) showing quartiles and (b) showing split based on difference performance levels	33
5.1	Chemical injection and sampling process	64
6.1	Some causes of dropped marks in an examination	69
11.1	RFT flow chart	120
11.2	Some aspects of an RFT form	121
11.3	Check sheet for days to process RFT	122
13.1	Bar chart of operator's drilling performance	143
17.1	Manufacturing, transportation and use process	187
18.1	Simplified process flow highlighting monitoring, adjustment and testing	196
20.1	Corporate organisation chart	233
21.1	The normal distribution	253
21.2	A normal distribution can only change location ... or spread	254
21.3	Some use facts about normal distributions	255
21.4	Probability of observing a point outside the control limits	256
21.5	The probability of a run of observations being on the same side of the average	257
21.6	The probability of alternating observations	258
21.7	Other indicators of special causes of variation	258

21.8(a) Rule: Are these indicators of the presence of special causes?   259
21.8(b) Random set of data: Are these indicators of the presence of
         special causes?   259
22.1   Selecting the appropriate control chart   264
22.2   Summary comparing variables and attributes data   265
22.3   Defectives vs. defects   267
22.4   Charts for defectives and defects   269
22.5   Comparison of X and $\bar{X}$ charts   271
24.1   Cusum mask   314
24.2   Cusum   314
25.1   The normal distribution   322
25.2   Process shift   323
25.3   Distribution $\bar{X}$ with sample size 9, standard deviation = $s/3$ 323
25.4   Process shift by $1s$ 324
25.5   ARL curves for $\bar{X}$ charts   325
25.6   OC curve for means   326
25.7   A typical process with units   331
26.1   Histogram of viscosity   338
26.2   Typical histogram shapes (a) Bell-shaped distribution,
         (b) double-peaked distribution, (c) multi-peaked distribution, (d) flat
         distribution, (e) toothed distribution, (f) separated distribution,
         (g) edged distribution, (h) skewed distribution and (i) truncated
         distribution   339
26.3(a) Histogram of a key measurement   340
26.3(b) Histogram of a key measurement with all three suppliers and overall
         histogram   341
26.3(c) Histogram of a key measurement with normal curve superimposed   342
26.4   (a) Bar chart showing the reason for emergency admission in hospital
         A and (b) grouped bar chart for five hospitals   344
26.4   Stacked bar chart of (c) the five hospitals showing the reason for emergency
         admission, (d) the reason for emergency admission for five hospitals,
         (e) five hospitals showing by percentage the reason for emergency
         admission   346
26.5(a) Ranked bar chart of the reason for emergency admission   347
26.5(b) Pareto chart of the reason for emergency admission   348
26.6   Check sheet for emergency hospital admissions   349
26.7(a) Scatter diagram of active ingredient vs. time since manufacture   349
26.7(b) Scatter diagram of active ingredient vs. time since manufacture with two
         suppliers   350
26.8   Patterns of relationship between two variables (a) No relationship,
         (b) positive relationship, (c) negative relationship and (d) complex
         relationship   351
27.1   Implementing control charting   354
27.2   A flow chart is a good source of ideas for what to measure   356
27.3   Establishing control chart average and limits   360
27.4   A typical monitoring process   362
27.5   Out-of-control (OOC) signal: action on the outputs   363

27.6	Typical process for process improvement	365
29.1	Rods Experiment: rods and ruler	378
32.1	The five phases of DMAIC	426
32.2	Process map inpatient ABX administration	431
32.3	Process capability analysis of TIME	432
32.4	Ishikawa diagram for administering ABX at the correct time	432
32.5	Ishikawa diagram for administering the correct ABX	433
32.6	Pie chart of the percentage of ABX administered	434
32.7	Frequency of incorrect drug or no ABX administered	435
32.8	Histogram and descriptive statistics showing that ABX is administered within specifications limits	435

List of Charts

The size of the charts within the book has been reduced for ease of reference. However, to view full-sized downloadable charts, please visit www.books. elsevier.com/companions/0750665297

2.1	Report for the 21 months up to and including September	26
2.2	Production history: (a) North, (b) South, (c) East and (d) West	28
2.3	NPT: (a) actual and (b) mis-reported data	31
3.1	Moving average for a process change. Plot of raw data and moving averages showing that the moving average responds poorly to changes in data	39
3.2	(a) Moving average for an out-of-control value. Plot of raw data and moving averages showing that after a process aberration the moving average takes a long time to reflect true process performance. (b) Moving average for two out-of-control values. Plot of raw data and moving averages showing that although a low observation has been recorded, the moving average may increase	41
3.3	Moving average for a trend. Plot of raw data and moving averages showing that the moving average lags behind the true trend, and that its slope varies	42
4.1	YTD expenditure vs. plan	48
4.2	This year's YTD values vs. last year's YTD figures	51
4.3	Control chart of monthly expenditure for the last 24 months	51
4.4	Grouped bar chart of this year's YTD values vs. last year's YTD figures	53
4.5	YTD average	53
4.6	Demonstrates that the control chart gives a better estimate of the true process average than the YTD	57
5.1	X/MR chart: chemical concentration	65
5.2	X/MR chart: chemical concentration highlighting differences by custodian	66
5.3	X/MR chart: chemical concentration for custodians 1 and 2	67
6.1	Examination results: X chart of % marks dropped	71
7.1	Moving average chart of the number of incidents	77
7.2	Moving average chart of the incidents frequency	77
7.3	Moving average of incident frequency	78
7.4	u chart of incident frequency	80
7.5	u chart of incident frequency with process change	81
7.6	Moving average chart of the incidents frequency for all months	82
8.1	c chart: number of accidents per month	86
8.2	X/MR chart of annual accident rate	87
8.3	X/MR chart of annual accident rate with process change	88
8.4	X/MR chart of annual accident rate with expanded scale	89
9.1	p chart: complications during surgery – all hospitals	92
9.2(a)	p chart: complications during surgery – hospital A	94
9.2(b)	p chart: complications during surgery – hospital B	95

9.2(c)	p chart: complications during surgery – hospital C	96
9.2(d)	p chart: complications during surgery – hospital D	97
9.3	Scatter diagram of the proportion of surgeries with complications vs. number of surgeries – hospital D	99
10.1(a)	p chart: medical errors – medical centre A	106
10.1(b)	p chart: medical errors – medical centre B	107
10.1(c)	p chart: medical errors – medical centre C	108
10.1(d)	p chart: medical errors – medical centre D	109
10.2(a)	p chart: medical errors – medical centre A	110
10.2(b)	p chart: medical errors – medical centre B	111
10.2(c)	p chart: medical errors – medical centre C	112
10.2(d)	p chart: medical errors – medical centre D	113
10.3	p chart: all medical centre errors	114
10.4	p chart: average error proportions for all centres	115
11.1	X/MR chart of days to process RFT	123
11.2	Multivariate c chart of RFT issues	126
11.3	Pareto chart of RFT issues arising	129
12.1	Scatter diagram of number of washouts vs. twist-offs	134
12.2	u chart: washouts and twist-offs	135
12.3	u chart: washouts and twist-offs with process change	137
13.1	$\bar{X}$/R chart: cost per foot drilled	144
13.2	X/MR chart for operator D showing two out-of-control (OOC) wells	145
13.3	X/MR chart for operator D: out-of-control points removed; calcs: calculations	146
13.4	$\bar{X}$/R: cost per foot drilled; two operator D wells removed	147
14.1	c chart: number of off-specification batches per month	155
14.2	c chart: number of off-specification batches per month with process change	156
14.3	p chart: off-specification tons	157
14.4	Scatter diagram of average batch size vs. proportion of off-specification tons	157
14.5	p chart: off-specification tons including process change	158
14.6	u chart: number of off-specification batches	159
14.7	X/MR chart: number of off-specification batches	161
14.8	X/MR chart: off-specification tons	162
15.1	X chart: cost per foot drilled for successive wells ordered by spud date	167
15.2	X chart: cost per foot drilled with process changes regression line	169
15.3	Comparison of moving average of span 12 and the X chart	170
16.1	X/MR chart: number of functions per month	176
16.2	X/MR chart: number of guests	177
16.3	Scatter diagram showing the close relationship between number of functions and number of guests	178
16.4	c chart: number of functions per month	179
16.5	c chart: number of guests:	180
16.6	np chart: number of functions per month	181
16.7	Histogram of the number of functions	182
16.8	Histogram of the number of functions after removing the nine values	182

16.9	$\bar{X}$/range chart of the number of guests per month	183
16.10	cumulative sum chart: number of functions per month	185
17.1	p chart: rejects at manufacture	189
17.2	Weighted cusum chart of rejects at manufacture	190
17.3	X/MR chart: rejects at manufacture	192
18.1	X/MR chart: bond sample values for data set A	197
18.2	X/MR chart: bond sample values for data set A showing process changes	199
18.3	X/MR chart: bond sample values for data set B	200
18.4	X/MR chart: bond sample for data values set C	201
18.5	$\bar{X}$/R chart: bond data set A averages plotted by shift	202
18.6	$\bar{X}$/R chart: bond value summary by analyst, data sets A, B and C	203
18.7	$\bar{X}$/R chart: bond summary by analyst, data sets A, B and C using unadjusted range	210
18.8	$\bar{X}$/R chart: Comparison of methods for calculating control limits	211
19.1	c chart: all injurious incidents	214
19.2	Cusum chart of all injurious incidents	216
19.3	$\bar{X}$/R chart: monthly incident analysis	217
19.4	Pareto charts of incident frequencies by (a) immediate cause and (b) part of body injured	219
19.5	c chart of hand tool incidents	220
19.6	$\bar{X}$/R chart: hand tool incidents	221
19.7	c chart: hand tool de-seasonalised incidents	222
19.8	c chart: hand tool de-seasonalised incidents with one process change identified	222
19.9	c chart: hand tool de-seasonalised incidents with process changes identified	223
19.10	Cusum chart: hand tool injuries chart of de-seasonalised data	223
19.11	Scatter diagram: relationship between eye incidents and wind blown incidents	225
19.12	c chart: hand tool de-seasonalised incidents with revised weights	229
20.1	X/MR chart: percentage of time-spent training ordered randomly within BU	235
20.2	X/MR chart: percentage of time-spent training ordered randomly within BU with separate averages and limits for each BU	236
20.3	$\bar{X}$/R and $\bar{X}$/S charts: training for each BU	238
20.4	X/MR chart: percentage of time-spent training ordered by group, out-of-control facilities removed	240
20.5	$\bar{X}$/R and $\bar{X}$/S chart: percentage of time spent training excluding out of control values	241
20.6	X/MR difference chart: percentage of time-spent training for each facility and the organisation average	242
20.7	X/MR chart: percentage of time-spent training ordered by group, data order changed	243
20.8	Histograms showing the non-normality of training percentage data: (a) all, (b) South and (c) North E	246
22.1	Run chart of the amount of active ingredient of a substance	266

23.1	$\overline{X}$/R chart: measured value charted by shift	281
23.2	$\overline{X}$/R chart: measured value charted by shift – all data	285
23.3	Median and range chart	288
23.4	X/MR chart: time to complete jobs	291
23.5	p chart: rejects on site	296
23.6	np chart: number of invoices paid late, sample size 50	299
23.7	c chart: incidents	301
23.8	u chart incidents	303
23.9	Multivariate np chart: rejected tenders	305
24.1	Run chart of golf scores	308
24.2	Run chart of par and actual golf scores	308
24.3	Run chart of differences (score – par)	309
24.4	Cusum chart of (score – par)	309
24.5	Cusum chart: downtime per week	310
24.6	The effect of changing the scale on a chart	311
24.7	Interpretation of a cusum chart	312
24.8	Interpretation of a cusum chart: with changes in slope	313
24.9	Cusum chart: downtime per week, using decision lines	316
24.10	Weighted cusum chart for incidents	319
25.1	Scatter diagram showing auto-correlation	328
25.2	Run chart of daily admissions over a 5-week period for casualty 1	332
25.3	Run chart of daily average admissions for casualty 1	333
25.4	Run chart of daily average admissions for all casualty departments	333
25.5	Run chart of daily average admissions for all casualty departments	334
25.6	Run chart of average admissions by casualty	335
26.1	Run chart: viscosity	343
29.1	Rods Experiment data recording sheet	379
29.2	Rods Experiment data recording sheet (completed)	380
29.3	Rods Experiment check sheet	381
29.4	Histogram of rod lengths	381
29.5	Rods run chart	382
29.6	X/MR chart: rods	383
29.7	X/MR chart: rods after process change	385
29.8	X/MR chart: rods with updated means and limits	386
29.9	Rods Experiment data recording sheet	387
29.10	$\overline{X}$/range chart of all rods	388
29.11	cusum chart: rods	390
29.12	Difference chart: rods	392
29.13	Z chart: rods	393
30.1	Out-of-control conditions	399
31.1	Out-of-control conditions	407
31.2	X/MR chart: downtime	408
31.3	u chart: repairs per 10,000 hours usage by task	409
31.4	$\overline{X}$/R chart: Days taken to raise invoices	411
31.5	np chart: off-specification blends	412
31.6	p chart: off-specification blends	414
31.7	np multi-characteristic chart: rejected tenders	415

31.8	p chart: complications during surgery – all hospitals	416
31.9(a)	p chart: complications during surgery – hospital A	417
31.9(b)	p chart: complications during surgery – hospital B	417
31.9(c)	p chart: complications during surgery – hospital C	418
31.9(d)	p chart: complications during surgery – hospital D	418
31.10	p chart: all hospitals: complications during surgery	419
32.1	X chart of the number of minutes for the ABX administration	433
32.2	X chart of the number of minutes before incision that drug is administered	436
32.3	c chart showing the decrease in of the number of infections	436

List of Case Studies and Examples

Chapter 2:	Non Productive Time	30
Chapter 3:	Exploring the mis-information in moving average charts	37
	How they fail to respond to process changes, out-of-control points, trends and seasonality	
Chapter 5:	Investigating variation in chemical concentration	63
	How control charts were used to identify, investigate and prove the cause of fluctuations in results	
Chapter 6:	Improving examination results by analysing past performance and changing teaching methods	69
Chapter 7:	Demonstration that moving averages are poor indicators of true process performance	75
	Monitoring the frequency of incidents	
Chapter 8:	Monitoring rare events	85
	How a sudden but uncertain change in safety record was shown to be significant	
Chapter 9:	Comparing surgical complication rates between hospitals	91
Chapter 10:	Comparing the frequency of rare medical errors between centres	103
Chapter 11:	Metrics proposal for a training administration process	119
Chapter 12:	Reducing problems during borehole drilling	133
	An example of monitoring two metrics on one chart	
Chapter 13:	Applying control charts to benchmarking in the drilling industry	141
Chapter 14:	Comparing the results of using different charts to analyse a set of data	153
	An application to a batch production process	
Chapter 15:	Using control charts to analyse data with a trend	165
	An application to cost management	
Chapter 16:	Identifying a decrease in the use of hospitality suites	175
Chapter 17:	Increase in reject rate at manufacture due to inspectors' fear of losing their jobs	187
Chapter 18:	Comparison of test results of production process	195
	From a batch production process to identify a key cause of variation and that the process is not capable of producing within specification	
Chapter 19:	Categorising, de-seasonalising and analysing incidents data using multivariate charts	213
Chapter 20:	Comparison of time-spent training across different facilities of an organisation	233

Chapter 23: Procedures and formula for drawing control charts   277
             Reject product   280
             Time to complete work packages   292
             Rejected pipes   295
             Invoices paid late   299
             Loss incidents   301
             Rejected tenders   304
Chapter 24: An introduction to cusum charts   307
             Downtime   309
             Loss incidents   317
Chapter 26: Data analysis tools   337
             Viscosity   338
             Railway supplies   340
             Emergency hospital admissions   344
             Active ingredient   349
Chapter 29: The Rods Experiment   377
             A practical case study that can be used for training
Chapter 30: Organisational review questions, workshops and exercises 395
             Downtime   399
             Repairs   400
             Days to raise invoices   401
             Blending test failures   401
             Rejected tenders   402
             Surgical complication (case study extension)   403
Chapter 31: Answers to exercises in Chapter 30   405
             Downtime   407
             Repairs   407
             Days to raise invoices   411
             Blending test failures   411
             Rejected tenders   414
             Surgical complication (case study extension)   415
Chapter 32: An introduction to Six Sigma   423
             (Pre-operative prophylactic antibiotic administration) 430

Reference of charts with part and chapter titles

Part and chapter titles	Control Charts														Other Charts			
	X	MR	X̄	Range	s	C	Median	n	u	np	p	cusum	Weighted cusum	Moving average	Histogram	Bar and/or Pareto	Scatter diagram	Check sheet
Part 2: Exploding Data Analysis Myths																		
2 Problems with monthly report tables, goals and myths	X														X	B		
3 Exploring the mis-information in moving average charts														X				
4 The problems with year-to-date figures	X															B		
Part 3: Putting SPC into Practice – the Cases																		
5 Investigating variation in chemical concentration	X	X													X			
6 Improving examination results by analysing past performance and changing teaching methods – monitoring incident frequency	X	X																
7 Demonstration that moving average are poor indicators								X						X				
8 Monitoring rare events	X	X				C												
9 Comparing surgical complication rates											X							
10 Comparing the frequency of rare medical errors: between errors											X							
11 Metrics proposal for a training administration process	X	X				M										P		X
12 Reducing problems during borehole drilling								X										
13 Applying control charts to benchmarking in the drilling industry	X	X	X	X												B	X	
14 Comparing the results of using different charts to analyse a set of data	X	X				C		X			X						X	
15 Using control charts to analyse data with trends	X													X				

Reference of charts with part and chapter titles

Part and chapter titles	Control Chart													Other Charts			
	X	MR	$\bar{X}$	Range	s	C	Median	n	np	p	cusum	Weighted cusum	Moving average	Histogram	Bar and/or Pareto	Scatter diagram	Check sheet
16 Identify a decrease in the use of hospitality suites	×	×	×	×		×			×		×			×		×	
17 Increase in reject rate at manufacture due to inspectors' fear of losing their jobs										×		×					
18 Comparison of test results of production process	×	×	×	×													
19 Categorising, de-seasonalising and analysing incident data using multivariate charts			×	×		M					×				P	×	
20 Comparison of time-spent training across different facilities of an organisation	×	×	×	×	×									×			
Part 4: Implementing and using SPC																	
23 Procedures and formula for drawing control charts	×	×	×	×		×	×	×	× M	×							
24 An introduction to cusum charts											×	×					
25 Issues for the more advanced SPC users																×	
26 Data analysis tools														×	BP	×	×
Part 5: Developing SPC Skills																	
29 The Rods experiment	X_{DZ}	×	×	×							×			×			×
31 Answers to exercise			×	×													
Downtime	×	×															
Repairs								×									
Days taken to raise invoices																	
Off-specification blends			×	×					×	×							
Rejected tenders									M								
Complications during surgery										×							
32 An Introduction to six sigma	×													×			

B = Barchart; P = Pareto chart; M = Multivariate; D = Difference.

Preface

"But SPC is for manufacturing, it won't work here". Comments like this were like a red flag to bull. Determined to prove "them" wrong was a task I set out to accomplish 15 years ago.

The first place to look, the literature, did not help much. There were few examples of statistical process control (SPC) outside manufacturing and, in any case, most examples used data to illustrate how to draw a control chart, rather than show how to investigate process performance using control charts, that is, they seemed to be technique rather than solution focused. Whilst this approach works well for teaching how to draw charts, it simplifies the often messy truth behind data analysis. With experience I also discovered that in many situations one chart did not give all the answers, nor were the charts always quite as simple to interpret as the literature would have me believe.

This led me to think that there was a gap in the information readily available, a gap that this book aims to bridge. Specifically my aims in writing this book are to demonstrate that:

- SPC is applicable in a wide range of organisations and applications, including non-manufacturing;
- control charts can be used for far more than just determining whether a process is in a state of control. They can often be used as an investigative tool to generate and test ideas as to what may be causing problems in processes;
- it is straightforward to begin using and benefiting from control charts. In addition, this book shows that the more one understands how to use them, the wider the applications – and the more information that can be gleaned from them – and so the greater the benefit to the organisation.

The benefit of using actual and sometimes less than optimal data as I do in this book is that the reader can be sure that these case studies have not been sanitised to gloss over or ignore some of the difficulties that may be encountered. Most case studies have been included not only to show a wide range of applications of SPC but also, for those more deeply involved in SPC, to illustrate how to overcome difficulties that may occur. The disadvantage of this approach is that those with knowledge in the application areas will realise that some issues covered in one case study have been ignored in others. Unfortunately it has not been possible to address every data issue in every case study. To do so would result in repeating similar methods and considerably lengthen each case. For example, seasonality is only addressed in one case study.

In addition, there are some more complex issues for which more advanced knowledge in the subject of the case study or statistical knowledge is required; these are and considered beyond the scope of the book. For example, in the drilling case studies I have not addressed the issue that if several wells are drilled in the same area, the cost and time to drill the wells may decrease due to a learning factor, and so the data values may not be independent of one another. Identifying the existence and effect of these trends, and then taking them into account, would require more statistical theory and techniques than this book covers. Despite the necessary limiting of the statistical aspects, these will often be secondary when compared to other effects.

It has been my intention to write this book for a wide readership; from those with little or no exposure to SPC, through those learning about SPC, to those who have been using SPC for some time. For managers who have not been involved in SPC my hope is that they will be able to appreciate its usefulness. To do so, it is not necessary to understand the mathematics behind the formulae. At the other extreme, those who have been involved with SPC for some time, may find ideas and methods that will help broaden the applications and uses to which SPC is put.

Finally, there is a web site where control charts as pdf and data files in Excel can be accessed and downloaded. Go to: www.books.elsevier.com/companions/0750665297

Tim Stapenhurst

Acknowledgements

The writing of a book is a team effort, although it is only the person who puts fingers to keypad who is credited with the final result. There have been many people who have contributed in different ways to this book and it would not be possible to mention them all, though they all have my thanks.

Those who have played a key part include Chris Saunders and Steve Spreckley who gave me the opportunity and support to experiment with applying SPC in a large non-manufacturing multinational organisation. At that time there was little help in the literature to applying SPC outside manufacturing and it was a challenge for us to see how to apply it to such areas as safety, environment, purchasing and projects. We took as our model ideas developed by Phil Cole, Will McNally and Guy Cochlan, who were applying SPC in other business streams and in areas such as deliveries, finance, and customer services. It was during this time that Dave Baldwin attended one of the one day 'Introduction to SPC' courses that we ran, and who a few days later handed me his proposal for the training administration process. His widow has kindly given me permission to use his proposal – not only as an example and SPC application – but also to demonstrate that even with a little training it is possible to start using the tools and ideas of SPC.

Writing a book is a daunting task, and although the idea of doing so had been at the back of my mind for sometime, it was a chance meeting with Len Airey, who had recently published his first book on his experiences in the Antarctic, that gave me the inspiration to start writing. It is, however, one thing to start a book and another to complete it. For his constant advice and support I am very grateful to Jonathan Simpson. Everything he advised later turned out to be correct.

Finally, even though I had a clear vision for the book and had much of the material prepared as case studies in the courses that I ran, it took many hours to complete the task and I thank my family for their patience and especially my wife pat for the long hours she spent reading through the draft.

Introduction

The aim of the book

This is a book about statistical process control (SPC). The aims of the book are to:

- encourage non-users of SPC from all types of organisations to use this powerful tool in order to understand, manage and improve process performance;
- demonstrate that some common methods used to understand process performance are misleading and may, for example, suggest that things are getting better when they are in fact getting worse;
- demonstrate that SPC, and control charts in particular, are applicable to all types of organisations, including service, governmental, charities and health care – NOT just manufacturing;
- demonstrate that control charts have a wide variety of uses including:
  - process monitoring,
  - prediction,
  - generating and testing theories for process performance upsets and problems,
  - benchmarking.
- provide enough information for the progressive manager to begin using these techniques.

To achieve this aim this book:

- is organised around typical organisational issues, rather than around a methodology;
- provides enough information up front to ensure that the reader can understand how and why these techniques and methods work;
- provides more information in **Part 4** for those wanting to increase their understanding and/or implement the ideas discussed;
- focuses on a wide range of situations.

This book is written for:

- managers who want to better understand what is going in their organisations, and identify correctly when intervention in the process is, and is not, required;
- improvement teams and individuals, that is, those involved in analysing process information in order to improve the process;
- quality, safety, production and other professionals who are responsible for monitoring performance on a regular (usually daily or weekly) basis.

Most books on SPC are technique oriented. They teach the statistical theory, then explain each of the SPC tools – mainly control charts – one by one, illustrating each with an example mainly geared towards manufacturing. These books are excellent for the SPC student.

This book is problem-, situation- and issue-based. SPC, and especially the use of control charts, is seen as the solution to a problem, not a technique. This approach shows how SPC can be used to gain insight into performance and requires a minimum of up-front theory. It complements the traditional, theoretical, SPC approach.

The structure of the book

Part 1 An introduction to SPC

It was the original intention to eliminate, or confine to an appendix, all statistical theory. However, it became apparent whilst writing this book that it would be helpful to include a short explanation of the fundamental concepts, applications and purpose of SPC before the case studies. These issues are addressed in *Part 1, An introduction to the theory of SPC*. Much of SPC concerns understanding process performance, and so it is necessary to explain firstly what is meant by the term *process* as used in this book. Secondly, it is important to understand some fundamental truths about *variation* in performance data, for example, that variation in data exists; that variation is of two types – frequently called common and special cause; that once we know what type of variation a process is exhibiting, we know the type of action to take.

At this point, it is useful to review some of the basic *statistical measures*. Whilst not vital for appreciating the benefits of SPC, most readers will probably want to review some of the measures used in SPC: the mean, range, standard deviation and variance.

Having covered the statistical theory, we discuss the fundamental *importance to management of understanding the information in variation* and we consider how the type of information in variation leads us to taking the appropriate action on a process. One of the most common failures in process management is the tendency to *tamper* or "over control" a process. We demonstrate the destabilising effect of tampering on process performance with a simple example.

This background has led us to appreciate that for effective management it is necessary to understand the information in process variation. In the section *control charts: the tool for understanding process performance* we show how control charts tell us what is happening in a process and suggest what management action is appropriate.

Finally, we dispel many of the common *myths about SPC* and outline some of the many uses of control charts – from monitoring processes, to testing theories, to prediction to benchmarking.

Part 2 Exploding data analysis myths

Many people and organisations will already have a variety of different methods of reporting and comparing performance and will wonder why they should change to control charts. In *Part 2, Exploding data analysis myths* we explore a variety of these methods and tools and expose some of their flaws. Open up any quarterly, monthly, weekly report, look on any a noticeboard and you are likely to be confronted with moving averages, year-to-date charts and tables of data. In *problems with monthly report tables, goals and quartiles* we address some of the issues of comparing pairs of numbers, trying to understand the information in tables of comparisons, and the potential risks of goal setting and using quartiles. In *exploring the mis-information in moving average charts* we demonstrate the high risks of trying to interpret moving averages, how they may mislead the reader into believing one thing – for example, that a process is improving – when the opposite is true. Finally, we look at some of *the problems with year-to-date figures*.

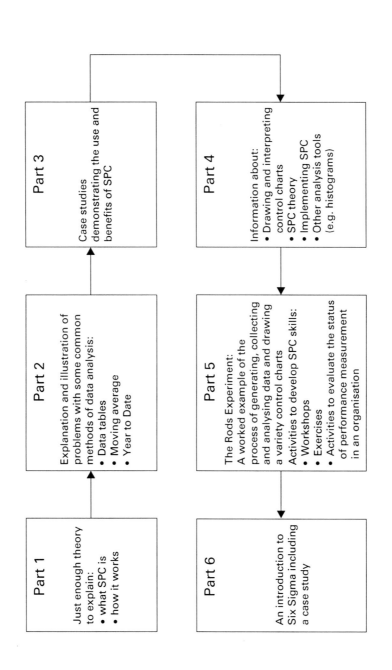

Part 1

Just enough theory to explain:
• what SPC is
• how it works

Part 2

Explanation and illustration of problems with some common methods of data analysis:
• Data tables
• Moving average
• Year to Date

Part 3

Case studies demonstrating the use and benefits of SPC

Part 4

Information about:
• Drawing and interpreting control charts
• SPC theory
• Implementing SPC
• Other analysis tools (e.g. histograms)

Part 5

The Rods Experiment: A worked example of the process of generating, collecting and analysing data and drawing a variety control charts

Activities to develop SPC skills:
• Workshops
• Exercises
• Activities to evaluate the status of performance measurement in an organisation

Part 6

An introduction to Six Sigma including a case study

In each of these chapters we include a comparison of the method under discussion with control charts so that the readers can judge for themselves which is the most useful.

Part 3 SPC case studies

The hope is that the reader will be motivated to discard the reporting methods of Part 2 and turn to *Part 3, SPC case studies* to see how SPC has been successfully applied in many different situations. The aim of this central part of the book is to encourage readers to apply SPC more widely than they are doing now by demonstrating how others have benefited.

In most case studies a variety of control charts and other tools have been used. The situations behind the case studies vary enormously, but in general each includes the background, the data, and the appropriate charts. Each chart is analysed in detail and points of interest arising are discussed either in the text or in the comments section towards the end of each case. The comments typically include, as appropriate, the effect of using different control charts and alternative analyses and answers to "what–if" questions. Details of calculations are often provided so that the interested reader can check his or her understanding of where the figures come from. However, it is not necessary to understand the calculations to benefit from this book. In an organisation it is only a few people who need to understand the detailed theory and formulae behind creating control charts – most people only need to understand how to use them.

Part 4 Implementing and using SPC

Having seen how useful SPC can be, *Part 4, Implementing and using SPC* provides much more information on how to select metrics, set up a framework for measurement, and draw and interpret control charts, as well as providing a more in depth understanding of some of the theory of SPC.

There is also a chapter on how to use and interpret other SPC tools such as histograms. The worked examples in Part 4 are also *case studies*, so even if you do not need to understand the theory of SPC, you will find more interesting applications.

Part 5 Organisational review questions, workshops, and exercises

Part 5, Organisational review questions, workshops, and exercises begins with the *Red Rods Experiment*. The red rods can be used as an excellent, if theoretical, case study taking the reader from data collection, through the use of various SPC tools including a variety of control charts. It can also be used as a training exercise.

The remainder of Part 5 begins with a series of questions aimed at helping individuals ascertain the status of data collection, reporting and analysis in their organisation. Questions on identifying appropriate metrics, selecting, drawing and interpreting control charts conclude this section.

Although you may think Part 5 is not for you, please glance through it. It contains typical exercises that a tutor may wish to use, or that will help a reader to test his or her understanding of the mechanics of control charting. It also contains exercises that will encourage the reader to gather information about an organisation and gain a deeper understanding of what is happening there. In addition, some of the control chart exercises are themselves further case studies.

Part 6 Six Sigma

Part 6, Six Sigma is devoted to a brief introduction to Six Sigma. Six Sigma is a process improvement methodology that has helped dramatically improve the performance of many organisations. The development of the methodology is ascribed to Motorola which, in the 1980s, embarked on a drastic reduction in defects in the products it manufactured. Their success is legendary and their methods have been adopted and adapted by many organisations since.

The methodology behind Six Sigma is a simple one – a step-by-step approach to improving performance by focusing on consistently producing better products and services that meet or exceed customers' needs more cheaply. The methodology is a disciplined project-by-project approach identifying the gap between desired and actual performance, identifying and testing potential improvements, selecting and implementing those that we have evidence will work, and monitoring the results to ensure that expected performance improvements have occurred and are sustained.

How to use this book

This book has been designed for the general reader interested in SPC to read from the beginning through to the end. It begins in Part 1 with some basic SPC theory. Part 2 explains the shortcomings of many of the reporting and analysis tools used in organisations today. Part 3 presents case studies demonstrating how SPC has been successfully used and Part 4 provides further information on implementing and using SPC.

This book has been written with the practitioner in mind, so that once the reader has found her or his way around, it can be used as a reference book for helping with issues that may arise when using SPC, such as what chart to use, how to use SPC in different situations, identifying applications for SPC, etc.

Some readers may have specific needs or reasons for picking up this book, and the route map here suggests ways in which you may wish to use the book.

With regard to the case studies, the reader can dip into the ones that seem most appropriate at the time. It is not necessary to read them all in the first reading of the book: they are intended as reference material. If you find you do not understand what SPC is trying to do, review Part 1 for an overview of the theory.

During the writing of this book some people queried the fact that the case studies (Part 3) are placed before the theory (Part 4) as it will be difficult to fully understand the case studies without first understanding the theory. This observation is correct and for this reason some readers may prefer to read Part 4 before Part 3. The case studies have been placed before the theory because the focus of this book is to demonstrate the application and use of control charts in practice. It is only necessary to understand the application and have a general idea of how to use and interpret a control chart, and this is covered in Part 1 of the book. The book has been written in modular form, and one option that may appeal to you is to read some case studies and then refer to Part 4 before returning to the case studies.

A route map/suggested itinerary

If this is you...	Try reading this...
I know very little about SPC. I want to know how it can help me, but at this time I do not want to go through all the theory.	This book was written in a specific sequence for you: • It begins by giving an overview of theory (Part 1), then • Demonstrates why many popular methods of analysing process data do not work (Part 2). Skip over the calculations and other aspects that do not interest you at this time. • Part 3 demonstrates with case studies that SPC helps us understand how an organisation is performing. Again skip the technical aspects.
We do not need SPC, we have charts and data analyses everywhere.	Take a look at Part 2. If you are using these types of analyses consider how they might be misleading and failing you.
I am involved in implementing SPC. I need more information on the theory, uses and implementation.	Start at the beginning and work through to the end.
We are using SPC, but have some difficulties in knowing how to use SPC in specific situations.	Look for case studies that are similar to your situation and see how others have tackled the situation.
I teach SPC and want some extra/new material	The myths may be interesting for you. I have used many of these case studies when teaching SPC, and found students interested in the approach. The case studies should provide a rich source of information, worked examples and/or exercises. There are also a large number of exercises at the back of the book.

How to I persuade others in my organisation to use SPC?	Always difficult. One of the things you will need to do is demonstrate why existing analysis methods do not work and that SPC will work. Review Part 2 and try selecting some case studies in Part 3 and applying the same techniques with data from your organisation.
How do I implement SPC?	Some ideas are given in Part 4. If overcoming resistance to change is an issue for you may need some help with organisational development which is not discussed in this book.
I am familiar with SPC, I have a reasonable understanding of the theory and have even tried a little charting. However, I think we should be getting more out of SPC.	Read the case studies (Part 2) to see the wide variety of applications and how charts can be used in practice to understand process performance. You can always return to the theory (Part 1) if you need.
We have been using SPC for some time. I am familiar with the standard theory and want some more advanced theory, applications and uses.	The case studies may give you some ideas on applications and some of the theory discussed in the case studies goes beyond the usual books on SPC.
I do not think SPC will work for us because…	Read the 'myths' in Part 1 and look through the case studies. If none of these bear any relation to what you do in your organisation then perhaps you are right.
We do not use SPC to monitor We use…	If you use any of the methods indicated in Part 2, read this section. Then read Part 3.

PART 1

An Introduction to the Theory of SPC

The aim of any type of data analysis is to gain understanding from data. When we collect process performance data we see that it varies. The information in this variation is important to the understanding of how the process is performing and statistical process control (SPC) is primarily the tool for understanding variation:

- *SPC* is the use of statistically based tools and techniques principally for the management and improvement of processes. The main tool associated with SPC is the control chart.
- A *control chart* is a plot of a process characteristic, usually through time with statistically determined limits. When used for process monitoring, it helps the user to determine the appropriate type of action to take on the process.

You may find these two definitions off-putting, and the purpose of this part of the book is to explain them, and also the basic concepts and ideas behind SPC as well as the importance and use of control charts.

First we explain, briefly, what is meant by the term 'process' as it is important to understand how the term is used in the book.

One of the crucial keys to understanding performance measurement, and hence statistical process control, is variation. If there were no variation there would be no problem: life would be much simpler and more boring. Much of a manager's work is given over to understanding, managing and controlling variation. This whole book deals with the analysis, understanding and management of variation.

Unfortunately statistics does come into SPC. Actually, statistics should come into all aspects of running an organisation because statistics is all about understanding data. There are only a few main statistical measures that need to be discussed here, namely the *mean*, the *standard deviation* and the *range*. You do not need to know the formula for these, but for those interested in doing so, the formulae are included.

Appreciating the existence of variation and something of the statistics for measuring it are building blocks. The next step is to realise the implications of variation for the understanding and management of processes. It is important to fully grasp the concepts outlined here, and particularly those on over-control (tampering).

So much for explaining the problems of variation and its impact on processes; the key tool for understanding variation is the control chart; therefore, we have summarised what a control chart is and how to use it.

Experience has shown that there are many popularly held erroneous beliefs about SPC. Unfortunately they have limited the application of this powerful tool. In the next section we dispel some of these myths, and by implication explain some of the many uses for SPC. If you have any pre-conceived ideas about SPC, or wonder if SPC is applicable to you, do read this section.

Finally, there have been many useful management tools, methodologies and philosophies aimed at management over the years. Each one has brought its benefits and left its mark. The current methodology that is being successfully implemented in many organisations is "Six Sigma". Like all the previous ones, Six Sigma does bring benefits, but it is not a panacea, and neither is SPC! However, each does have its place, application and use, and we summarise the use of Six Sigma briefly in anticipation of a later chapter devoted to it.

1 Statistical process control

A word on processes …

As statistical process control (SPC) is used for analysing process data, it is pertinent to explain what we mean by a process. There are many good definitions of a process. One very simple definition with a wide application is:

A process is everything required to turn an input into an output for a customer.

This definition can be applied to a spectrum of processes, from small tasks (such as filling in a form) to a complete business system (such as order fulfilment). Processes may involve just one person at one end of the spectrum to complete departments including suppliers at the other. The concept can be applied to the design, development and manufacturing of goods and/or services, such as patient care, government or legal processes.

Let us consider this definition in a little more detail. We start with an "input", for example, a raw material, a sick patient or a blank form, and we "do something" with it to produce an output – a finished product, a "well" person or a completed form. The "something" which we "do" to turn the input into the output is the process. We need "things" to carry out the process. These "things" can be grouped into: equipment, people, materials, procedures and environment. For a hospital, the "people" include the nursing, administration and other staff as well as any contractors; equipment includes beds, monitors, testing equipment; procedures are the (usually documented) steps that the people follow to do the job; "materials" include medication and dressings; and "the environment" includes both the physical environment (such as temperature) and more abstract aspects (such as culture) (see Figure 1.1).

Clearly there must be a customer for the process, and the customer may be internal or external to the organisation.

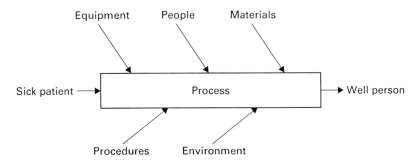

Figure 1.1 A process: everything required to turn an input into an output for a customer

... And a word on variation

All the (non-trivial) process outputs, inputs, methods of working, etc. vary. This is where many management problems begin. For example:

● The time taken to process invoices, treat a patient, answer an enquiry, etc. will vary from occasion to occasion.
● The number of safety/breakdown/interrupt incidents in a time period will vary.
● Performance of different work groups, departments and people will vary.
● The performance of a particular work group, department or person will vary over time.

Sometimes this variation may be considered relatively unimportant, such as variations in the amount of light transmitted through a pane of ordinary window glass. Other instances of variation, for example flight arrival times, may be much more serious.

As SPC is fundamentally about understanding and managing variation, we need to spend some time considering some key aspects of it. As a focus for this discussion, see Figure 1.2.

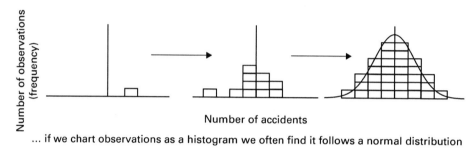

... if we chart observations as a histogram we often find it follows a normal distribution

Figure 1.2 Observations vary from one another, and build into a distribution

Suppose we measure some attribute of a process on a regular basis. For example, it may be the number of accidents per month. When we have taken our first observation we could plot the value on a histogram, as shown in Figure 1.2. The next observation is likely to be different. As we continue taking and plotting observations we would gradually see the distribution of observations take on a pattern. Frequently this distribution will take the shape of the bell curve, known by statisticians as the normal distribution (see Figure 1.3).

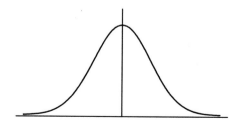

Figure 1.3 The normal distribution

Distributions vary in one or more of three ways (Figure 1.4):

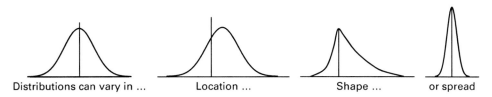

Distributions can vary in ...   Location ...   Shape ...   or spread

Figure 1.4 Distributions can vary in only three ways

- the central location (i.e. a "typical" value);
- the shape (e.g. one peak or two, symmetrical or skewed);
- the spread or variability (how different the values are: e.g. the span from the maximum to the minimum).

We use the "normal" distribution as the basis for discussions throughout as it is the most common distribution that we are likely to come across. However, the ideas, methods and charts that we discuss were developed by Shewhart not because of the statistical theory but because they worked in practical situations. More recent analysis has shown that these ideas work for many different distribution shapes, such as those in Figure 1.5.

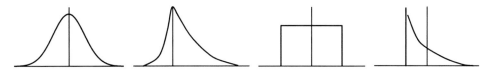

Figure 1.5 In practice, control charts work with many different shapes of distribution ... including these

Some statistical measures

There are a few statistical terms relating to location and variation that are used when discussing SPC. The calculations are given at this point for reference only and it is not necessary to understand the calculations in order to understand SPC and control charts.

When collecting data, statisticians refer to the first value collected as x_1, the second as x_2, the ith as x_i and the last value, called the nth as x_n.

As an example, we use the following set of data values (e.g. these values could be the number of patients admitted each shift for the last nine shifts): 3, 6, 5, 6, 4, 7, 2, 6, 4.

For this set of values $n = 9$ and $x_1 = 3$, $x_2 = 6$, etc., and x_n, that is $x_9 = 4$.

There are two main descriptions that we are interested in: the location and the spread (Figure 1.6).

Measures of location

There are three commonly used measures of location, namely the *mean*, the *median* and the *mode*.

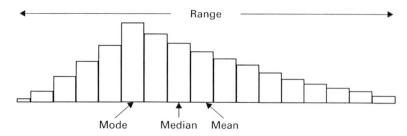

Figure 1.6 Common measures of location and variability

The mean

Also known as the average, denoted by statisticians as $\bar{x}$, which is calculated as:

$$\bar{x} = \sum_{i=1}^{n} \frac{x_i}{n}.$$

The Greek letter sigma, Σ, is used to show that we add together all the values of the x_i from $i = 1$ to n; that is, $x_1 + x_2 + x_3 + \cdots + x_n$.

Using the above data:

$$\bar{x} = \sum_{i=1}^{n} \frac{x_i}{n} = \frac{3 + 6 + 5 + 6 + 4 + 7 + 2 + 6 + 4}{9} = \frac{43}{9} = 4.78.$$

The mean is by far the most common statistic used for describing the location of a set of data.

The median

The value at which half of the observations fall above and half fall below. It is found by ordering the values in ascending or descending order: 2, 3, 4, 4, 5, 6, 6, 6, 7; and picking the middle one. The median of our data set is **5**. (If there is an even number of values, the median is the average of the middle two values.)

The mode

The most frequently occurring number observed. In our set of data the mode is **6**.

Measures of variability

Common measures of spread or variability are the *range*, the *standard deviation* and the *variance*.

The range

The difference between the maximum and minimum values observed. In our example the range is $(7 - 2) = 5$. The range is simple to calculate but is greatly influenced by single outlying values.

The standard deviation

Denoted as s, is calculated using the formula:

$$s = \sqrt{\frac{\sum_{i=1}^{n} (x_i - \bar{x})^2}{(n-1)}}.$$

Using the above data:

$$s = \sqrt{\frac{(3 - 4.78)^2 + (6 - 4.78)^2 + \cdots + (7 - 4.78)^2}{(9 - 1)}} = 1.64.$$

As this calculation uses all the observed values it is less influenced by a single extreme value, unlike the range.

The variance

The square of the standard deviation. In our example the variance is $1.64^2 = 2.69$.

Why is understanding variation important to management?

There are two situations which may occur when we repeatedly take measures from a process:

1. The distribution of data is much the same with each set of measurements.
2. The distribution changes with each set of measurements.

1. The distribution is much the same each time (Figure 1.7)

This is what management and operators alike would like to see: predictable results, with a minimum of variation centred on a target value. The key advantage is that in this situation we know what will happen next. Things are running smoothly, we can plan,

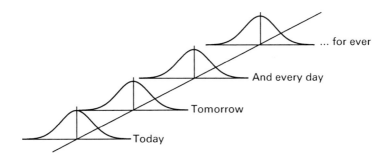

Figure 1.7 A process in control. What management likes: boring predictability. The same today, tomorrow and every day

we can estimate. Even if the process is delivering some unacceptable results (e.g. 6% rejects, 30% late orders, 9% over-budget) we know that these unacceptable results will continue at these rates.

In this situation the process is said to be *in a state of statistical control* (often shortened to *in control*). Processes in a state of control are said to be subject ONLY to *common cause variation*; that is, the random fluctuations which we expect in any set of measurements.

When a process is in control it does NOT necessarily mean that it is working well. It means that it is stable and predictable. Improvement typically comes about by working on procedures and methods, for example, by re-engineering all or part of the process. It requires process analysis (e.g. using flow charts to analyse workflows) and usually results in changes to working practices, new equipment or training. These improvement activities are the responsibility of management.

2. The distribution changes with each set of measurements (Figure 1.8)
Many processes are in this situation. When we work under these conditions life is exciting, interesting and unpredictable (others may see it as stressful, frustrating and worrying), and we spend time fire fighting. Many troubleshooters have built their reputations on "solving" these problems.

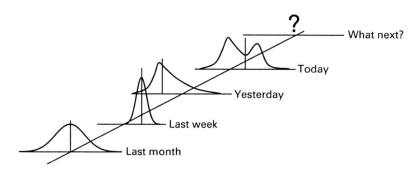

Figure 1.8 A process out of control. It is interesting, exciting, unpredictable and great for fire fighting. Not so good for planning though

In this situation the process is said to be *not in a state of statistical control* (often shortened to *out of control*). Processes which are out of control are said to be subject to BOTH *common cause variation* and *special cause variation*. That is, in addition to the inevitable random fluctuations, other specific factor(s) are affecting the result.

Special causes of variation are not always present on the process. When they occur, they change the location, the spread or the shape of the distribution of the process outputs. Some typical causes of special cause variation are: a machine set up incorrectly, untrained staff used as a stand in, a new supplier, new methods of working, holidays, flu epidemic. Special cause variation is also called "assignable cause variation" because the variation can be assigned to a particular cause. Often, special cause variation can be resolved locally (e.g. re-set the machine, maintain equipment). In other cases where we may not be able to stop a recurrence, we may be able to take action to mitigate the

Common cause variation	Special cause variation
Measurable	Measurable
Always present	Sometimes present
Many causes	Few causes
Part of the process	Not part of the process
Predictable	Not predictable
May be a problem	Usually a problem
Reduced by analysing and improving the process	Removed by identifying and removing the cause OR if the cause cannot be removed, mitigating the effects

Figure 1.9 Common and special causes of variation

effects (e.g. power cut, lightening strike, illness). The key is to identify what changed at the time the out-of-control condition occurred and take appropriate action. Figure 1.9 summarises attributes of common and special causes of variation.

Example: Journey to work

A simple example may help to illustrate the difference between special and common causes of variation. It takes me about 35 minutes to drive to work every day. Some days I may have to stop at more traffic lights than others, some days I wait longer at road junctions and there are many other factors that change my travel time. I know that normally I will arrive at work in 35 minutes plus or minus about 5 minutes; it is predictable, and the process of driving to work is said to be in a state of control. Occasionally things go wrong. There is a snowstorm, or an accident and it takes may be 45, 50 minutes or even longer. These events, "special causes", are generally unpredictable. Some special causes I can prevent: filling the car with petrol ensures that I will not run out of fuel; maintaining the car helps prevent breakdowns. Sometimes I cannot prevent or deal with a special cause. If there is a traffic accident I may be able to take an alternative route, but I cannot stop the snow.

Conversely, one day it may only take me 25 minutes to get to work, perhaps because I left half-an-hour earlier and the traffic was lighter. This is a special cause of variation, and I may choose to change my process to always leave half-an-hour early. Alternatively, it may be due to a public holiday, and it is probably outside my control to make more public holidays!

Other ways in which we could change the process include using a different route, using a different mode of transport, driving faster or getting a job at a different office.

One way of summarising common and special causes of variation is: ALL data contain "noise" (common cause variation). Some data contain "signals" (special cause variation).

To detect a signal you must know what the noise is, and then the remainder is the signal. The job of the control chart is to distinguish and quantify signal and noise.

Summary of the implications of process variation

If a process is NOT in a state of control:

- Prediction of the future will be of minimal practical value as we do not know when and how special cause variation will affect the process.
- We cannot manage it because we do not know what will happen next.
- We do not know the capability of the process; that is, the limits within which it can perform.
- Improving the process will be difficult, as we first have to ignore the causes and effects of special cause variation.
- Trying to improve the process will have minimal effect, as special cause variation will still affect process outcomes.
- As prediction is of little practical use, customers who are aware that the process is not in control will be sceptical of our ability to produce within specification, on time, every time.

For these reasons the FIRST task is to eliminate special cause variation to bring a process in control (Figure 1.10). When this is achieved and we have an in-control process:

- The ability of a process to meet requirements (specifications) can be assessed (where appropriate), for example, by calculating capability indices. Capability indices are meaningless if the process is not in a state of control.
- Work can begin on improving the process, by reducing common cause variation.
- Process performance can be predicted.
- Process performance can be benchmarked.

It is also important to be aware that:

- In general, individuals working *in* the process can resolve special cause variation.
- Reducing common cause variation is primarily the task of management, though they will use the expertise and knowledge of those working in the process.
- Some special cause variation may be beneficial. Occasionally special cause variation produces improved results. In these cases we need to identify the special cause and make it part of the process.
- At most 20% (Deming suggests only 3%) of problems in a process are worker controllable. This implies that blaming or rewarding the workforce for process outcomes is often inappropriate.

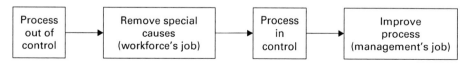

Figure 1.10 Process improvement process

Tampering (over-control) and its effect on performance

What happens if we treat common cause variation as if it were special cause? The answer is that things may get worse! Returning to the example of driving to work, let us suppose that one day the journey takes 38 minutes. Believing this to be special cause variation, I look for the cause and realise that on this day I listened to the radio more intently because there was some interesting news. I decide that leaving the radio switched off will help me to drive faster. The next day my journey time is 34 minutes, and seeing the improvement I conclude that I was correct. I have proof: I kept the radio off and my travel time to work dropped by 4 minutes. The following day the journey takes 33 minutes; the next day it takes 37 minutes. Why did it jump? Again I investigate and I make more futile changes.

In business these erroneous changes, or knee jerk reactions, often include attaching blame to one or more people and taking, or threatening, punitive action. Other changes include issuing edicts such as a new restriction or an extra check in the process.

Common examples of tampering include:
- Every morning, the gunnery officer instructed his gunners to fire one shot at the target. The gun was then adjusted to compensate for any error.
- Last time we carried out a project like this, we over-ran by 15% on time and 5% budget. So this time we will add 15% to our estimated time and 5% to the budget.
- Last year my budget of £100,000 was overspent by £10,000:
  - Do I have ask for £110,000 this year to cover the extra expenditure?
  - Do I tell my staff they only have £90,000, so that if they overspend by £10,000, I will still within budget?
  Both the actions are tampering.
- In month 11 of the budgeting year, I am £15,000 under spent. In order not to get my budget cut next year, I will use the £15,000 to buy computer equipment.
- Fred, the new buyer was not given any formal training – the manager thought he would pick it up as he went along, just like everybody else did. Fortunately, the new buyer was befriended by Freda, a buyer who joined a couple of months previously and who had not made many friends. So Freda helped Fred, and romance flourished.
- Whenever non-productive time increased, the manager had to explain to the board what had caused the increase, and what action he is going to take to reduce it.
- The break-even point had been calculated as 30,000 tons. Whenever production was lower, the manager had been instructed to write a report explaining why.
- It was the organisation's intention to always purchase the best valves. To this end, whenever a new potential valve came to market, three were bought and tried for 6 months. If none of them failed, they were considered suitable. If one or more failed, a note was sent to the supplier explaining that no further valves of that type would be purchased until the problems were fixed and another trial could be carried out.
- ... the five groups will be measured. The best performer will be rewarded, the worst will be retrained.
- ... the records of each department will be compared. There will be an investigation into those departments below average.
- It is imperative that the value does not fall below 3.5. Therefore, daily readings will be taken and as soon as there is a result below 3.5, the item will be re-treated within a day.

Figure 1.11 Examples of tampering

In a typical example a customer complains about a late delivery. In response, the manager speaks to the "appropriate people" and issues a note stating that orders for this particular client must be given priority. Ensuring that the clients' orders by-pass the system causes extra work for the staff. Three weeks later a different customer complains about a late delivery and more changes are made.

Such actions will occasionally be appropriate, but more by luck than judgement; in other words, if you change enough things one of them will be right. More often, however, the changes demoralise the workforce, and they always take effort to implement. One department I worked with made around 90 changes after a particularly poor month. Undoubtedly some of these changes were good (it is difficult to imagine all 90 could be bad!), and things did improve for a while. But which of the actions were effective? Which did more harm than good? Nobody knew. Was the apparent improvement just a result of the workforce not daring to report the truth? Tampering is such an important issue because we do it so often without realising what we are doing. There is a list of typical tampering scenarios in Figure 1.11.

Worked example: Golf practice

To give a simple numerical example, consider a golfer who can drive a ball within 30 yards of the target distance, and on average hits the target distance. The process is centred because the average drive equals target distance. If we were to plot the result, 10 such drives might be positioned as in Figure 1.12(a). The distance from the target in yards is given on the figure. This is a stable in-control process.

Suppose the golfer takes the "corrective" action by comparing the position of the last drive with the target distance and adjusting for the difference. Taking the values plotted in Figure 1.12(a) and "correcting" them we get:

● Drive 1 is 12 yards too long, so the golfer adjusts her drive 12 yards less, the result is that:

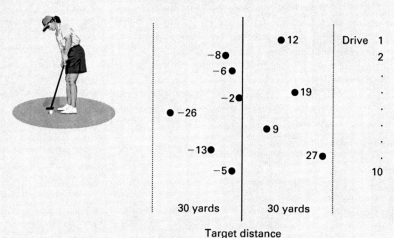

Figure 1.12(a) Golf-driving practice without tampering

(continued)

Worked example – *continued*

- Drive 2 is 20 yards too short (−8 − 12). The golfer adjusts her drive 20 yards longer, the result is that:
- Drive 3 is 14 yards too long (−6 + 20). The golfer adjusts her drive 14 yards shorter, the result is that:
- Drive 4 is 5 yards too long (19 − 14). The golfer adjusts her drive 5 yards less, the result is that:
- Drive 5 is 7 yards too short (−2 − 5). The golfer adjusts her drive 7 yards longer, the result is that:
- Drive 6 is 19 yards too short (−26 + 7). The golfer adjusts her drive 19 yards longer, the result is that:
- Drive 7 is 28 yards too long (9 + 19). The golfer adjusts her drive 28 yards less, the result is that:
- Drive 8 is 41 yards too short (−13 − 28). The golfer adjusts her drive 41 yards longer, the result is that:
- Drive 9 is 68 yards too long (27 + 41). The golfer adjusts her drive 68 yards less, the result is that:
- Drive 10 is 73 yards too short (−5 − 68).

These new values are plotted in Figure 1.12(b).

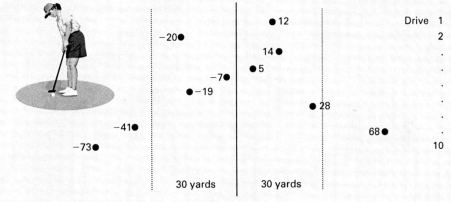

Figure 1.12(b) Golf-driving practice with tampering

Figure 1.12(c) shows the results of both the raw golf drives and those achieved after correcting the drives. The increasing variability is obvious and undesirable. The same happens all too often in organisations, though it is seldom as blatant as in this chart.

Hopefully common sense would soon tell the golfer not to keep adjusting. However, it is important to remember that ANY adjustment to a process average increases variability, as is illustrated by Figure 1.12(b), at least in the short term. This will become clearer as we read through the case studies in Part 3 of this book. Before we adjust a process we should know what the effect of the changes will be, not just adjust and hope! Note that strictly speaking, removing a special cause of variation is NOT a process adjustment, as a special cause is not normally considered as part of the process, but something external acting on it.

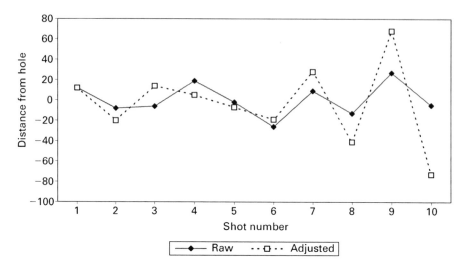

Figure 1.12(c) Run chart of the effect of tampering on golfing scores

There are other standard methods for over-control. The one explained above is to "adjust for the mean". Others are:

- adjust the process only when we obtain a value a certain distance away from the mean (e.g. a value outside specification);
- only adjust by half the distance from the last value and the mean.

However, ANY automatic change is tampering! In particular, automatic adjustments by computers, operators or control systems are tampering. It may be that this is appropriate, but then again, it may be that it is not. The important thing is to understand the process and the effects of the automatic control before deciding whether to use it or not. The seriousness of tampering cannot be underestimated.

If a process is in a state of statistical control, there is no point setting a target or goal: we will get what the process is set up to deliver. If we want to get something else we need to change the process. There are a variety of proven methods for improving processes: plucking numbers and wishes out of the air are not included.

To reiterate, the first task is to identify and remove special causes of variation. Once the process is in a state of control, the second step is to improve the process.

Only once a process is in a state of (statistical) control can we begin predicting future performance and estimating process capability. The question is "how do we know if a process is in a state of control?"; and this is what we address next.

Control charts: the tool for understanding process performance

The control chart was developed specifically to determine whether process outputs exhibit common cause variation only, or whether, and when, special cause variation is occurring.

In the discussion on distributions we plotted consecutive observations on a histogram. If instead we plot them as a run chart the chart would look similar to the top chart in Figure 1.13. Such a chart shows that the data are randomly scattered around the mean and are within limits. We could guess at the limits within which the data are likely to lie, but

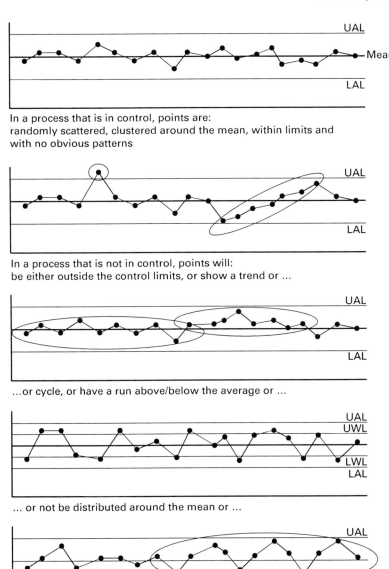

In a process that is in control, points are:
randomly scattered, clustered around the mean, within limits and
with no obvious patterns

In a process that is not in control, points will:
be either outside the control limits, or show a trend or ...

...or cycle, or have a run above/below the average or ...

... or not be distributed around the mean or ...

... or show some other pattern

Figure 1.13 Control chart signals; UAL: upper action limit; LAL: lower action limit; UWL: upper warning limit; LWL: lower warning limit

statistics give us a tool for calculating them. The details are explained in Part 4 of the book, but for now it is enough to know that they are placed at three times the standard deviation on either side of the mean. It is also important to realise that these limits, known as control limits or action limits, are estimated from the data. They have no connection with specification limits, which are the limits within which we would like the process to operate. Specification limits are a statement of what we (or the customers) want; control limits are what the process is telling us we will get. Although the terms action limits and control limits do have a slightly different meaning, in a practical sense they are interchangeable.

When the data are randomly scattered and concentrated around the mean within these limits the process is in a state of statistical control.

If the process is not in a state of statistical control the chart will exhibit data that are not randomly scattered around the mean or within limits, as in the other charts in Figure 1.13.

For example, there may be a point outside the control limit; or a run of 7 or more points (a trend) as shown in the second of the charts.

The third chart shows two other common non-random patterns. The first is known as cycling, where the points alternate between high and low. The second is a run of 7 or more points above the average. Conversely, a run of 7 or more points below the average is also an out-of-control indicator.

Some people include warning limits at two standard deviations on either side of the mean, as in the fourth chart of Figure 1.13. In this chart there are a high proportion of points away from the mean (outside the warning limits), and not many close to it. This is another signal suggesting that a process is not in a state of statistical control. In most examples in this book I omit warning limits, as a matter of personal preference.

There may be other patterns that appear in a control chart. An example is given in the last chart of Figure 1.13. For a complete discussion on interpreting control charts, refer to Part 4.

Dispelling some myths of SPC

There are several myths and misunderstandings about SPC that have led to it being applied almost exclusively to production lines. This is a great shame as this very well established and successful tool has applications far wider than on the production line. I dispel some of those myths here.

Myth: SPC is only for manufacturing

Perhaps the most pervasive myth regarding SPC is that it only applies to production line manufacturing. This myth exists because SPC was developed to solve a manufacturing problem, which it did very well. It was therefore very easy for other manufacturing organisations to adopt the ideas and tools of SPC. Until recently books on SPC mainly related it to manufacturing, as these were the only areas where it was known to be used successfully. In recent years, non-manufacturing examples have appeared, but still the myth persists that SPC applies only to manufacturing.

The truth is very different. SPC is about monitoring process performance and if you can measure performance, the likelihood is that SPC is the tool for analysing it. The number of "incidents" per month may well refer to the number of rejected items on a production line; equally well it can refer to the number of late flight arrivals, mis-diagnoses, traffic accidents, injuries, system downtime events and so on. Similarly, the length of items from a production line could equally be the length of time waiting in a queue, order-processing time, time to complete a project, etc.

SPC and the use of control charts can be applied to health care, travel, education and training, oil and gas, distribution, public services, government, information technology (IT), construction and many other sectors. It can be used in finance, health and safety, planning, projects, design and most other areas of an organisation.

Examples in this book alone cover chemical monitoring, health and safety, education (HSE)/health care, drilling, training, facility usage, continuous manufacturing, projects and finance, amongst others.

Myth: SPC is only for engineers and statisticians

If the waitresses in the *Esquire Nightclub* in Tokyo can apply SPC, why not you?

In 1985 the *Philadelphia Area Council for Excellence* organised a tour to a variety of Japanese organisations in Japan. The tour only included those companies that were involved in the Total Quality Program and so those on the tour became highly sensitised to its Q Mark logo. One evening some of the delegates on the tour decided to visit a nightclub called the *Esquire Club* and were surprised to see the waitresses wearing the Q Mark logo. Intrigued, they spent some time talking to the staff and finding out about their quality improvement activities.

A group of seven waitresses and their supervisor regularly identify sources of waste in the club. They use standard SPC tools to collect and analyse relevant data, then identify, select and implement solutions. A project, explained to the delegates, included copies of the team's charts and data, and showed how the team reduced losses on beer and sake sales by 90%.

The astonishing story of how these women are using SPC to improve processes at a nightclub can be found in Donald J. Wheelers book, *SPC at the Esquire Club*.

In manufacturing organisations it is not unusual to see control charts being used on the shop floor by the operatives, inspectors and other front-line staff. With appropriate training most non-technical people can pick up the skill of keeping charts up to date.

Myth: Control charts are only used for monitoring performance

Another common myth is that control charts are used only for monitoring. In much literature, control charts are presented as a tool for monitoring a process to determine whether it is "in control" or "out of control". Certainly this is a key use of control charts, the aim being to reduce the number of occasions managers waste time, effort and resources inappropriately intervening in a process (called *tampering*). However, control charts can also be used for other applications including:

- *Generating and testing theories* as to causes of problems or process upsets. In the examples in this book, many use control charts in this way. On some occasions, the control chart may be all that is needed to identify that there is a problem, generate theories as to causes of the problem and test the theories.
- *Predicting future performance*: Many a manager would pay clearly to be able to predict what will happen tomorrow, next week and next month. Indeed many build prediction models, employ analysts and theorise as to what may happen next. One key use of control charts is that they do predict future performance. The form of the

prediction is: "if the process does not change, it will deliver an average of X and each individual value will lie between Y and Z".

I worked with one manager who signed a contract which penalised him financially whenever he failed to deliver a service in a specified time. He paid heavily for this. The control chart which was drawn after the contract was signed not only showed him that his process was not capable of regularly meeting the specification, but also estimated the penalties that he would incur.

Similarly, control charts can also indicate where it is *not* possible to predict future performance.

Note that prediction using a control chart is very different to some other common methods of "prediction" such as wishes, plans, targets, hopes or edicts from on high. Prediction is also different to setting goals. To achieve a new goal when we have a stable process requires improvement: the control chart can help us plan what we need to do to achieve the goal and monitor our progress towards it.

● *Assessing performance*: Control charts can be used to assess performance against a target and suggest when to adjust the process to bring performance closer to the target.

Myth: Data must be time dependent to be plotted on a control chart

Whilst it is true that most control charts are time related (i.e. data are plotted in chronological order), it is not unusual to use control charts to analyse data in other ways. In this book there are case studies comparing performances for different months of the year, and, from benchmarking studies, comparing performances between different departments and organisations.

Myth: Control charts are used singly and not as part of a suite of tools

In most published case studies it seems that only one type of control chart is used. This is a great shame as sometimes groups of charts can help us better understand how a process is performing. In some situations we identify that the process is changing with one chart and use other charts to investigate the causes of changes.

When monitoring accidents we may use one chart to monitor the total number of accidents. This will tell us how we are performing from a *safety point of view* overall and reflect the overall safety culture. However, subtle changes in accident causes would be better identified by monitoring causes of accidents separately. We could use another chart to investigate the theory that accident rates are seasonal, and if they are we may need to adapt the way we manage each season.

We can use charts for comparing performances between groups using a control chart, but before we do so we need to ensure that the performance of each group is itself consistent, and we do this with a control chart.

These are just a few of the situations where using groups of charts help us to get a valuable insight into how our processes are performing. This book contains many more examples in the case studies.

Myth: Control charts are not appropriate when a process is continually being re-set

If a process only produces a few readings before it is re-set (e.g. re-setting a machine tool) it is still possible to apply SPC analysis methods to the data. There are some SPC charts aimed specifically at what are called *short runs*. In the extreme situation, the process may be adjusted after each measurement has been taken. Such an example is in project work where no two projects are the same. Other examples include production runs where machinery is regularly adjusted to produce a different item. Difference and Z charts specifically address these issues.

Myth: Control charts are not appropriate when a process is being improved

When improving a process, it is important:

- to know what the previous performance levels were during periods of stability;
- to know if there were any aberrations in the process so that the causes can be identified and (hopefully) eliminated;
- to be able to quantify the size of process change so that we know the effects of our actions;
- to know what levels of performance the new process should achieve. If we do not know this it suggests that our process changes are based on hopes rather than solid process knowledge.

Control charts do this for us quickly and easily.

Are there situations where SPC is not appropriate?

The simple answer is yes, but in the normal running of an organisation the first tool to consider using when managing a process is a control chart.

Like other tools, techniques and methodologies, such as Six Sigma, Total Quality Management (TQM), ISO 9000 and many others, control charts and SPC alone will not transform an organisation. However, in the successful running of organisations it is necessary to understand how well the organisation and its processes are performing, and control charts are the tool for doing so. One advantage of SPC is that it is a good starting point. It can be adopted by individuals, departments or the whole organisation. It is easy to begin. For someone who is numerate, all that is needed is a book on SPC and preferably a computer with a standard spreadsheet package. SPC also leads naturally into other aspects of quality as people endeavour to improve their processes.

The relationship between SPC and Six Sigma

SPC is primarily a methodology for monitoring process performance. It is used for assessing performance levels and hence indicates when it is appropriate to consider embarking on a process improvement project. There are a variety of tools that may be used to aid process improvement including TQM, Business Process Re-engineering and more recently Six Sigma.

The concept of Six Sigma is to improve process performance to where error rates are less than 4 per million. It does this by selecting specific processes whose performance (typically measured using control charts) needs to be improved and then following a disciplined approach to process improvement based on data collection and analysis.

Once the process has been improved, monitoring continues as before by using control charts. So whilst control charting is continually used to monitor a process, Six Sigma projects are used as and when required to improve the process to new levels of performance.

Finally

- The process target is what we want the process to deliver.
- The process average (or mean) is what the process is delivering.
- The process aim is what we have set the process to deliver (that is right; the process does not always deliver to the settings because it is not working properly!).
- The specification limits are the limits within which we want the process to deliver.
- The control limits are the limits within which the process is delivering.

Summary

- All work is a process. A process turns an input into an output for a customer who may be internal or external to the organisation.
- To understand how a process is performing and help manage it we collect data.
- Data vary, and frequently follow what statisticians call a normal distribution.
- Distributions can vary in location (measured by the average), shape (e.g. number of peaks, skewness) and/or variability (usually measured by the range and standard deviation).
- There is important information in variation. It can tells us whether a process is subject only to random variation ("in control"), whether the process has changed or whether there are occasional external influences (special causes) acting on the process.
- If a process is subject to special causes these should be investigated.
- An "in control" process is predictable and ready for improvement if desired.
- Control charts are the tools used to analyse and quantify variation.
- There are many myths about SPC and control charts which have resulted in them being ignored by many organisations. The truth is that SPC has many applications in many different types of organisations.
- Control charts have many applications apart from monitoring. They can be used to predict performance, generate and test theories about causes of performance levels and benchmarking amongst others.
- Six Sigma and SPC work hand in hand. SPC is the tool for continuous monitoring and Six Sigma is the tool used to improve process performance when improvement is required.

PART 2

Exploding Data Analysis Myths

In this part we examine some of the more common and often less than useful methods that many organisations try to use to glean information from data.

The purpose of the most performance monitoring reports and charts is to show how the process is performing: are we getting better, worse or staying the same? Did everything go smoothly? or was there a problem? (i.e. a special cause of variation). This is information. Unfortunately, many reports and charts provide masses of data but only little information.

In Chapter 2, Problems with Monthly Report Tables, Goals and Quartiles, we:

- Review how comparisons between pairs of numbers can lead to erroneous conclusions. As people realise the risks of comparing two numbers, some begin to search for more comparisons as they grapple to understand what is happening in their organisation, and so we next …
- Eavesdrop as a Director tries to get to grips with how well (or otherwise) his/her manager is performing. He/she compares the latest result with previous performances, averages, plans, year-to-date (YTD) and other individual pieces of data. See what can you make of the data!
- Compare performances between different regions using raw data. Simple and common enough, how easy will you find it to spot the laggard and pick the prizewinner?
- Blow the whistle as people try to ensure they meet targets. If you are a manager and achieving targets is important, you may well have resorted to the same tactics as revealed in these cases. Somehow, seeing the ruse printed in black and white brings home the futility of it all. If people are reporting to you and you set the targets, do you know whether you are getting a true picture?
- Investigate the meaning of "top quartile", the cry of the moment in some organisations. But if everyone you are comparing against are actually similar in performance levels, then the chance of being in the top quartile is 25%. So what does being in top quartile mean, if anything? Do we really want to be there? … or is there a better target?

In Chapter 3, Exploring the Mis-information in Moving Average Charts, we demonstrate that a moving average:

- Can show a process improving when it is actually getting worse (and vice versa);
- Lags behind the true process performance when the data are trending, and will still be showing a trend once the trend has stopped;
- Mainly compares the relative performance of two data values.

What we explain in theory here is demonstrated with live data in Part 3.

In Chapter 4, The Problems with Year-to-Date Figures, we invite you to see if you can tell: what is happening to expenditure; when would you panic and when would you party; does YTD analysis tell you anything useful? We demonstrate that YTD charts are very difficult to interpret, for example:

- A YTD chart shows an increase (decrease) if the last data value was higher (lower) than the last YTD figure, and does not depend on whether the data value is above, below or equal to the plan (or any other value).
- A YTD value will usually change even if the last two data values were the same.
- There is no simple method for determining whether a process has changed, or if there is an extreme value when using YTD figures.

In all these cases, we show how a simple control chart explains quickly and easily what is actually happening, and what targets are likely to be appropriate.

2 Problems with monthly report tables, goals and quartiles

Introduction

Most organisations produce daily, weekly, monthly, project, department and a whole host of other reports. In this chapter we investigate some of the typical tools that organisations try to use to make sense of the data they collect.

One common pitfall we often fall into is comparing two numbers and drawing conclusions. Whilst comparing two numbers is simple to do; it usually over-simplifies a more complex situation, and the fact that we try to do so suppresses a basic truth of process life: that process outputs vary randomly.

Some organisations have realised the difficulty of drawing conclusions from using two numbers and so developed tables of numbers so that they can understand how performance has changed. For example, they may compare the latest figure with the previous one, and with the last year's corresponding figure, or may compare this year's year-to-date (YTD) figure with last years'. All these comparisons tend to add to the confusion rather than clarifying the situation.

We also discuss some of the possible negative effects on the organisation of setting targets, and investigate the appropriateness and possible mis-interpretation inherent in the common aim to be top quartile.

Comparing pairs of numbers: a trap for the unwary

For many organisations, having equipment and plant available for use is important, be it information technology (IT) equipment/systems, drilling rigs, processing plants or any of hundreds of other items and systems. One key measure of the availability of equipment is the percentage of time it is available for use, and many organisations monitor availability, reliability, downtime or other similar measures. One such organisation included in its monthly report a chart similar to Figure 2.1(a), a three-dimensional (3D) bar chart of the quality of the service they provided to the organisation, measured in downtime. Knowing that one figure on its own does not mean much, and wanting to show how well they had done, they included the previous month's downtime figure and concluded that things were getting better. Unavailability had decreased from 21% to 19%, a decrease of around 10%. The chart looks dramatic. Take a few minutes to draw your own conclusions.

There are a number of issues with the chart. The purpose of a chart should be to impart information as accurately and impartially as possible, clearly and concisely. How easy is it to read off the values? Do we need 3D charts. If so, why? What does the 3D add? What about the scale? If we want to dramatise differences, then expanding the scale is one way

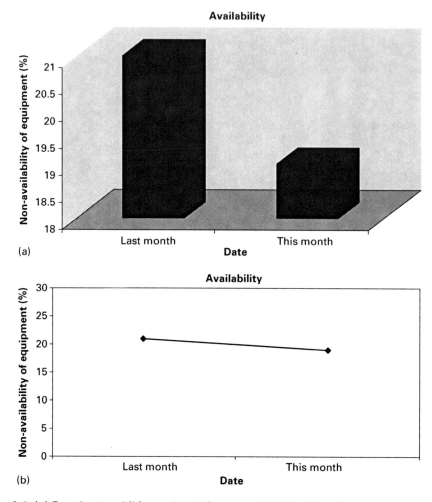

Figure 2.1 (a) Bar chart and (b) run chart of non-availability of equipment

of doing it. Good disciplined chart drawing is important and, though not a key aspect in this book, is a theme that crops up occasionally.

A more "truthful" way of presenting this data is with a simple run chart as shown in Figure 2.1(b). Since we are monitoring through time, it makes sense to join the points to lead the eye through the chart. Scaling is also important as we can appreciate more easily the difference between the two figures.

Once we have a well-produced, unbiased chart, we need to be aware of the message in the chart. What is the message in this chart? Have things improved? Actually, we do not know. This month is lower than last month, but then just because it takes 21 minutes to drive to work, 1 day and 19 minutes the next, we would not conclude that driving to work is getting faster.

So, we need more information. Perhaps we should compare this month with the same month a year ago? Maybe against plan? As we add more and more numbers, we might end up in the following situation:

Death by numbers: the Saga of the monthly report (Table 2.1)

Manager (M): September's performance is 12.
Director (D): Is that good or bad? Do we applaud or boo?
M: Well, if we compare it to the historic average of 9, it's up by 33% which is excellent.
D: Sounds good. What was it last September?
M: 5. So it's up by 240%.
D: Sounds like you've done a great job this month.
M: Well, I had a word with the team and things have got better.
D: So what's the average monthly figure this year?
M: The YTD average is 9.7.
D: And I guess that is up from last months YTD figure?
M: Yes, by just over 3%.
D: And the planned YTD average is 9, isn't it?
M: Yes, so we are 7.4% above that. Also, if we look at the YTD average for last September, we are 13% above that.
D: That's good. Well done. It seems as if that performance bonus may be coming your way.

Perhaps not quite the way the conversation would go, but the gist is common in organisations today.

Table 2.1 reproduces the figures as a table. You might like to review the table and answer the following questions:

● What conclusions would you draw?
● Is the Director right to congratulate the Manager?
● If the performance is not quite so good next month, do you think the Manager will have his reasons to hand when he goes to see the Director?
● What is likely to happen next month?
● Was this month a fluke?
● Are the data seasonal?
● Is there a trend in the data?

Table 2.1 September monthly report

Item	Actual	Historic average	Difference from historic average (%)	Difference from September last year (%)	YTD average	Difference from last months YTD (%)	Planned YTD	Difference from planned YTD (%)	YTD change from last September YTD
Cost	·	·	·	·	·	·	·	·	·
Failures	·	·	·	·	·	·	·	·	·
Performance	12	9	33	240	9.7	3.1	9	7.4	13
	·	·	·	·	·	·	·	·	·
	·	·	·	·	·	·	·	·	·

It is not easy to interpret tables like Table 2.1. Often we do not even try; we just produce the data and let the reader enjoy the challenge of interpreting them. Why do we make life so difficult for ourselves?

The truth could be easily discovered and reported by using a control chart. Chart 2.1 shows the data for September report. The dialogue above refers to last year's YTD, and so we have assumed the data goes back as far as the beginning of last year, and so there are 21 monthly values.

Chart 2.1 is in a state of control (i.e. the data are randomly scattered around the mean with fewer points; the further away we move from the mean, there is no trend or seasonality). The monthly average is just over 9. This September's value of 12 is high, but not especially high; in April there was a value of 13. The variability in the data is such that we could receive a value of anywhere between just below 2 and just over 16 in any one month without being concerned that there was anything abnormal about the figure. If there is no change to the process, then next month's result will be between 2 and 16 (for simplicity all data have been reported as integers), and it is very likely that it will be less than 12, and a 50% chance that it will be less than the average of 9.

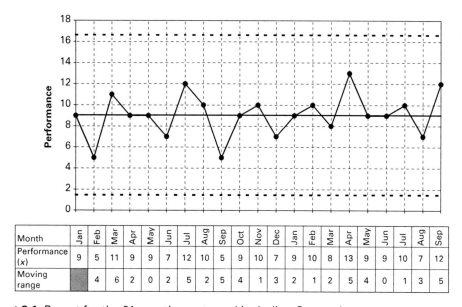

Month	Jan	Feb	Mar	Apr	May	Jun	Jul	Aug	Sep	Oct	Nov	Dec	Jan	Feb	Mar	Apr	May	Jun	Jul	Aug	Sep
Performance (x)	9	5	11	9	9	7	12	10	5	9	10	7	9	10	8	13	9	9	10	7	12
Moving range		4	6	2	0	2	5	2	5	4	1	3	2	1	2	5	4	0	1	3	5

Chart 2.1 Report for the 21 months up to and including September

Now, let us answer the questions:

- What conclusions would you draw?
  Process is in a state of control. Nothing unusual has happened.
- Is the Director right to congratulate the Manager?
  No. Process is running smoothly (or Yes. Because the process is running smoothly, depending on your management style)

- If the performance is not quite so good next month would the Manager have his reasons to hand when he goes to see the Director?
  He probably would, because he previously claimed that he "had a word with the team" and things got better. However, the control chart shows that any value between 2 and 16 is expected with no explanation needed, so why waste time inventing one?
- What is likely to happen next month?
  If the process does not change, we will receive a value between 2 and 16.
- Was this month a fluke?
  No.
- Are the data seasonal?
  Unlikely; no obvious pattern, though we could investigate.
- Is there a trend in the data?
  No.

Using charts like these make reporting very simple to do and to understand. Keeping charts like these up to date can be just a matter of plotting a point on a wall chart or entering a value into a spreadsheet. Any process aberrations, comments, investigations or other information pertinent to the process can be logged on the chart for quick and easy reference. In this way the chart contains a potted history of the process. Collating the monthly report is a matter of copying (or printing if computerised) another copy for inclusion in the monthly report. Very little commentary is then required as the chart has all the key information on it.

Who wins the prize? How not to compare regional performance statistics

Of course, there are many ways that production, sales, costs or anything else are compared between groups. Table 2.2 is one typical, if simplified, scenario for comparing production rates between regions. Take a few minutes to review the table. What conclusions would your draw? For example, Which region is the best performer? Which is the worst? What action should be taken in each region?

In some organisations, "best" performers are rewarded, perhaps by employee or group of the month or similar awards, and "worst" performers are penalised. This, of course, only encourages data to be "massaged".

Now look at Chart 2.2(a–d). They are run charts of the historic production of each region. Below each chart is a table which includes the period number, the target

Table 2.2 Period production report

Region	Production	Change (%)	Target	Difference from target (%)
North	120	15	110	9
South	318	23	330	−4
East	256	−8	220	16
West	86	−7	93	−8

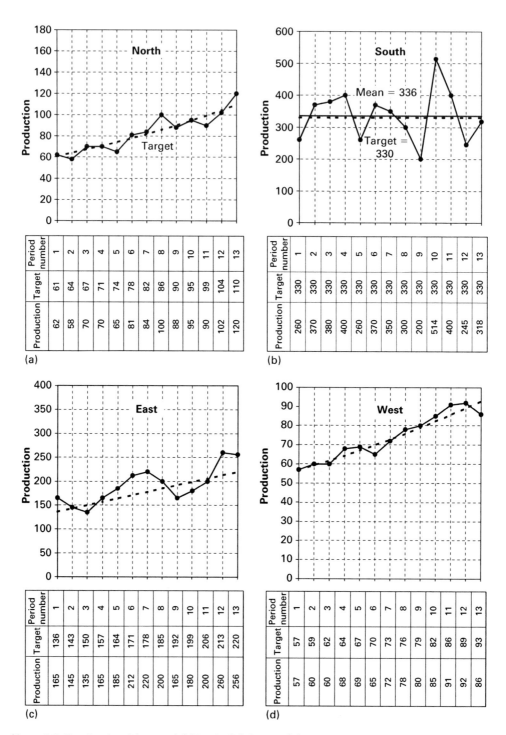

Chart 2.2 Production history: (a) North, (b) South, (c) East and (d) West

production and the actual production figures. The dotted lines on the charts represent the targets and the solid line with points the actual results. Apart from South, which is a stable region, all other regions are showing steady growth and the production follows the targets quite well. For no facility there is any other conclusion except that production is following the target and is subject only to random variations from it. No one region is performing particularly well or badly against target. Region South shows that the production average at 336 is slightly above the planned production rate of 330. However, with such large background (common cause) variation in the difference is not significant.

Falsifying the data (and how to spot it): one result of setting targets

At one time I worked for a large company as the Manager of the internal consultancy group. All groups had to report, on a monthly basis, the percentage of time spent on projects for clients. At one time an edict came down to us that we had to report at least 70% of out time to these projects. Like everyone else, I didn't want management questioning our figures; so I did what everyone else did: the months where we had spent, say 80% of our time on projects I would only report perhaps 75% or 76% and "keep" the other hours for when we had a lean month. When the project work only accounted for, say 67%, I would put back in the hours I had held back.

If a process is in a state of statistical control, it will follow one of a number of distributions, most often the normal curve (see Figure 2.1(a)). Sometimes, especially with counts data (counting occurrence of events), or data where there is a limiting value (e.g. a count can never be negative), the distribution may be skewed as in Figure 2.2(b).

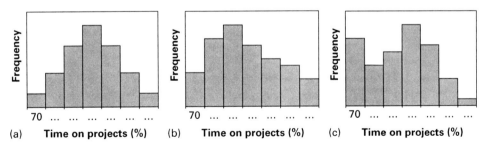

Figure 2.2 (a) One of the most common shapes for a histogram of data in a state of control; (b) another common shapes for a histogram of data in a state of control and (c) histogram of percentage of time spent on projects

What we expect to see on the control chart is a set of data randomly distributed around the average, with fewer points further away from the average. When data are being falsified it is usually to hide extreme values, or to keep values within a limit, as with the time writing situation. Whilst we can learn to spot these situations with a control chart, they are more readily identified using a histogram. Had my Manager plotted a histogram of my time recording he would have seen a shape similar to that in Figure 2.2(c).

The giveaway is the left-hand bar. Histograms like these are often seen if there is a limit on the process and the person submitting the data does not want to report data below the limit and so reports them as just above the limit.

Case study

One month the Manager of a particular business unit was called before the Directors to explain why non-productive time (NPT) was so high (27% compared with an average of around 18%, but with huge variability) and what he was going to do about it. He duly gave his reasons and explained what was he going to do to improve the situation the following month. Unfortunately for him, the following month the NPT was even higher, 35%. The Manager told the analyst who calculated the figure to check it, but it was correct. The Manager did what most of us would do to avoid a particularly unpleasant interview with his managing team: he told the analyst to reduce the figure to 20%. The Directors were happy, the Manager was relieved, but the analyst was troubled. When we produced a control chart of the data, we discovered that the variability in the data was such that in any 1 month we could expect downtime of up to 40%. Had the Directors understood, they would not have called the Manager to account every month when they considered the NPT to be too high, and the ensuing tension would not have occurred. Rather, they would have focused on the real issues – analysing the process to understand the underlying causes of high NPT and then improving the process.

In situations like this it is far more difficult to identify that the data is being falsified. There is no natural cut-off value, such as a specification limit, and the occasional high value would probably be tolerated. Extremely high values would probably have been reduced a little, and only reduced by a lot where two high values occurred consecutively, as in the (true) story above. The difficulty of spotting the falsified data in these situations is demonstrated in Chart 2.3. Chart 2.3(a) is the control chart of the actual data and Chart 2.3(b) is the control chart of the mis-reported data. The only difference is the penultimate point which was mis-reported as 20%, when the actual figure was 35%. The last month, the month after the data had been falsified, NPT was back below the average of 18%.

These two examples illustrate the way in which setting targets, paying for performance, or similar attempts at motivation can be counter-productive. If data is falsified it cannot be an accurate reflection of what is happening in the organisation. If management do not know what is happening, how can they manage effectively?

In both cases, it is important to realise that the problem lies in the fear culture fostered by management, and it is they who need to change their management styles to foster openness and honesty.

(continued)

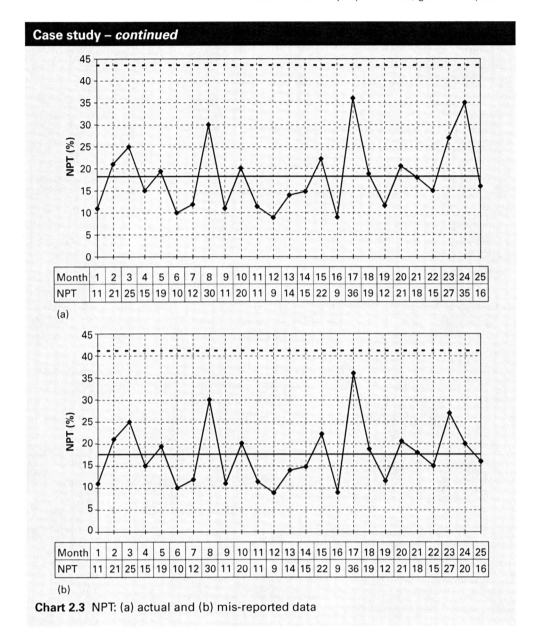

(a)

(b)

Chart 2.3 NPT: (a) actual and (b) mis-reported data

There are three ways of meeting a target:

- *Falsify the data*: Simple, quick but can be easy for management to identify.
- *Manipulate the process*: For example, change definitions, exclude some data giving reasons. Common methods include bringing forward/delaying orders to meet targets/ budgets. More difficult to achieve and often harder to spot.
- *Improve the process*: Much more difficult to achieve, but clearly the desired response to setting targets.

The culture in the organisation needs to be such that only the last of these methods is used.

Querying the top quartile: Does it mean anything?

One of the problems with goals is deciding how to set them. Should they be based on last year's performance? A technical limit perhaps? $X\%$ less/more than last time? Another issue is that goals set in one area may be achieved at the expense of loss in performance elsewhere. In order to know that we are incurring that loss we need to measure it. For example, a sales department may meet its sales goals without regard to whether the manufacturing department is able to meet delivery dates (with the result that the percentage of ontime deliveries may drop), or without regard for the profitability of each sale. Add to this, the difficulties of interpreting any figures using tables such as those above, and it is easy to see how difficult it is for the management to steer an organisation.

Some organisations have attempted to overcome these and other performance-related problems by setting a clear target: to be top quartile in all that they do. The concept is simple: benchmark all (key) activities and require that manager's progress to being top quartile. Such an approach seems to have many advantages. For example:

- It matters little how well we perform against internal targets and goals. What is important is that we are competitive in the marketplace, and this is precisely what benchmarking can achieve: comparison with others carrying out similar activities.
- By being involved in a multi-company benchmarking study, a good set of performance metrics will be developed which will cover all key areas of the business being benchmarked.
- Many benchmarking studies are facilitated by independent consultants resulting in an independent report, covering all the key metrics of the organisation.
- If the benchmarking study is carried out regularly, an independent view of changes in performance over time will be seen.
- In addition, there may be *Best Practice Forums* or other opportunities to identify how and where to target improvement activities.

Before discussing the goal to be top quartile, we need to explain what quartiles are. The concept is quite simple:

- the best 25% of performers are deemed to be top quartile;
- the second 25% of performers are in the second quartile;
- the third 25% of performers are in the third quartile;
- the remaining 25% of performers are in the fourth quartile.

Worked Example 1

Consider the following set of failure rates submitted to a benchmarking study by 20 different companies:

7 8 12 13 16 17 18 18 22 24 25 25 26 27 30 30 31 33 34 35

(continued)

Worked Example 1 – *continued*

As there are 20 numbers, there will be five in each quartile (Figure 2.3(a)). Therefore, the companies with failure rates of:

- 7 8 12 13 16 will be in the top quartile;
- 17 18 18 22 24 will be in the second quartile;
- 25 25 26 27 30 will be in the third quartile;
- 30 31 33 34 35 will be in the fourth quartile.

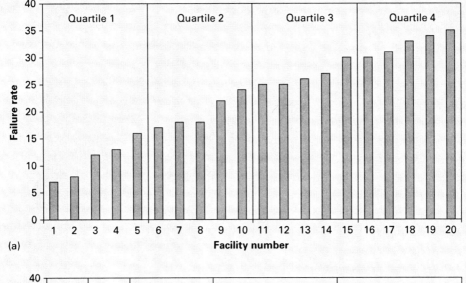

(a)

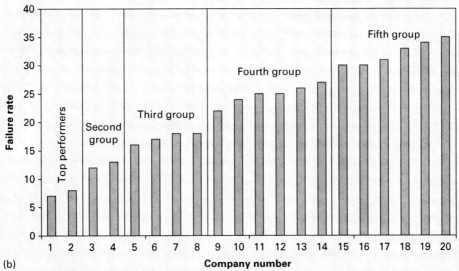

(b)

Figure 2.3 Bar chart of the performances of 20 facilities (a) showing quartiles and (b) showing split based on difference performance levels

(continued)

Worked Example 1 – *continued*

There are a number of practical difficulties with quartile calculations, for example:

- Notice that some of the failure rates are the same (e.g. there are two values of 18 and 25). As Figure 2.3(a) shows, one of the companies with failure rate 30 is in quartile 3 and the other in quartile 4. It seems inappropriate to place companies with very similar performances in different quartiles.
- When the number of facilities points is divisible by 4, as in the illustration, we can always put the same number of facilities into each quartile. Where the number of data points is not divisible by 4, it is not possible.

There are also some dubious conclusions that are likely to be drawn by simply referring to quartile position, and we may also miss some opportunities for learning. For example:

- The top quartile figures are 7, 8, 12, 13, 16. If all that is required is that we be in the top quartile, achieving a 16 is as good as a 7 as both are in the top quartile. The company reporting 16 can conclude that as they are top quartile, there is little opportunity for improvement at this time. Not only it is misleading to imply that these two performances are in some sense equally good, but we also miss the opportunity to learn. The top two companies with values 7 and 8 are likely to be significantly better performers (or at least in some way different) than the next companies at 12 and 13, and these in turn may well be significantly better performers than the other quartile 1 company, reporting 16, etc. If the main purpose for comparison is learning and improvement, then a better approach would be to group companies according to performance levels; for example, as shown in Figure 2.3(b), and then aim to look for reasons for lower/higher failure rates. In Figure 2.3(b), the companies with the lowest two failure rates, 7 and 8 are grouped together. The next band also consists of two companies, 12 and 13. Then follows a group of four companies with values between 16 and 18, and so on.
- There is another question we may need to ask. This example is about failure rates. However, if related to maintenance effort, it could be that the companies reporting 7 and 8, the lowest amount of maintenance effort, are not doing enough maintenance and risking the integrity of the equipment: it is often dangerous to look at one figure by itself.
- It is not unusual to see one performance much better than all the others. When this happens, it is more likely that this is due to mis-reporting rather than genuine high level of performance and should always be investigated.

For these reasons, it is more appropriate to attempt to group performances together (as illustrated in Figure 2.3(b)), rather than arbitrarily grouping facilities that may really have little or no performance similarity. Perhaps, better still, is to identify a good performance level, somewhere around 7 or 8 in this example, and consider setting that as the target.

There is another concern with the goal of being top quartile. It is important because most people do not recognise the implication. With ANY set of data, 25% of the data will lie in the top quartile, 25% in the second quartile, etc., BY DEFINITION. This DOES NOT mean that the top-quartile performances are significantly different than the second- or other-quartile performances. The following simple example explains.

Worked Example 2

Suppose we initiate a dice throwing competition in which we want to throw high values. We have one six-sided dice and ask each of 40 people to throw the dice five times. The highest possible is $5 \times 6 = 30$ and the lowest $5 \times 1 = 5$. The top quartile will be those 10 people with the highest total. We would not conclude that they are "better" at throwing dice than the remaining 30 people. If the experiment is repeated with the same people, we would expect two or three of the 10 people in the top quartile in the first experiment to be in the top quartile in the second experiment. If the experiment were carried out a third time, it would not be surprising if one person was still in the top quartile. It is only if the same person is in the top quartile for perhaps a fifth and sixth time that we would suspect that they are either cheating or have a method of throwing a dice that is likely to result in a high number.

We can generalise this argument and state that if there is no underlying difference in performance and that differences in results are purely random, then the probability of being in the top (or any other) quartile consecutively is given below:

Number of times in top quartile	1	2	3	4	5
Probability		1 in 4	1 in 16	1 in 64	1 in 256	1 in 1024

This is a very simple example, if only, because there is no skill involved. However, it does illustrate that being in a certain quartile of itself may have little meaning and we need to look beyond that.

Investment Example

In recent years the performance of unit, investment and similar funds have been reported in financial magazines. Performances are frequently presented as tables giving performances of various funds, grouped by fund type. One such set of tables includes the ranked performance over the last 3 months. For example, fund A may be ranked 15 out of 90 funds in the same group, and would therefore fall into quartile 1. Twelve consecutive quartile results for one such randomly selected fund reporting every 3 months are: 1 4 1 1 1 3 3 4 1 4 1 1. The seeming randomness of these results adds weight to the theoretical argument above.

Summary

- Comparing individual performance values, for example, current value vs. previous value, gives the illusion of being meaningful. Often the conclusions are erroneous.
- In an effort to overcome the shortcomings of comparing only two performance figures, organisations have turned to many different types of comparisons and often produce tables of numbers to help draw conclusions. The result is usually confusing.
- The tool for monitoring performance is the control chart, which is much easier to understand and interpret than tables of numbers.
- Using comparison tables to compare performance across regions (or other organisations groupings) can lead to erroneous conclusions in many situations as it is necessary to understand performance over time.
- There are three ways to meet a target: falsify the data, manipulate the system or improve the process. Organisations need to ensure that only the last of these methods is used.
- Whilst it is easy to set goals to be "top quartile", measuring against and understanding their severe limitations should be understood. It is more important to identify the gaps between the best performer(s) and the other performers, and then investigate the reasons for these gaps. Only once we understand the reason for the performance gaps and we can evaluate whether a particular performance was "good" or not. Understanding the reason for the gap is also fundamental in beginning the improvement process.

3

Exploring the mis-information in moving average charts
How they fail to respond to process changes, out-of-control points, trends and seasonality

Introduction

The moving average is best explained by an example. When monitoring monthly sales figures, for example, one common practice is to monitor the average of the last 12 monthly figures. For example, in March the reported sales would be the average of the sales from the previous April up to and including the current month, March. In the following month, April, we drop off the previous April's sales and add in the current April sales and recalculate the moving average.

The reason often given for using the moving average is that it smoothes out the variation and removes the effects of seasonality. Not only is this correct, it is also part of the problem. There is information in variation that tells us how the process is performing.

In the case studies in Part 3 we will see clearly mis-information inherent in moving average charts. This chapter explains *why* moving averages suffer from these problems and in particular shows that:

- The moving average may imply an increase (decrease) in process average when the actual average has decreased (increased).
- When trends are present in the data, the moving average lags behind the true process performance, and will still be showing a trend even after it has stopped.

We also investigate the response of moving averages to process changes and extreme values. Finally, we contrast the use of moving averages and control charts in typical monitoring objectives.

Example

Data and background

Consider a process in which monthly observations are recorded. In this example, the observed values have historically always been 10, and so the monthly moving average, calculated as the average of the last 12 monthly values, has also been 10. The data in Table 3.1 show the last 12 monthly values along with the moving average. In this chapter we consider three scenarios that begin at month 13:

1. *Process change*: A sudden drop in process average to 4, which lasts for 8 months before another sudden change in the process causes the process average to jump to 8.
2. *Single out-of-control point*: A single very low value of 5, followed by a series of 10s.
3. *Trend*: In which each value successively decreases by 0.5.

Before reading on, the reader might like to make a note of how they would want a chart to respond to these three different situations.

(continued)

Example – *continued*

Analysis

Responding to a process change

Table 3.1 shows the observed values and the resulting moving averages. Chart 3.1 is a plot of the moving averages and raw data. A simple analysis of Chart 3.1 and data shows that:

- Although the process made a simple step change in month 13 from 10 to 4, with no other variation in the data, the moving average shows only a gradual decline from a value of 10.
- Before the moving average can flatten out at the new process average of 4, another change occurs at month 21. The moving average never does accurately reflect the true process value of 4 during this period.
- After the process average *increases* from 4 to 8 in month 21, the moving average continues to show a *decrease*.

In order to understand why the moving average does not reflect actual process performance, we need to examine the calculations behind the moving average.

The moving average is calculated as the average of the last 12 values. When the process average changes from 10 to 4, the new moving average is calculated as:

$$\frac{10 + 10 + 10 + 10 + 10 + 10 + 10 + 10 + 10 + 10 + 10 + 4}{12} = 9.5.$$

The important factor is that a value of 10 has been dropped off and a value of 4 added.

Table 3.1 Raw data and moving averages

Month	Observed value	Moving average	Month	Observed value	Moving average
1	10	10.0	17	4	7.5
2	10	10.0	18	4	7.0
3	10	10.0	19	4	6.5
4	10	10.0	20	4	6.0
5	10	10.0	21	8	5.8
6	10	10.0	22	8	5.7
7	10	10.0	23	8	5.5
8	10	10.0	24	8	5.3
9	10	10.0	25	8	5.7
10	10	10.0	26	8	6.0
11	10	10.0	27	8	6.3
12	10	10.0	28	8	6.7
13	4	9.5	29	8	7.0
14	4	9.0	30	8	7.3
15	4	8.5	31	8	7.7
16	4	8.0	32	8	8.0

(continued)

Example – *continued*

The following month a similar change occurs, and the average is:

$$\frac{10 + 10 + 10 + 10 + 10 + 10 + 10 + 10 + 10 + 10 + 4 + 4}{12} = 9.0.$$

In the following months the average continues to decline by 0.5 as shown in the Table 3.1. It is only once all the values of 10 have been replaced by values of 4 that the true process average will finally be reached. However, before that happens, the process changes again. In month 20, when the last 4 is recoded, the moving average is:

$$\frac{10 + 10 + 10 + 10 + 4 + 4 + 4 + 4 + 4 + 4 + 4 + 4}{12} = 6.0.$$

The next month a 10 is dropped off and an 8 added, so that the average will continue to fall:

$$\frac{10 + 10 + 10 + 4 + 4 + 4 + 4 + 4 + 4 + 4 + 4 + 8}{12} = 5.8.$$

Chart 3.1 clearly shows that the step changes in the process are represented in the moving average as continually changing values.

These findings may be generalised and summarised as:

- For a moving average span of n ($n = 12$ in our example) values, it is not until n values, after a process change, that the moving average correctly reflects the new process average.
- If spans are longer than the number of observations between process changes, the moving average will NEVER accurately reflect the true process average.
- When the process average changes in one direction, the moving average may show a process change in the opposite direction. This will always happen when the second change occurs within n observations of the first and the second change moves the average back towards the original average (in this case, the second change was to a value of 8, which is between the previous process average of 4 and the original value of 10). Thus, we may believe a process is improving when it is actually getting worse, and *vice versa*.

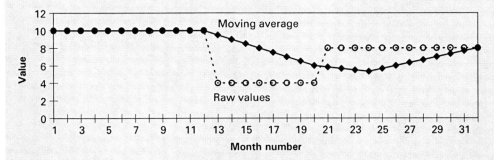

Chart 3.1 Moving average for a process change. Plot of raw data and moving averages showing that the moving average responds poorly to changes in data

(continued)

Responding to a single out-of-control point

Occasionally something happens to our process that results in a single value which is not within the range of values that we expect. For example, when driving to work the journey may take us between 25 and 35 minutes day after day. If one day it takes 50 minutes and the next it is back within the range 25–35 minutes, we conclude that something special happened on the day the journey took 50 minutes. That something special could have been an accident, or that we drove a different route or we had a flat tyre. These incidents are special causes of variation and are identified by what we call a single out-of-control point (Part 1 gives a more complete introduction to variation).

The second scenario that we investigate is how moving averages react to a single out-of-control point in an otherwise stable process. Again suppose the background observed values are 10 followed in month 13 by a single unusually low value of 5, after which the observed values return to 10, as shown in Table 3.2.

At month 13 where the 5 occurs, the moving average drops, but only from 10 to 9.6. This is calculated as:

$$\frac{10 + 10 + 10 + 10 + 10 + 10 + 10 + 10 + 10 + 10 + 10 + 5}{12} = 9.6.$$

Since the following values are all 10, the values being added and dropped off each month are the same, keeping the moving average the same. This continues until, 12 months later, the 5 is dropped off and a 10 is added at which time the moving average increases to 10.

Table 3.2 Single out-of-control point

Month	Observed value	Moving average	Month	Observed value	Moving average
1	10	10	17	10	9.6
2	10	10	18	10	9.6
3	10	10	19	10	9.6
4	10	10	20	10	9.6
5	10	10	21	10	9.6
6	10	10	22	10	9.6
7	10	10	23	10	9.6
8	10	10	24	10	9.6
9	10	10	25	10	10
10	10	10	26	10	10
11	10	10	27	10	10
12	10	10	28	10	10
13	5	9.6	29	10	10
14	10	9.6	30	10	10
15	10	9.6	31	10	10
16	10	9.6	32	10	10

(continued)

Example – *continued*

Chart 3.2(a) shows both the moving averages and the raw data values. A brief review of the chart and table shows that in general terms, when a high (or low) value is observed:

- For a moving average span of n values, the moving average will increase (or decrease) by an amount equal to (observed value − value dropped off)/n.
- Although only one outlying value has occurred, the moving average continues to be affected until that value is dropped off n observations later.

Note that in this case the value dropped off is the process average, 10, but in the general case, the change in moving average depends *not* on the process average, but on the difference between the new value and the value dropped off. This has an interesting and disturbing implication:

- If the value added is low but not as low as the value dropped off the moving average will *increase*. Using our simple data set, in month 25 when the value of 5 is dropped off, if it is replaced by another low value, 8, the moving average increases from 9.6 to 9.8 (Chart 3.2(b)).

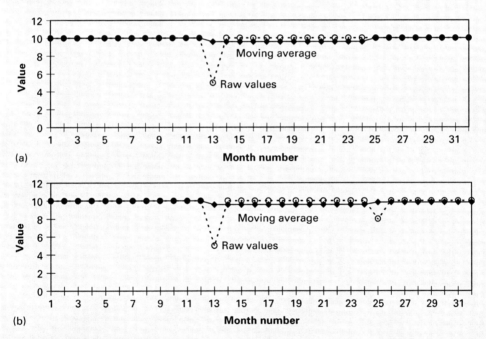

(a)

(b)

Chart 3.2 (a) Moving average for an out-of-control value. Plot of raw data and moving averages showing that after a process aberration the moving average takes a long time to reflect true process performance. (b) Moving average for two out-of-control values. Plot of raw data and moving averages showing that although a low observation has been recorded, the moving average may increase

(continued)

Example – *continued*

Responding to a trend

The third scenario is that of a steady trend. From month 13 onwards, each successive value is 0.5 less than the previous, as shown in Table 3.3.

Chart 3.3 shows both the moving averages and the raw data values.

Beginning at month 13, the slope of the graph gradually increases for the next 12 months, until all the values of 10 have dropped off. Only after this point does the slope of the moving average correspond with the slope of the raw data. Again we see that the moving average is slow to respond to changes in the process.

Table 3.3 Trend

Month	Observed value	Moving average	Month	Observed value	Moving average
1	10	10	17	7.5	9.4
2	10	10	18	7.0	9.1
3	10	10	19	6.5	8.8
4	10	10	20	6.0	8.5
5	10	10	21	5.5	8.1
6	10	10	22	5.0	7.7
7	10	10	23	4.5	7.3
8	10	10	24	4.0	6.8
9	10	10	25	3.5	6.3
10	10	10	26	3.0	5.8
11	10	10	27	2.5	5.3
12	10	10	28	2.0	4.8
13	9.5	9.96	29	1.5	4.3
14	9.0	9.9	30	1.0	3.8
15	8.5	9.8	31	0.5	3.3
16	8.0	9.6	32	0.0	2.8

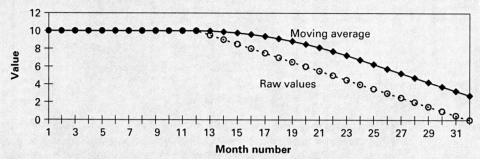

Chart 3.3 Moving average for a trend. Plot of raw data and moving averages showing that the moving average lags behind the true trend, and that its slope varies

(continued)

Example – *continued*

The findings may be summarised as follows:

- For a moving average span of n values, the slope of the moving average will take n values before the trend is fully recognised.
- From the onset of the trend until a span of n values AFTER a steady state as been reached the moving average will be a meaningless figure.

In general we may add the disturbing observations that:

- For a moving average span of n values, any upset in the process (e.g. change in process average, out-of-control condition or onset of a trend) the moving average will not reflect the current status of the process for n observations.

Coping with seasonality and trends

Seasonality

The argument for using monthly moving averages as explained in these case studies is based on the theory that it is not appropriate to compare consecutive months as this does not take account of seasonality. If seasonality is suspected, control charts can be used to test the theory that seasonality has a significant influence, and to help determine which months fall into which seasons. The method for doing this is demonstrated in the case studies in Part 3. Having identified that seasonality does exist there are a number of methods for dealing with it. The simplest and the most obvious is to use a control chart to monitor the variation between the observed value and the seasonally adjusted expected values. Using seasonal adjustments is a well-established technique and used, for example, by governments to analyse certain fiscal measures.

Trends

Another argument put forward for using moving averages is based on the idea that this is the best method for dealing with data that shows a trend. Control charts are well able to take account of trends, as demonstrated in the case studies in Part 3. One method of coping with trends is to identify the trend and monitor the variation between the actual value and that expected from the trend. As illustrated above, the moving average does not cope well with trends.

Seasonality and trends

A more complicated scenario is that of seasonality imposed upon a trend. Since moving averages do not cope well with either seasonality or trends, it is unlikely that they can cope with both at once. However, this complicated scenario can be catered for by combining the above methods in control charts. The effect of the trend and seasonality would be used to generate an expected value which is subtracted from the actual value:

$$\text{variance} = \left(\begin{array}{l}\text{observed} \\ \text{value}\end{array}\right) - \left(\begin{array}{l}\text{predicted value based on} \\ \text{trend and seasonality}\end{array}\right).$$

The result is plotted on a difference chart.

It is worth noting that this method of calculating and monitoring the variance from an estimated (or planned or predicted) value can be applied to many situations.

What moving averages actually monitor

Clearly the moving average monitors the average of the last n values. It includes the effects of any process aberrations such as process changes and out-of-control values. A cursory review of the calculations above shows that the change in moving average from one value to the next is dependent only on the difference between the value added and the value dropped off. Analysis is further complicated by the fact that plotted values are auto-correlated; that is, each value is related to the previous one.

The failing with moving averages is that they aim to suppress variation in data. In contrast, processes speak to managers through variation and control charts are the tool for interpreting variation.

The two main objectives when monitoring process outputs are usually to identify:

● *process changes* (usually identified by a change in average);
● *special causes that need investigation* (usually identified by an unusually high or low value).

Table 3.4 contrasts the way in which moving averages and control charts fulfil these and several other typical objectives of monitoring a process.

Despite the problems of using moving averages as a monitoring tool, there are occasional applications where moving averages may be appropriate. In these situations they are drawn as control charts of moving averages, often with the raw data superimposed. As moving averages smooth the data, the interpretation guidelines for moving average control charts are not the same as for other charts. For further information on these charts see standard texts such as Oakland (2003) or Wheeler (2003).

Table 3.4 Contrasting moving averages with control charts as tools for monitoring processes

Monitoring objective	Moving average	Control chart
To identify process changes	Not an objective, and difficult to do	A key objective of the control chart, and the only tool in common use aimed at doing so. Identifying process changes is well researched and understood
To identify current process average	Often not reported, and generally not trustworthy. The last point plotted gives the average for the last span of n values. However, if this span includes process changes or outlying points, the moving average will not be trustworthy. Even if there are no outlying values or process changes, the average is only based on the last n values. The smaller n is the worse the estimate, the larger n is the more likely it is to include process changes and out-of-control points	Estimated from all data since the last process change and outlying (out-of-control) values correctly excluded from the average
To identify outliers (out-of-control values) that warrant investigation	No guidelines	Well-known criteria for signalling out-of-control conditions
To remove out-of-control values from calculations	Not usually done	Standard procedures exclude out-of-control points from calculations, but include them in charts for reference
To identify trends	Identifies the existence of trends, but quantifying the trend is difficult	Trends can be identified and if appropriate, incorporated into the charting process
Memory of process information	Process comments not usually included on chart (though they could be)	Standard practice is to include significant events and other information on charts to help with analysis and investigations

Summary

Note: n is the number of observations being used to calculate the moving average. For example, the frequently used monthly moving averages has a value of $n = 12$.

- Moving averages are not very good at identifying process changes or outlying (out-of-control) values.
- The moving average may imply an increase (decrease) in process average when the actual average has decreased (increased).
- For a moving average spanning n values, after a change in process average it will take the moving average n observations to provide an unbiased estimate of the new average.
- If there are process changes within n values of each other, the moving average will never accurately reflect the true process average.
- The moving average only uses the last n values to estimate the true process average. If n is small, the estimate will not be stable, if n is large it is more likely to include the effects of process changes and outlying values.
- An outlying value will increase or decrease the following n moving averages, leading to an incorrect estimate of process average.
- A trend in the data will take n values after the trend has stabilised to be correctly tracked by a moving average.
- Control charts are a far better tool to use for monitoring process performance than moving averages.
- Despite the serious concerns about moving averages, there is a place for *moving average charts*, albeit a very minor role, in the armoury of the manager. A good introduction to *moving average and exponentially weighted moving average control charts* can be found in standard statistical process control (SPC) texts such as Oakland (2003).

4

The problems with year-to-date figures

Introduction

Many organisations try to use year-to-date (YTD) figures to understand how well they are performing. In this chapter we challenge you to make sense out of some YTD figures, explore the difficulties in interpreting them and investigate alternative methods that are easier to interpret.

The YTD value is the cumulated total from the beginning of the year to the current date. They are intended to be a simple method of representing the performance since the beginning of the year. They are usually applied to monthly figures starting in January of each year, and for this reason we use the example of monthly data here. However, they can be generalised and exactly the same concept of totalling figures from a start point to date can be applied in many situations, usually with the same difficulties as illustrated here.

The hope in using YTD figures is that we can somehow look at one value (the total from the beginning of the year) and that it will tell us something useful. Unfortunately, it does not, but in a belief that it should, we compare the YTD actual figure with a planned figure, or with last year's figures.

In this chapter we investigate the shortcomings of monitoring actual YTD against planned YTD, this year's YTD against last year's YTD, and the average YTD figures. We also demonstrate that control charts are a far superior tool for monitoring performance.

Example

Analysing YTD against plan

As an example, Table 4.1 gives the raw data and the YTD figure for a complete year of actual expenditure.

The data is usually plotted as a chart, often with a plot of the previous year's data or the plan YTD.

Table 4.1 YTD figures for raw data

Month	Jan	Feb	Mar	Apr	May	Jun	Jul	Aug	Sep	Oct	Nov	Dec
Expenditure	7	12	9	8	8	10	16	5	7	11	8	7
YTD	7	19	28	36	44	54	70	75	82	93	101	108

(continued)

48

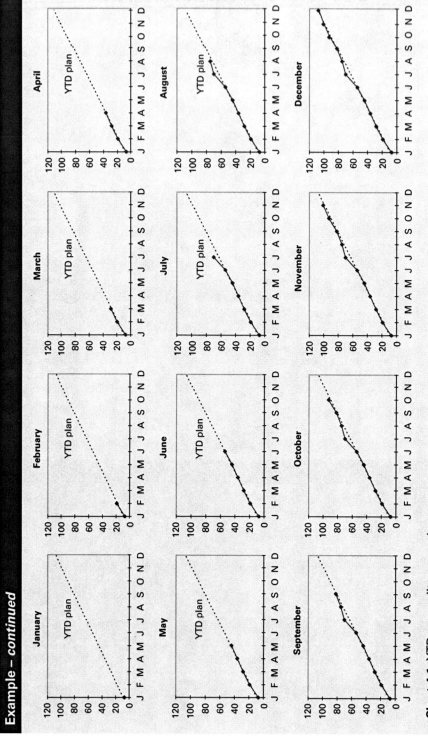

Chart 4.1 YTD expenditure vs. plan

(continued)

Chart 4.1 gives the series of 12 monthly charts that would result when the data are plotted along with a target expenditure of 9 units per month (108 units per year).

To simulate how this would work in practice, you may like to look at the charts in month order and before moving onto the next month, consider:

● Has the process improved or worsened? If so, when? If not, how do you know?
● Are there any months that were abnormally high/low when action should have been taken? If so, which ones and how do you know?

Table 4.2 provides interpretation of the data.

This is a simple example. The planned figure frequently changes monthly, making interpretation a little more difficult.

Table 4.2 An interpretation of the YTD data

Month	Expen-diture	YTD	YTD plan	Comment
January	7	7	9	YTD 2 less than the plan. It this a good start? Or, with only 1 month gone, is there too little data to draw any conclusions?
February	12	19	18	YTD 1 greater than the plan. Should we get concerned, after all, last month we were two below the plan.
March	9	28	27	YTD 1 greater than the plan. Should we get concerned? Or, with only 3 months gone, is it too early to draw conclusions?
April	8	36	36	YTD equals the plan. Presumably, there are no concerns.
May	8	44	45	YTD is 1 less than the plan. After 5 months, can we conclude that we are doing OK?
June	10	54	54	YTD equals the plan. Presumably, there are no concerns.
July	16	70	63	A very expensive month. Should we take action? We are over-half-way through the year, and well above the plan.
August	5	75	72	Better, but still above the plan, but much closer than last month. Should we take action?
September	7	82	81	Even better, even closer to the plan, but still too high. Only 3 months to go. Should we take action or not?
October	11	93	90	Worse again; 10 months gone and above the plan.
November	8	101	99	Better. Still above the plan, but much closer than last month. However, only 1 month to go. Should we take drastic action to try to meet the plan, after all we have been above the plan for 4 months in a row.
December	7	108	108	Great, we hit the plan. Now we start again next year.

(continued)

Example – *continued*

Analysing this year's YTD against last year's YTD

Another frequent ploy is for organisations to compare this year's YTD with last year's YTD. Chart 4.2 shows the results month by month over the year. You may like to review the chart, month-by-month commenting on what information it provides and what action you might take on the process. Table 4.3 provides one typical interpretation.

Telling the truth: the control chart

Chart 4.3 is a control chart over the 24 months for which data are available. The chart shows that the process we are monitoring has an average of 9 and that in any

Table 4.3 An interpretation of Chart 4.2

Month	Expenditure	YTD	Last year's YTD	Comment
January	7	7	12	Expenditure well down on last year. Good start.
February	12	19	20	Expenditure still down on last year and we must be doing well.
March	9	28	26	Above last year. And things must be bad, because if we look we see that after January, we have caught up and now overtaken last year's figure. If this trend continues we will be well above last year's figure by the end of the year. Better do something.
April	8	36	35	Still higher than last year, but not by much. Maybe things are not too bad.
May	8	44	49	Whatever we did in March is certainly kicking in now. Well down on last year.
June	10	54	59	Half-way through the year and nearly 10% down on last year. Congratulate the team. Where is the champagne?
July	16	70	70	An awful month. All our gains gone. Should not have let up after recent successes. Better take some more action.
August	5	75	78	Good month. The action taken last month worked.
September	7	82	85	Nine months gone. Still about 3.5% down on last year. Pity about July or we had have been well down.
October	11	93	95	Still holding. Looks like being a good year, if only we can keep this up.
November	8	101	99	Awful. And after things were going so well. Need an extra effort in December.
December	7	108	108	Relief! We met the plan!

(continued)

Example – *continued*

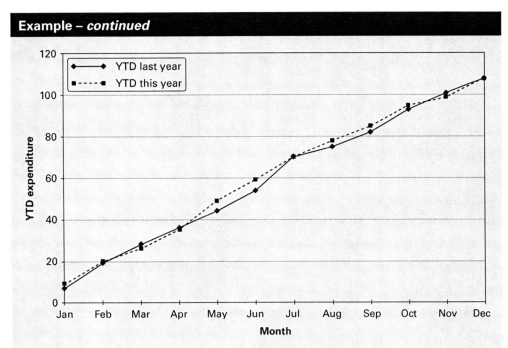

Chart 4.2 This year's YTD values vs. last year's YTD figures

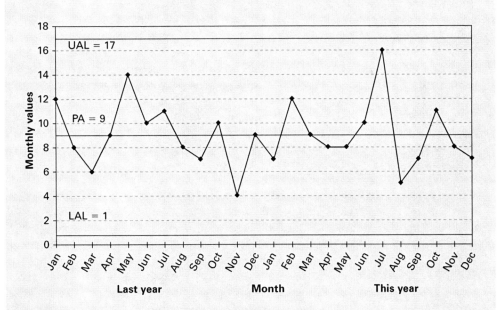

Chart 4.3 Control chart of monthly expenditure for the last 24 months; UAL: upper action limit; PA: process average

(continued)

Example – *continued*

one month the process may yield a figure as high as 17 without anything in the process having changed. The interpretation for each month is simple: "process continues in a state of control with an average of 9". We then have the choice of taking steps to improve the process (by following a process improvement methodology) or leaving the process to continue operating at an average of 9. The time spent in other interpretations are at best a waste of time, and if they lead to knee jerk reactions could degrade the process.

Why YTD charts do not work

There are a variety of reasons why YTD charts are not appropriate for process management:

- They encourage naive point-by-point analysis (e.g. comparing actual value against planned value). They totally ignore the truth behind all (useful) process-monitoring data: that values vary due to natural (common cause) variation. The result of this is that they encourage action on often irrelevant point-by-point comparisons. Such action is likely to be erroneous and degrade the process, resulting in losses to the organisation.
- Comparing this year's figure against last year's assumes that last year's figure was in some sense "normal".
- They do not provide a clear picture of how the process is behaving. Ideally we need more than 15 data points as a minimum to determine process behaviour. Particularly in the early months we have far too little information in a YTD chart. It is only by the end of the year at best that we begin to have enough data to draw reasonable conclusions; so for most of the year the data is of little use however it is analysed. They do not provide an easy way of estimating the process average (the YTD figure would need to be divided by the number of months to get an monthly average and multiplied by 12 to get the annual average).
- Even after a year, we only have 12 values from which to calculate the process average, and that value would only be correct if the process is in a state of control for the whole year. Control charts use all the relevant data to determine current process performance (i.e. since the last process change, excluding out-of-control values).
- Frequently YTD charts are drawn as in Charts 4.1 and 4.2. It is difficult to interpret slopes of lines, especially when the scale is condensed. On these examples, the largest figure in any 1 month is 16, and yet the scale has to extend to 108. The result is that relatively large and important swings show up as relatively small variations on the chart.
- A difference of a certain value between this year's YTD and the plan, or last year's YTD will be of different significance depending on where it occurs. For example, in Table 4.3, January, there is a difference of 5 units between this year's YTD and last year's YTD, a percentage difference from last year of $100 \times (5/12) = 43\%$. In June, the difference is also 5 units but the percentage difference is $100 \times (5/59) = 8.5\%$.
- There are no guidelines for determining whether the process has changed, is on target, or is subject to out-of-control conditions.
- Often YTD charts are plotted as bar charts (see Chart 4.4 for an example). These suffer from the same difficulties of interpretation as the line charts already described.

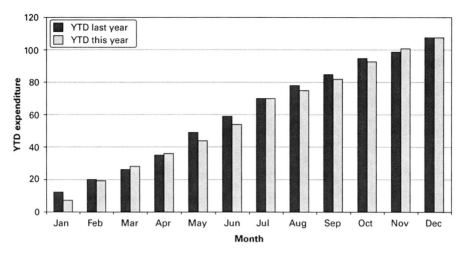

Chart 4.4 Grouped bar chart of this year's YTD values vs. last year's YTD figures

Example

Analysing the YTD average

To overcome some of the above problems, some organisations use the YTD average. This is simply calculated as:

$$\text{YTD average} = \frac{\textit{YTD value}}{\textit{number of values in the YTD calculation}}.$$

Table 4.4 gives the resulting table for the same YTD data, and Chart 4.5 is the resulting chart.

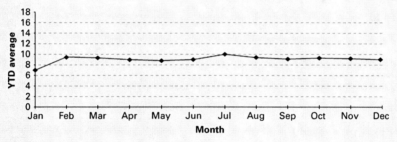

Chart 4.5 YTD average

Table 4.4 Calculations for the YTD average

Month	Jan	Feb	Mar	Apr	May	Jun	Jul	Aug	Sep	Oct	Nov	Dec
Expenditure	7	12	9	8	8	10	16	5	7	11	8	7
YTD	7	19	28	36	44	54	70	75	82	93	101	108
YTD average	7.0	9.5	9.3	9.0	8.8	9.0	10.0	9.4	9.1	9.3	9.2	9.0

(continued)

Example – *continued*

Again the reader is encouraged to move from month-to-month interpreting the chart and deciding what action he would take as each point is added. Table 4.5 contrasts what the YTD average chart is telling us with what is actually happening.

Table 4.5 An interpretation of Chart 4.5

Month	Raw data	YTD average	Observations
January	7	7.0	YTD average is 7, less than process average of 9. False impression that things have got off to a good start.
February	12	9.5	Bad month. The YTD average is now above the true process average of 9.
March	9	9.3	The actual result for the month is equal to the average, but the YTD average shows a decrease.
April	8	9.0	
May	8	8.8	Raw data does not change; however, the YTD average decreases.
June	10	9.0	
July	16	10.0	
August	5	9.4	Raw data decreases by 11, YTD average by 0.6. Compare, for example, with month 2 where raw data increases by only 5 and YTD average by 2.5. Changes in raw data have a progressively smaller effect on the YTD average as the year progresses.
September	7	9.1	Raw data increases but YTD average decreases. We cannot use YTD average to determine whether the raw data increased or decreased. The YTD average only tells us whether the current month is higher or lower than the last months' YTD average. The YTD does "home in" on the true process average but will take 20–30 values to do so, which we never reach if we use monthly YTD charts data!
October	11	9.3	
November	8	9.2	
December	7	9.0	YTD average equals the process average for only the third time (see months 4 and 6).

The above analysis makes it clear that the YTD average chart has drawbacks and is not generally suitable as a tool for interpreting process performance.

Comparing YTD and YTD average charts with control charts

Table 4.6 summarises and generalises the conclusions of this chapter. For each criteria as listed on the left-hand side of the chart, note what you would like the chart to tell

Table 4.6 Responses of different analyses methods to new values

Criteria and chart reference (All references to this year)	Response for each analysis method			
	YTD	YTD vs. last year's YTD*	YTD average	Control chart
Current value equals PA (see March)	Will always increase. The slope of the YTD line will increase, decrease or stay the same depending on the previous YTD value	Will move towards last year's YTD if the current value is between the last year's YTD and last month's YTD figure Otherwise, will move away	As pervious column, BUT the amount by which it moves depends on how many values are in the YTD calculation There will be less movement in the YTD average as more values are included	Value will lie on the PA line
Current value above (below) PA (above: February, June, July, October; below: other months)	As above in both cases	As above in both cases	As above in both cases	Value will lie above (below) the PA
Current value represents an extreme above (below) PA. Nearest is July which is the highest value	As above in both cases (but will an probably result in high increase in slope for values, and decrease for low values)	As above in both cases (but will probably result in an increase in slope for high values, and decrease for low values)	As above in both cases (but will probably diverge from the average)	Value will lie outside the control limits signalling a special cause of variation
Increasing (decreasing) trend (No defined trends, but see August to October for a run of three increasing values.)	The slope will increase (decrease), but since the YTD is by definition a slope, identifying a change in slope will be difficult, and identifying the magnitude of the change even more so	As the previous column However, the trend should be easier to identify as the last year's YTD acts a reference	A trend will appear, BUT for a constant trend the slope will gradually decrease For the slope to remain the same, the trend must increase by an amount related to the number of values in the YTD	Trend will be obvious and its slope will directly reflect its magnitude

* The same response holds for YTD vs. planned YTD.
PA: process average.

you before reading the comments under the analysis method. If you find the actions for the YTD options difficult to understand compared to the control chart, that is due to the shortcomings of the methods: they are difficult to interpret, which is why we do not want to use them!

Finally, Table 4.7 compares the estimate of the process average from both the YTD average and the corresponding control chart average.

For last year both the YTD average and the control chart have the same estimate for the process average. However, for this year, the calculation estimate for the YTD average begins again, which is why the YTD average, which was settling down to be around 9.0, falls dramatically to 7.0, and varies between 7.0 and 10.0 during the year. In contrast, the control chart estimate varies between 8.8 and 9.4. Chart 4.6 shows the convergence of the two estimates.

Table 4.7 Comparison of process average estimates

Month	Value	YTD average	Control chart estimate
Last year			
January	9	9.0	9.0
February	10	9.5	9.5
March	5	8.0	8.0
April	9	8.3	8.3
May	8	8.2	8.2
June	9	8.3	8.3
July	10	8.6	8.6
August	9	8.6	8.6
September	7	8.4	8.4
October	14	9.0	9.0
November	8	8.9	8.9
December	9	8.9	8.9
This year			
January	7	7.0	8.8
February	12	9.5	9.0
March	9	9.3	9.0
April	8	9.0	8.9
May	8	8.8	8.9
June	10	9.0	8.9
July	16	10.0	9.3
August	5	9.4	9.1
September	7	9.1	9.0
October	11	9.3	9.1
November	8	9.2	9.0
December	7	9.0	9.0

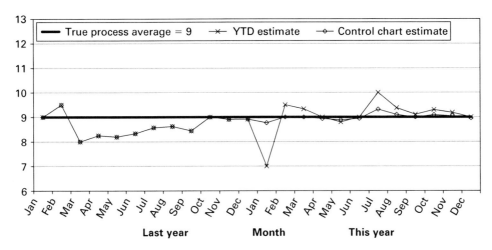

Chart 4.6 Demonstrates that the control chart gives a better estimate of the true process average than the YTD

Summary

We have demonstrated in this chapter the serious difficulties in attempting to interpret YTD tables and charts. Apart from the practical difficulties, the philosophy behind the YTD figure is insidious. For example, the YTD analysis implies that:

- You can tell something useful from one single value.
- Every year all previous data is irrelevant, or at best included in value-by-value comparison with the current year.
- There is no awareness of the possibility or attempt to identify process changes or out-of-control conditions.

PART 3

Putting SPC into Practice – The Cases

During the writing of this book I was privileged to be present at the 100th birthday celebrations of Dr J. Juran, the well-known management guru and author. One of the many speakers remembered how in the 1970s one company explained to Dr Juran a problem they had which was costing them money and clients. When they had finished explaining the problem Dr Juran fell silent, deep in thought. After a minute or so he told them that he had solved that problem in 1941. He sent them his report and they saw that he did indeed have the solution.

Many of us have problems in assessing how our processes are performing. Many of these problems have been solved by using the statistical process control (SPC), and these case studies are part of the evidence.

The sources of the case studies

The industries and companies which provide the setting for these case studies have been anonymised. The reason for this derives from a lesson I learned from a Quality Manager in a large oil company. He was trying to instil into managers an enthusiasm for quality improvement. They asked for evidence that the ideas worked, so he presented numerous published case studies demonstrating that quality improvement programmes could indeed significantly improve performance. "OK" his fellow managers said, "but show us where they have worked in our industry". After some further research he was able to supply one or two more case studies in the oil industry to demonstrate that pursuing a quality improvement programme had resulted in significant benefits. "But these companies are not based in our country…". Some people will find any excuse to avoid change.

If SPC works for a safety department in the airline industry, it will work for a safety department in the oil, manufacturing, health care, public service, and other industries and organisations. The origin of the case studies is frequently irrelevant, and so where possible and appropriate I have removed it.

Similarly, when analysing incident rates, the type of incident (safety, breakdown, etc.) is not important – the methodology still applies, and you can interpret terms such as "incident" in whatever way is appropriate to your situation.

Control charts in the real worlds are not always so clear

In many books on SPC the control charts presented are usually easy to interpret. When teaching the theory of SPC this is understandable. In reality, neither control charts nor any other technique is "magic". Organisations and processes are complex living and therefore changing things. We do not know exactly what is happening in a process or organisation, and progress lies in obtaining better and better approximations. This is reflected in the case studies and control charts in this book where interpretation is not always clear. The charts give us hints which we need to investigate further.

A word on chart formats

There are a variety of formats that can be used when drawing control charts. Each software package has developed its own format, and each person may develop formats to suit their own particular needs. This book deliberately reproduces charts in several formats with the intention of highlighting the fact that there is no single best way. We suggest you try different ideas out in different situations and see what works best. For example, if you are expecting people who are not well trained in the use of control charts to enter data on them, you will need to keep them simple and easy to use, preferably with no calculations required. Alternatively, for analytical purposes you may choose to keep details of calculations on the sheet. Always consider the needs of the user of the chart before designing it.

In the workplace it may often be more appropriate for people to add hand written comments – perhaps on what investigations were carried out and what they showed, adjustments, or other process changes. One of the functions of the control chart is to act as the process memory so that when investigating special causes of variation or improving the process, pertinent information is already at hand on the chart.

Charts in this book consist of up to four sections:

1. The plot (or plots) of the data. In the case of X–MR chart, the X chart will be above the MR chart (see below for a brief description of different chart types).
2. The raw data required to plot the chart.
3. On some occasions a "calculation" section is provided above or below the chart. Whilst items 1 and 2 would frequently be distributed with the chart, Section 3 seldom is because it is not necessary to do so. It is included here only to aid the understanding of calculations necessary for control limits, etc.
4. Occasionally there will be a histogram of the data turned through 90 degrees.

Using Chapter 5 as an example of an X/MR chart, the X chart is above the MR chart. The histograms (for Charts 5.2 and 5.3 two histograms) are shown to the right of the

X chart (on the top right-hand side of the chart). These histograms are turned through 90 degrees and show the distribution of the raw data. Below the histograms is a calculation section. For further details on histograms and chart calculations see Part 4. To the left of the calculations section are the raw data including an index (in this case, observation number) and comment section.

Layout of and information in the case studies

The primary purpose of this book is to demonstrate the practical benefits and uses of SPC and control charts. For this reason the case studies have been placed before the detailed theory, which is in Part 4, and a minimum of theory has been provided in Part 1. These case studies contain both the story of how SPC was applied to data with the benefits that resulted and often in depth complex data analysis. If your main reason for reading this book is to gain insight into the many benefits of SPC and to read about how others have gained from them, it is not necessary to understand the calculations and you may choose to skip over these aspects. If you want to understand the mechanics of how each chart was developed, what analyses may be carried out in slightly different circumstances, etc., you will need to be highly numerate and have a good understanding of the detailed statistical theory. All the required information is provided in Part 4.

Since the case studies cover a wide range of applications the formats are not always the same, but in general case studies are split into the following sections.

Charts used: This lists of the different charts used in the case study. Different chart types are used for different types of data, but it is not necessary to understand the differences between charts to follow the case study. The different chart types are:

- X also known as individuals charts, are generally used for charting measurements such as lengths. Their primary purpose is to monitor the average performance of a process, but they are also the key charts for identifying special causes of variation.
- MR (moving range) charts are used for monitoring variability. X and MR charts are usually used together.
- The c charts are used to monitor counts, such as number of accidents per month.
- The u charts are used in place of c charts when the opportunity for counts varies, for example, number of incidents per million man hours worked per month, and the number of hours worked each month varies.
- The np charts are used to monitor proportions of non-conforming items, for example, the proportion of customers that complain per month.
- The p charts are used in place of np charts when the number of units inspected varies, for example, if the number of customers per month varies.
- $\overline{X}$ (pronounced X bar) charts are used to monitor averages. For example, on a production line, if four samples are measured we may plot the average of the four.
- R (range) charts are used to monitor the range (i.e. maximum–minimum) in the same situation as the $\overline{X}$ chart, and are used in conjunction with it.
- s (standard deviation) charts are an alternative to the R chart and are used as the sample size increases because they provide a better estimate of the variability of a set of data than the R chart. R charts are commonly used because the calculations are simpler than those for s charts.

- *Introduction*: This provides an overview of the key aspects of the case study.
- *Background*: This provides background information to the data and the situation in which it was reported.
- *Analysis*: This section explains the process and details of analysing the data.
- *Comments*: This section provides further information, suggestions, observations and theories drawn from the case study.
- *Calculations*: This provides detailed workings of calculations.
- *Summary*: This section reviews the main learning points from the case study.

Further details of these and other charts are in Part 4. One common chart type not yet mentioned is the cusum (cumulative sum) and the weighted cusum chart. These specialist rather complex charts are extremely powerful at identifying small changes in process average. Part 4 provides the details.

How to use the case studies?

There are a variety of ways in which you may choose to read these case studies. If you are mainly interested in gaining an appreciation for what SPC can do for you, start at the beginning and read through the case studies. Ignore the calculations and any technical terms and comments that you do not understand (you can always return to these once you have read Part 4) and just focus on the broad application areas. In general, the shorter and/or simpler case studies are first, with the more complex cases later.

If you want to know how to analyse a particular situation, perhaps one that you have in your own organisation, search through for those case studies which relate to your situation and concentrate on them.

If you want to see how to use a particular type of chart, select those studies with the particular chart you are looking for. These are listed at the front of each case.

If you are involved in a particular industry or sector (such as health care or education), or have an interest in a particular application (such as benchmarking), read through the titles and introductions to select those that are of interest to you.

Finally, if you are deeply involved in SPC and know the basic ideas and theory, you will find many cases that explore in depth some of the subtleties of charting; for example, the effect of using different types of chart to analyse the same set of data.

5

Investigating variation in chemical concentration
How control charts were used to identify, investigate and prove the cause of fluctuations in results

Charts used: X and MR

Introduction

This example is taken from the continuous processing industry. It demonstrates how one control chart was used to:

- identify a performance problem,
- generate theories as to the causes,
- test and ultimately "prove" the theory that differences in measurements were due to different chemical analysts and not actual process performance.

No other analytical method was needed to bring about a significant improvement in the process.

The whole analysis took place at the data manager's computer using a standard spreadsheet package and took approximately 1 hour to complete. By the end of the session all that was required was to confirm with the chemical analysts what the differences in sampling and analysis method were. As a result, sampling methods were standardised and streamlined. The control charts presented here were produced later as part of the report.

Background

In order for a particular continuous process to run smoothly and maintain the life of the equipment, it is necessary to inject a chemical at the feed every few months. To add the chemical, the process is stopped, typically for a couple of days. The chemical is added and the process restarted.

The chemical injection is expensive not only because of its chemical and manpower cost, but also because production is temporarily halted.

In order to minimise these costs, the chemical concentration is measured every 2 weeks at the output of the process. Once chemical concentration levels fall below a pre-determined threshold more chemical is added (Figure 5.1).

The measured concentrations are stored in a database.

Figure 5.1 Chemical injection and sampling process

Analysis

The first step was the usual one of plotting the raw data on a control chart. Chart 5.1 shows the un-interpreted X–MR chart of the chemical concentration:

- The X chart (top part of Chart 5.1) plots the individual chemical concentration values. There are several instances of cycling (alternate high and low values). See, for example, observations 21 to 28.
- When data are normally distributed, which is what we expect with this type of data, most values will be close to the average with fewer values, the further we move from the average. The histogram clearly shows that there is very little data near the average, and that they are grouped into high (above 20) and low (below 10) values.
- The moving range chart shows similar out of control signals. Note particularly the large number of values above the mean.

The above points clearly suggest that the process is not in a state of statistical control.

The cycling and double-peaked histogram give a clue that there may be two sub-processes involved: one process yielding the high values and the other the low values. A brief discussion with the data custodian led to the theory that the suspected sub-processes may be explained by the existence of two chemical custodians responsible for sampling and analysis. Fortunately the identity of the chemical custodian was recorded along with the chemical concentration.

To test the theory that the variation is largely due to chemical custodian, the chart is redrawn with the chemical custodian identified (Chart 5.2). Custodian 1, identified by the open circles produces a consistent result between 22 and 26 with two notable exceptions. Custodian 2 produces results with much higher variability. On several occasions stand-in custodians are identified by hatched circles. Chart 5.2 is not easy to interpret as the data from different custodians are intermingled and so the data have been re-charted (Chart 5.3) with each of the two main custodian's data grouped together.

Chart 5.3 clearly shows that the results from the two main custodians are completely different. Since there is no other known cause of such large changes on chemical concentration it was decided to investigate the actions of the two custodians. Not surprisingly, it was found that the sampling and analysis methods used by the custodians were different, and that these caused the large variation between analysts.

These findings were presented at the following chemical custodian meeting and a trainee engineer who had worked with both custodians 1 and 2 was able to explain some of the differences she had noticed in the sampling and analysis methods.

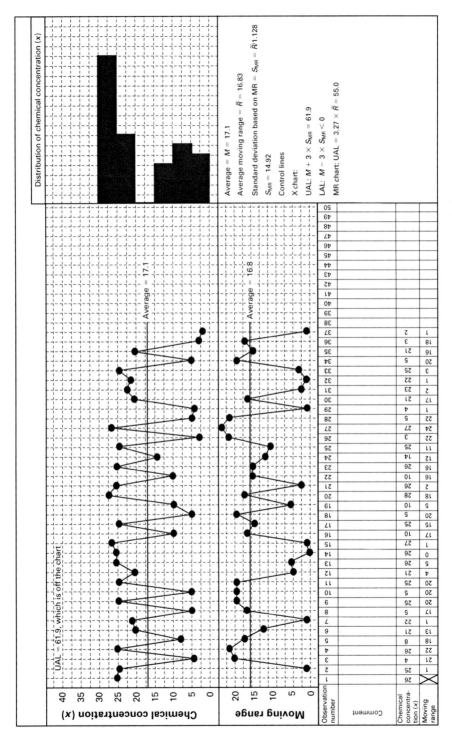

Chart 5.1 X/MR chart: chemical concentration; UAL: upper action limit; LAL: lower action limit

Distribution of chemical concentration (x)

Average = M = 17.1
Average moving range = $\bar{R}$ = 16.83
Standard deviation based on MR = $S_{MR} = \bar{R}/1.128$
S_{MR} = 14.92
Control lines
X chart:
UAL: $M + 3 \times S_{MR}$ = 61.9
LAL: $M - 3 \times S_{MR} < 0$
MR chart: UAL = $3.27 \times \bar{R}$ = 55.0

Chemical concentration (x)

UAL = 61.9, which is off the chart

Average = 17.1

Moving range

Average = 16.8

Observation number	1	2	3	4	5	6	7	8	9	10	11	12	13	14	15	16	17	18	19	20	21	22	23	24	25	26	27	28	29	30	31	32	33	34	35	36	37
Comment																																					
Chemical concentration (x)	26	25	4	26	18	13	22	17	25	20	25	21	26	0	27	10	25	20	10	28	26	10	26	12	25	3	24	22	4	17	23	22	3	20	16	18	1
Moving range	✕	1	21	22	8	5	1	5	20	5	20	4	5	26	1	10	15	5	18	28	2	16	16	14	25	3	22	5	4	1	23	1	25	5	21	3	2

66

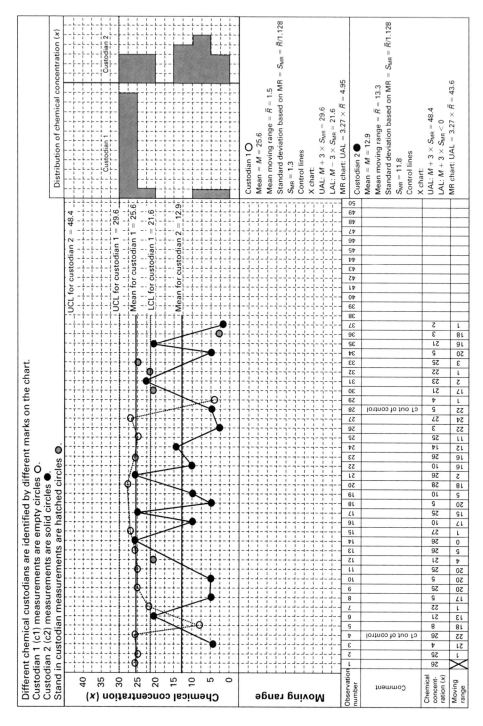

Chart 5.2 X/MR chart: chemical concentration highlighting differences by custodian; UCL: upper control limit; LCL: lower control limit; UAL: upper action limit; LAL: lower action limit

Custodian 1 exhibits a much more consistent measurement than custodian 2.
Custodian 1 also has two out-of-control measurements.
Custodian 2 is more erratic.

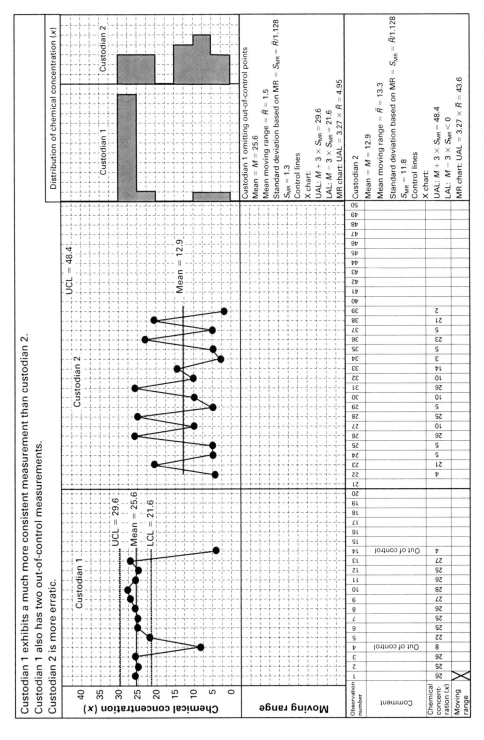

Chart 5.3 X/MR chart: chemical concentration for custodians 1 and 2. UCL: upper control limit; LCL: lower control limit; UAL: upper action limit; LAL: lower action limit

As a result, a streamlined sampling and analysis procedure was developed and implemented. The old data was archived as being of little practical use.

Comments

- Data collected over the period of this chart was all but useless as the key determinant of the reported values is the chemical custodian. Past decisions based on these measurements are likely to have been erroneous and incurred unnecessary cost if chemical was added prematurely, or impaired equipment, if concentration levels fell too low. These costs were not estimated.
- If values are known to come from different processes we should not plot them as if they came from the same process. For example, if it is know that different suppliers, shifts, groups or people, machines give different results we should monitor them using separately calculated averages and control limits…
- To test the theory in Chart 5.2, the moving range chart is not required and so is omitted. It is standard practice to include the moving range chart with an individuals chart when beginning the analysis as the moving range monitors variability.

Calculations

The calculations are shown on the charts.

Summary

- This case study demonstrates how a simple control chart and an hour of time was all that was required to:
  - identify the existence of a performance measurement problem,
  - test theories,
  - identify the root cause of the problem.
- Comparing results from different shifts, analysts, teams or other groupings is often useful as they are often a significant cause of variation. This variation is usually due to different methods of working.
- Patterns of results, such as cycling, are frequently indicative of measurements from alternating sources such as shifts or machines.
- The charting and analysis took a little over 1 hour and resulted in implementing procedures that would ensure high data quality which would allow significant savings in reduced downtime of equipment and lower chemical cost.

6

Improving examination results by analysing past performance and changing teaching methods

Charts used: X

Introduction

There has been much debate over the last few years about the publishing of examination results. Do good results reflect the ability of the students, the school, tutors, the environment or other factors, or a combination of these?

For some years I facilitated a short intensive professional course which concluded with an examination that delegates had to pass in order to progress towards gaining a professional qualification. In this case study we look at how using a control chart to analyse examination results helped tutors improve their teaching and hence increase the pass rate.

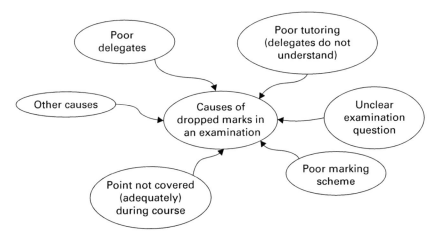

Figure 6.1 Some causes of dropped marks in an examination

Background

In order to progress to a certain professional qualification, one requirement is attendance and passing of an intensive short course. To pass the course delegates are assessed on all aspects of the course including attendance, teamwork, written work, role plays and an examination. The 2 hour examination is held at the end of the course and consists of:

- Ten multiple choice questions each with four possible answers: A, B, C or D – each question is worth 1 mark.
- Nine true/false questions – each question is worth 1 mark.
- Eight short essay questions – each worth 4 marks.
- Four short case studies – each with two 3-mark questions.

This gives a total of 75 marks.

There is a marking scheme against which all papers are marked. For the case studies and essay questions this consists of a list of points that the delegates should mention in their reply.

There were usually around 20 participants on a course and two lecturers selected from a pool. If:

- the teaching is consistent throughout the week,
- the examination questions are equally difficult,
- the marking scheme is fair,

then we expect that on average each question will be equally well answered.

We can measure the average "correctness" of answers by counting the percentage of marks "dropped" by the delegates. For example, if a question is allocated 4 marks and there are 20 delegates, the total possible mark for the class is $4 \times 20 = 80$ marks. If the marks for each delegate are added together and total 60 then $80 - 60 = 20$ marks have been dropped which is equal to $(100 \times 20)/80 = 25\%$.

The aim of this analysis is to identify which questions, if any, have a higher than expected level of dropped marks.

After all the papers had been marked, the number of marks dropped for each question was entered directly into an X chart (Chart 6.1). The percentage of marks dropped was then calculated and plotted on the chart. Finally the average and control limits were calculated and drawn. Note that the order of plotting the data does not affect the interpretation as we will not be looking for trends, and therefore joining the points with a line is not necessary. Strictly speaking, a line joining the points is only drawn when the data are plotted in some sequence where looking for trends is appropriate.

Those questions dropping the highest percentage of marks were reviewed first for clarity of the question and then for the possibility of the prescribed answer being wrong. Once these two potential sources were eliminated, we reviewed how well or otherwise the tutors had covered the topic on the course.

Analysis

The only question which is above the control limit is case study 1b (see Chart 6.1). Reviewing previous results, this question does not normally lose many marks (hence it is unlikely to be a systematic problem). The two tutors reflected on the question and reviewed the answers. The question seems to be understood by the delegates, but the tutors remembered that most of the teaching points being examined in this question were covered late one evening, and because we were running a little late, we were in a hurry to wrap up the day so that the delegates could carry out their evening assignments.

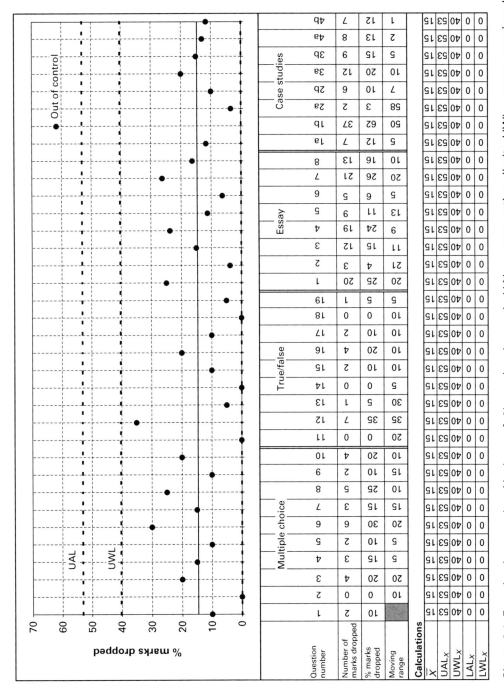

	Multiple choice										True/false									Essay								Case studies							
Question number	1	2	3	4	5	6	7	8	9	10	11	12	13	14	15	16	17	18	19	1	2	3	4	5	6	7	8	1a	1b	2a	2b	3a	3b	4a	4b
Number of marks dropped	2	0	4	3	2	6	3	5	2	4	0	7	1	0	2	4	2	0	1	20	3	12	19	9	5	21	13	7	37	2	6	12	9	8	7
% marks dropped	10	0	20	15	10	30	15	25	10	20	0	35	5	0	10	20	10	0	5	25	4	15	24	11	6	26	16	12	62	3	10	20	15	13	12
Moving range		10	20	5	5	20	15	10	15	10	20	35	30	5	10	10	10	10	5	20	21	11	9	13	5	20	10	5	50	58	7	10	5	2	1

Calculations

X̄	15	15	15	15	15	15	15	15	15	15	15	15	15	15	15	15	15	15	15	15	15	15	15	15	15	15	15	15	15	15	15	15	15	15	15
UAL_X	40	40	40	40	40	40	40	40	40	40	40	40	40	40	40	40	40	40	40	40	40	40	40	40	40	40	40	40	40	40	40	40	40	40	40
UWL_X	53	53	53	53	53	53	53	53	53	53	53	53	53	53	53	53	53	53	53	53	53	53	53	53	53	53	53	53	53	53	53	53	53	53	53
LAL_X	0	0	0	0	0	0	0	0	0	0	0	0	0	0	0	0	0	0	0	0	0	0	0	0	0	0	0	0	0	0	0	0	0	0	0
LWL_X	0	0	0	0	0	0	0	0	0	0	0	0	0	0	0	0	0	0	0	0	0	0	0	0	0	0	0	0	0	0	0	0	0	0	0

Chart 6.1 Examination results: X chart of % marks dropped; UAL: upper action limit; UWL: upper warning limit; LAL: lower action limit; LWL: lower warning limit

In addition, neither tutor raised the learning points during the review session held the following morning.

The tutors learned that if key learning points are being taught in any situation where the delegates may not be giving their full attention, the points should be reviewed by the tutors at the next review session. This became the standard practice and the number of occasional out-of-control questions has all but disappeared (Figure 6.1).

In question true/false 12 is not above the control limit, but after reviewing previous examination results, it is invariably one of the highest for marks dropped in the "true/false" section. Unfortunately, the question is not clearly worded, and whilst the tutors are able to suggest that the wording be changed, this will take months to effect, if accepted. The dropped mark percentage on this question does appear to have dropped as tutors clarify during the course exactly what is required to a question such as that being asked, but of course, they cannot compromise the independence of the examination.

Question 4a of the case studies (see Chart 6.1) is also not above the control limit, but it used to be much higher. After reviewing these charts for some time, tutors realised that this case study was frequently being incorrectly answered. Being a case study, tutors are allowed to award marks for the thought process even if the answer does not conform to the prescribed answer. In tutor meetings this question has been discussed at length and it was agreed that the answers being submitted by a minority of delegates should be accepted if the correct reasoning was given.

Using this approach to analysing examination results, the tutors improved the way they covered course material, challenged and changed marking schemes and, over longer periods of time, changed the examination. As a result of these actions the marks achieved by delegates gradually increased and we believe that occasionally people just passed who would have just failed the examination.

Calculations

The chart and table shows the result of a typical examination.

Along with the question number the type of question of each section is annotated. The next row gives the total number of marks dropped by all delegates on that question. For example, for Question 14 (a true/false question) no marks were dropped, whilst for Essay Question 6, 5 marks were dropped.

The "% marks dropped" is calculated in the following ways:

- For questions with only 1 mark (i.e. the first 19 questions):

$$\text{\% marks dropped} = \frac{\text{number of marks dropped} \times 100}{\text{number of delegates}}.$$

For Question 1, 2 marks were dropped, and so:

$$\text{\% marks dropped} = \frac{2 \times 100}{20 \text{ delegates}} = 10\%.$$

- For questions with more than 1 mark:

$$\text{\% marks dropped} = \frac{\text{number of marks dropped} \times 100}{\text{number of delegates} \times \text{number of possible marks}}.$$

For the case study Question 3a, which has 3 marks, 12 marks were dropped giving:

$$\% \text{ marks dropped} = \frac{12 \times 100}{20 \times 3} = 20\%.$$

The next row gives the moving range which is calculated in the usual way, and used to calculate the control limits.

The calculation table below the chart gives the average, $\overline{X}$, and the warning, and control limits. The standard deviation is calculated from the moving range in the usual way as:

$$S_{MR} = \frac{\overline{MR}}{1.128} = \frac{14.46}{1.128} = 12.8.$$

The control and warning limits are calculated in the usual way at 2 and 3 standard deviations, respectively, above and below the average.

Note that the lower limits are both 0, implying that sometimes there will be zero dropped marks.

Comments

- This analysis provided a very interesting lesson for the author. One of the purposes of training courses is to impart information to the delegates which the delegates are expected to understand, remember and apply. The examination is one way of testing to see whether this has happened. It is the tutors and course materials job to ensure that the appropriate information is imparted and explained. It is also a requirement that the examination is unambiguous. Where a high number of people give the wrong answer, it is likely that the problem lies not with the delegate but somewhere else – as illustrated in this case study. Tutors owe it to their students and employers to analyse examination results with the aim of improving the examination, their teaching methods and the course.
- It would be possible to keep a separate control chart of the percentage of marks dropped for each question which would be added to every time a course was run. This would allow tutors to monitor changes over time. On occasion, a chart may be drawn for one or two questions to test a theory; for example, that a new approach has reduced the marks dropped in that section.
- It would be possible for each tutor to keep a chart of the percentage marks dropped on the courses that they are involved in. This may help tutors monitor and improve their own performance.
- The points on the control chart have not been joined by a line. This is because the order of questions is not relevant.
- The moving range chart is not important for this analysis because it monitors the variability from one point to the next where the points are in logical order. In the case of examination results (or any other data) where the order of the data is random, monitoring the relationship between consecutive points is meaningless.

Summary

We have shown how control charts and simple analyses helped tutors improve the coverage of course material, marking schemes and examination questions which resulted in higher marks, and as far as can be judged by examination results, a more complete understand of course material.

Similar methodologies are widely applicable to examination and other assessment tools.

7

Demonstration that moving averages are poor indicators of true process performance
Monitoring the frequency of incidents

Charts used: Moving average and u

Introduction

In Chapter 3 we explained how moving averages can give the wrong messages to the unwary user. We showed, for example, that the moving average may increase when the process values were declining, that single out-of-control points are difficult to identify and that the actual moving average value may be of little use.

People often use moving averages to analyse a set of data because it smoothes out fluctuations. The claim is that they are interested for identifying and monitoring trends, and that control charts cannot do this. On one occasion having received a series of comments along these lines, we analysed safety incidents data using a control chart and compared the results with the moving average charts that were currently in use. The results were quite startling and support the theory of Chapter 3.

The case study shows that:

- Moving averages may incorrectly suggest a process is improving.
- Moving averages do not reflect the true process average for some time after a process change.
- Control charts accurately reflect process changes and can be used for predicting future performance.

Background

Every month the number of (safety) incidents is recorded. Table 7.1 gives the month number and the number of incidents in that month. The moving average of the number of incidents is calculated and also provided in the table. As an example, the first moving average of 2.0 is calculated as:

$$\frac{\text{total number of incidents in the previous 12 months}}{12} = \frac{24}{12} = 2.0.$$

Chart 7.1 shows the first 12 moving averages.

The number of hours worked each month varies from a minimum of 0.12 to 0.36 million hours in the first 12 months. We would expect more incidents where more hours

Table 7.1 Incident data and calculations

Month number	Month	Number of incidents	Moving average	Hours worked (millions)	u = incident frequency = number of incidents / number of hours worked	Moving average
	Year 1					
1	Jan	3		0.23	13.0	
2	Feb	1		0.25	4.0	
3	Mar	1		0.24	4.2	
4	Apr	1		0.24	4.2	
5	May	2		0.20	10.0	
6	Jun	1		0.25	4.0	
7	Jul	2		0.12	16.7	
8	Aug	1		0.32	3.1	
9	Sep	1		0.35	2.9	
10	Oct	3		0.34	8.8	
11	Nov	6		0.35	17.1	
12	Dec	2	2.0	0.36	5.6	7.38
	Year 2					
13	Jan	1	1.8	0.37	2.7	6.49
14	Feb	2	1.9	0.41	4.9	6.48
15	Mar	1	1.9	0.41	2.4	6.18
16	Apr	6	2.3	0.39	15.4	7.24
17	May	2	2.3	0.40	5.0	6.88
18	Jun	2	2.4	0.69	2.9	6.43
19	Jul	1	2.3	0.53	1.9	5.69
20	Aug	1	2.3	0.41	2.4	5.59
21	Sep	3	2.5	0.39	7.7	5.94
22	Oct	1	2.3	0.49	2.0	5.38
23	Nov	2	2.0	0.55	3.6	4.44
24	Dec	1		0.55	1.8	4.11
	Year 3					
25	Jan	1		0.51	2.0	4.01
26	Feb	2		0.34	5.9	4.06
27	Mar	1		0.49	2.0	4.01
28	Apr	1		0.61	1.6	3.02
29	May	2		0.55	3.6	2.95
30	Jun	0		0.50	0.0	2.70
31	Jul	0		0.46	0.0	2.56
32	Aug	2		0.43	4.7	2.73
33	Sep	1		0.41	2.4	2.38
34	Oct	3		0.40	7.5	2.76
35	Nov	0		0.44	0.0	2.46

are worked and so comparing the number of incidents unadjusted by the hours worked is not appropriate.

The usual approach in these situations is to calculate the incident frequency as the number of incidents divided by the number of hours worked. In this case the result will be the number of incidents per million man hours (mmh). The incident frequency is

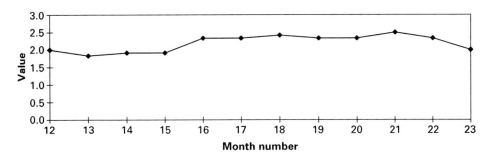

Chart 7.1 Moving average chart of the number of incidents

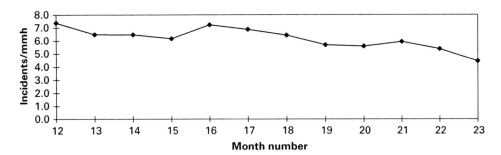

Chart 7.2 Moving average chart of the incidents frequency

also provided in the table. The moving averages for the incident frequency are calculated as:

$$\frac{\text{total number of incidents in the first 12 months}}{\text{total number of hours worked in the first 12 months}}$$

$$= \frac{24}{3.25} = 7.38 \text{ incidents/mmh.}$$

Chart 7.2 shows the moving average for the incident frequency, and it is clear that the chart is quite different from Chart 7.1.

Continuing with the moving average analysis of the incident frequency, every month a new moving average is calculated based on the previous 12 months and the new point added to the chart as the oldest point is dropped off.

In order to investigate likely management thinking 12 consecutive moving average charts have been reproduced in Chart 7.3 to simulate a year's analysis, beginning in month 23, November, which is the first month that a full set of 12 moving averages is available for plotting.

Before continuing, you might like to take some time to review these charts and note your own conclusions about how the process is performing.

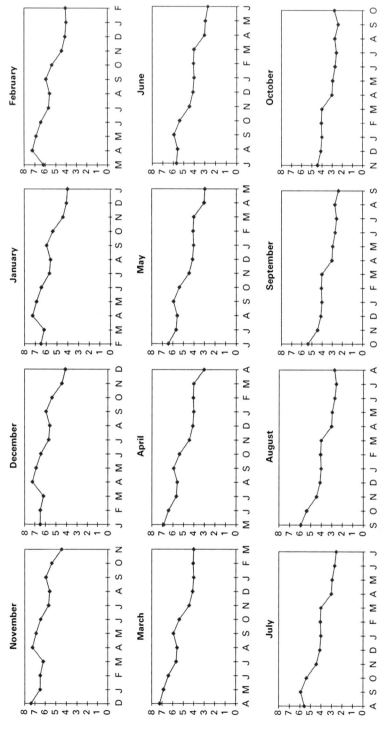

Chart 7.3 Moving average of incident frequency

Analysis

Analysis of moving averages is notoriously difficult because the raw data values are masked. One interpretation is offered below, you may have other interpretations:

November There has been a steady decline over the last 12 months, despite occasional hiccups – especially in April. Is the April figure a cause of concern or not? Is September, another month in which the moving average rose, a cause of concern? How would we decide?

December Decline continues.

January Decline continues, but rate of decline may be slowing down.

February Steady decline over the first 9 months, may be it has now stabilised, or rate of decline slowed. What is the current rate? What is the rate of decline?

March Shows decline over the last 12 months (especially since the previous increase from March to April has now dropped off). Process was now stable, or in gentle decline.

April New decline. Perhaps we have started a new era of reduction, or the resumption in incident decline. Let's hope this was not just a flash in the pan.

May–July Further small decline. Overall trend still down.

August Slight increase, may be nothing to worry about. Overall trend still down.

September Decline continues. August obviously an aberration. Overall trend still down.

October Slight increase, should we worry? Overall trend still down.

Having analysed the moving average charts, the next step is to review the control chart. Unlike the usual implementation of moving average charts where points are dropped off the chart, the control chart will plot as many values as is reasonable, usually all of them, or, if there are too many, all values since the last process change.

In this case all values easily fit onto one chart (Chart 7.4). Control charts give two immediate advantages over moving average charts:

1. We do not need to wait 12 months before we can calculate the first moving average. We can cautiously begin plotting charts and calculating averages with as few as half a dozen data points. By the time the first moving average point is plotted we will already have 12 points on the control chart.
2. We do not need to wait a further 12 months (i.e. a total of 24 months) whilst we wait for a full year's worth of moving averages to become available for plotting.

In Chart 7.4 we see that up to month 17 the data appears very variable – two points out of control, and several other very high (months 1 and 7, and to a lesser extent 5 and 10). From month 17 the process appears to have settled down to a lower incident rate.

Assuming that there is a process change in month 17, we re-draw the control chart with the process change included, (see Chart 7.5). This chart shows that the average before the assumed process change is 7 incidents/mmh and that no months are above the control limits. After the process change, the average incident rate reduces to 3 incidents/mmh and once again no months are above the control limit. We can now state that unless the process changes, the incident frequency will continue at the current level of just 3 incidents/mmh.

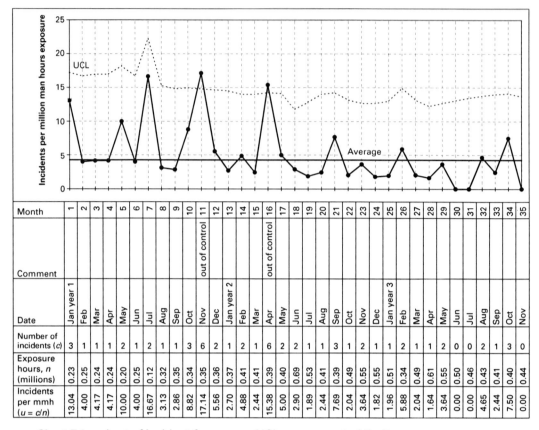

Month	1	2	3	4	5	6	7	8	9	10	11	12	13	14	15	16	17	18	19	20	21	22	23	24	25	26	27	28	29	30	31	32	33	34	35
Comment											out of control					out of control																			
Date	Jan year 1	Feb	Mar	Apr	May	Jun	Jul	Aug	Sep	Oct	Nov	Dec	Jan year 2	Feb	Mar	Apr	May	Jun	Jul	Aug	Sep	Oct	Nov	Dec	Jan year 3	Feb	Mar	Apr	May	Jun	Jul	Aug	Sep	Oct	Nov
Number of incidents (c)	3	1	1	1	2	1	2	1	1	3	6	2	1	2	1	6	2	2	1	1	3	1	2	1	1	2	1	1	2	0	0	2	1	3	0
Exposure hours, n (millions)	0.23	0.25	0.24	0.24	0.20	0.25	0.12	0.32	0.35	0.34	0.35	0.36	0.37	0.41	0.41	0.39	0.40	0.69	0.53	0.41	0.39	0.49	0.55	0.55	0.51	0.34	0.49	0.61	0.55	0.50	0.46	0.43	0.41	0.40	0.44
Incidents per mmh ($u = c/n$)	13.04	4.00	4.17	4.17	10.00	4.00	16.67	3.13	2.86	8.82	17.14	5.56	2.70	4.88	2.44	15.38	5.00	2.90	1.89	2.44	7.69	2.04	3.64	1.82	1.96	5.88	2.04	1.64	3.64	0.00	0.00	4.65	2.44	7.50	0.00

Chart 7.4 u chart of incident frequency; UCL: upper control limit

Comparing the conclusions of the moving average and control charts:

- The moving average suggests that the incident rate was gradually falling.
- The control chart shows that the process improved at around month 17 and there is little evidence of a sustained gradual improvement.

The reason that the moving average misled us is because after month 17 we were dropping off the higher values before the process change and replacing them with lower values after the process change.

If the control chart had been used for monitoring performance, management would have known:

- In the first 17 months there was an average of around 7 incidents/mmh.
- In any one month, they could expect an incident rate as high as 20–25 incidents/ mmh, depending on the number of hours worked (the actual value is given by the control limit).
- From month 17 there was an average of around 3.5 incidents/mmh.

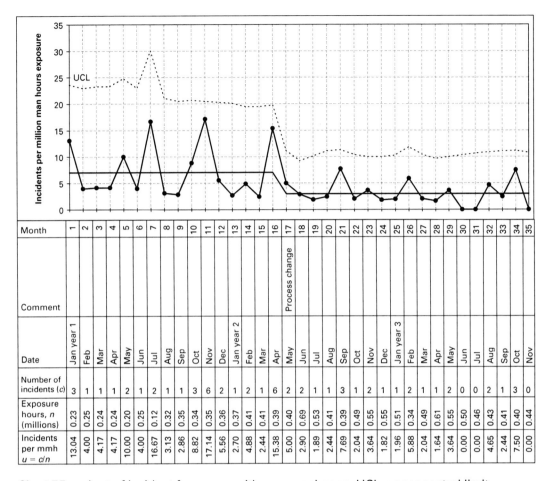

Chart 7.5 u chart of incident frequency with process change; UCL: upper control limit

- In any one month, they could expect an incident rate as high as 10–12 incidents/mmh, depending on the number of hours worked (the actual value is given by the control limit).
- There was not a gradual improvement as suggested by the moving average.

None of this information is available from the moving average chart.

This analysis was carried out retrospectively and we should investigate the cause of the apparent process change at month 17. Knowing that the safety department carried out major changes in safety procedures, training and awareness, it is very likely that the improvement coincided with one such initiative.

Finally, it is of interest to see whether including all the moving averages on one chart changes the moving average interpretation (Chart 7.6). The chart strengthens the impression that the process is improving, and therefore is not considered appropriate for this type of analysis.

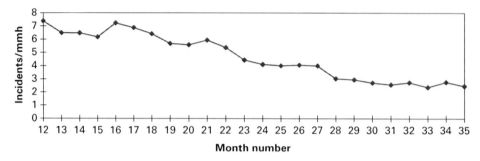

Chart 7.6 Moving average chart of the incidents frequency for all months

Comments

- In a typical monthly moving average chart, almost 2 years of data are required before the first chart is complete. As shown in Table 7.1, 12 data points are required before the first point can be plotted and a further 11 before a full set of 12 points can be plotted. In contrast, by the time the first moving average point is drawn, a control chart can already begin providing useful information.
- Moving averages may occasionally be an appropriate tool to use in analysing data. However, they should be selected for exceptional circumstances, and should themselves be drawn as control charts.
- It is important when dealing with data that are counted (such as incident, errors and rejects) to consider whether the variability in the counts are related to opportunity. If they are, we need to normalise the data by analysing, for example, incidents per hour, errors per thousand units produced, errors per square meter inspected, rejects per number inspected.
- The reason that the moving average misleads us is explained in Part 2 of this book. In this particular situation it plots the difference between the value dropped off and the new value added. This tacitly assumes that the value dropped off is in some sense "average". Where a process change occurs, it will take a full year for the moving average to accurately reflect the current process average, and that is assuming there are no out-of-control points or process changes.
- We assumed that a process change occurred at about month 17. In practice we would need to try to identify when the change occurred and what caused it.

Calculations and data

Moving average chart

Some calculations are shown in Table 7.1 and in the text.

The calculations for the first 16 months of Chart 7.5 are as follows:

The calculation for u, the incident frequency, is given in Table 7.1.

$$\bar{u} = \text{average incident frequency for the first 16 months}$$
$$= \frac{\text{total number of incidents}}{\text{total hours}}$$
$$= \frac{34 \text{ incidents}}{4.83 \text{ mmh}}$$
$$= 7.0 \text{ incidents/mmh}.$$

$$s = \sqrt{\frac{\bar{u}}{n}}, \text{ which for the first month} = \sqrt{\frac{7}{0.23}} = 5.5.$$

Upper action limit (UAL) for month $1 = \bar{u} + 3s = 7.0 + 3 \times 5.5 = 23.5$.
Lower action limit (LAL) for month $1 = \bar{u} - 3s = 7.0 - 3 \times 5.5 < 0$, so that it is likely that sometimes we will have no incidents.

Summary

- Chapter 3 explained theoretically why moving averages were inferior to control charts for monitoring process performance. In this case study, we have used live data as an example to confirm and illustrate the mis-information in moving averages.
- In this case study, the moving average suggested that the incident rate was in gentle decline whereas the data showed a sudden process change, which was correctly identified by the control chart.
- Moving averages are aimed at removing the variation in individual data points by smoothing the data. Control charts use the information in this variation to help us understand how the process is performing.
- The control chart was used to predict future incident rate. Such predictions are generally much more difficult to make with a moving average chart.
- The above analysis shows how moving averages can blind us to the important information in variation. It also shows that the information in control charts can be used to determine likely process changes and aberrations, and hence be a driver for investigation and improvement.
- On presenting these results to management, one person commented that sometimes it was convenient to let management believe things were running smoothly, so that they did not interfere.

8 Monitoring rare events
How a sudden but uncertain change in safety record was shown to be significant

Charts used: c and X/MR

Introduction

Sometimes events occur only rarely. Typical examples of "rare events" include natural disasters, air and rail accidents, and serious injuries/death in the workplace. If we were to monitor the number of rare events per week or per year, etc., we would obtain a chart with many zeros and the occasional non-zero value. In extreme situations the resultant control chart would signal every event as an out-of-control condition.

In deriving an effective solution to overcome this problem, we show how control charts can be used to monitor the time between events.

Background

The occurrence of serious accidents in an industrial manufacturing sector is recorded in a national database and there has been regulatory concern that the number of accidents increased suddenly, after a protracted period with only occasional accidents. The apparent increase coincided with a period of political change and economic uncertainty, following many years of stable government led by one political party.

Table 8.1 lists the dates of the accidents. The first accident was recorded in March 1995. The start date of monitoring is not known. Data is available up to and including March 2004. In most months no accidents were recorded.

Table 8.1 Accident dates

Number	Date	Number	Date	Number	Date
1	13 Mar 1995	9	2 Jul 2001	17	16 Jan 2003
2	28 Feb 1996	10	11 Apr 2002	18	14 Mar 2003
3	28 Jan 1997	11	13 May 2002	19	18 Mar 2003
4	3 Jun 1998	12	7 Nov 2002	20	30 Mar 2003
5	29 Jul 1999	13	12 Dec 2002	21	1 Apr 2003
6	12 May 2000	14	13 Dec 2002	22	9 May 2003
7	13 May 2000	15	17 Dec 2002	23	5 Jan 2004
8	25 Sep 2000	16	7 Jan 2003		

Analysis

The usual choice of chart when monitoring events such as accidents is a c chart (if the opportunity of an event occurring remains the same) or a u chart (if the opportunity of an event occurring is variable).

In this case the opportunity for an event occurring remained constant over a number of years and so a c chart would be the usual choice. Chart 8.1 is the resulting chart. For much of the chart there are a large number of 0s with an occasional 1. However, towards the end of the chart there appears to be a flurry of accidents.

The upper action limit (UAL) for Chart 8.1 is 1.59, so every month that two accidents occur, a special cause of variation is indicated. However, if there is a process change, for example at April 2000, the recalculated UAL up to April 2000 would be 0.93 and so every accident would be a special cause. This is of no practical use for identifying and investigating process changes or special causes of variation.

The problem is that with such rare events the average number of events per month is very low, and so the UAL is low (see calculations below). To overcome this problem, when working with frequencies of less than 1, rather than monitor the number of

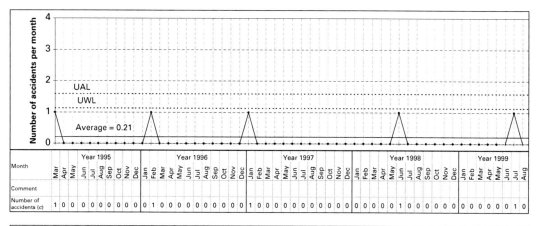

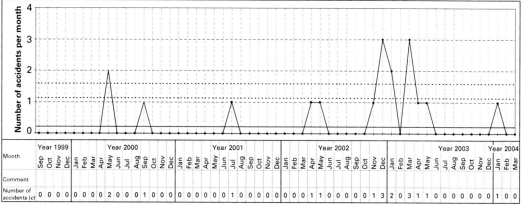

Chart 8.1 c chart: number of accidents per month; UAL and UWL: upper action and warning limits, respectively

events per month, we monitor the time between events. We count the number of days between events and convert this into a number of accidents per year (or other convenient unit).

For example, there are 352 days between the first accident on 13 March 1995 and the second on 28 February 1996. This equates to $365/352 = 1.04$ accidents per year. Chart 8.2 shows the results when using this approach. Since we do not know when

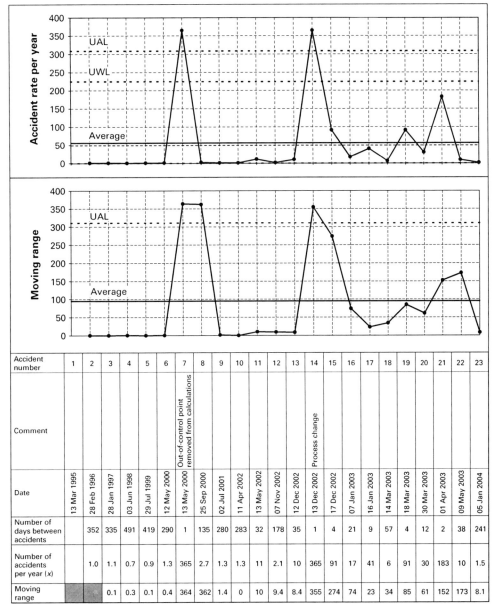

Chart 8.2 X/MR chart of annual accident rate; UAL and UWL: upper action and warning limits, respectively

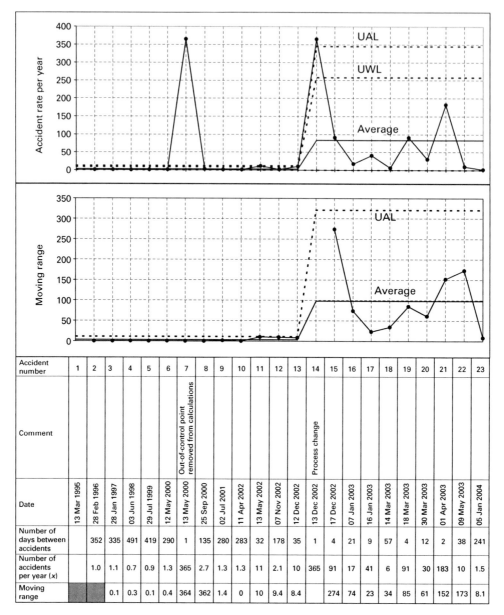

Chart 8.3 X/MR chart of annual accident rate with process change; UAL and UWL: upper action and warning limits, respectively

monitoring began it is not possible to calculate the number of days before the first accident occurred.

The table in Chart 8.2 includes the date of the accident, the number of days since the last accident and the calculated annual accident rate.

Chart 8.2 shows clearly that there is an out-of-control condition for accident 7. In addition, there is clear evidence of a process change commencing round about accident 14,

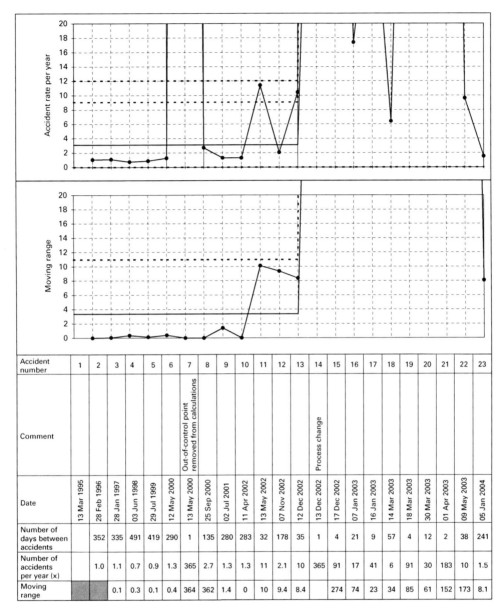

Accident number	1	2	3	4	5	6	7	8	9	10	11	12	13	14	15	16	17	18	19	20	21	22	23
Comment							Out-of-control point removed from calculations							Process change									
Date	13 Mar 1995	28 Feb 1996	28 Jan 1997	03 Jun 1998	29 Jul 1999	12 May 2000	13 May 2000	25 Sep 2000	02 Jul 2001	11 Apr 2002	13 May 2002	07 Nov 2002	12 Dec 2002	13 Dec 2002	17 Dec 2002	07 Jan 2003	16 Jan 2003	14 Mar 2003	18 Mar 2003	30 Mar 2003	01 Apr 2003	09 May 2003	05 Jan 2004
Number of days between accidents		352	335	491	419	290	1	135	280	283	32	178	35	1	4	21	9	57	4	12	2	38	241
Number of accidents per year (x)		1.0	1.1	0.7	0.9	1.3	365	2.7	1.3	1.3	11	2.1	10	365	91	17	41	6	91	30	183	10	1.5
Moving range			0.1	0.3	0.1	0.4	364	362	1.4	0	10	9.4	8.4		274	74	23	34	85	61	152	173	8.1

Chart 8.4 X/MR chart of annual accident rate with expanded scale

in December 2002, with a point above the control limit and a group of much higher values than generally seen earlier.

Removing accident 7 from the calculations and imposing a process change at accident 14 results in Chart 8.3.

Chart 8.3 still shows accident 14 as being above the UAL.

Chart 8.4 shows Chart 8.3 on an expanded scale so that we can examine the first 13 accidents in more detail. It is interesting to note that accident 11 is only just below the

UAL. Accident 12, although below the average, is still high compared to all the previous accident rates, and accident 13 is also only just below the UAL. This suggests the process change may have occurred earlier, and should be placed at 11, or that two changes may have occurred, one in the spring of 2002 and the second in December 2002.

The next step is to investigate what was happening at these times in order to deduce what caused the increase in accident frequency. However, there is no doubt that since December 2002 the rate has markedly increased, as initially suspected.

Calculations

The calculations for the c chart (Chart 8.1) are:

$$\bar{c} = \frac{\text{total number of accidents}}{\text{total number of months}} = \frac{23}{109} = 0.21$$

$$s = \sqrt{\bar{c}} = 0.46$$

$$\text{UAL} = \bar{c} + 3s = 0.21 + 3 \times 0.46 = 1.59.$$

For the period up to April 2000, $\bar{c}$ = 5 incidents/61 months = 0.081 and the UAL is 0.93.

The X/MR chart calculations follow the usual formulae, and the charts show the calculation for turning the days between accidents into accidents per year.

Summary

In this case study, we have shown that the usual attributes charts are not appropriate for monitoring rare events. Although this case study concerns accident rates, the same methodology applies to many types of rare event including, for example, spills, disasters and breakdowns. One guideline suggests that attribute charts should not be used when the average event rate falls below 1. In these situations, we can convert the data into a number of events per year (or other appropriate base) and chart the results using an X/MR chart. See also Part 4 for more information.

9

Comparing surgical complication rates between hospitals

Charts used: p

Introduction

One of the purposes of this book is to demonstrate that statistical process control (SPC) is applicable to a wide range of applications in a wide range of business sectors. This example illustrates the use of p charts in the health care industry.

The percentage of surgical interventions that result in complications for each of four hospitals are combined and charted on a single chart. The resulting control chart suggests that the process is in a state of control. However, further investigation reveals that three of the four hospitals have sudden changes or trends in the data which are hidden when the data are combined. This illustrates the importance of ensuring that results from sub-processes (hospitals in this case) are analysed separately with a control chart to ensure that the sub-process is in a state of control before combining outputs.

We also examine the problem of and propose solutions for "chunkiness" in charted values whereby the values fall into "bands" with no data in-between the bands.

This case study is one of the few not based on actual data. The data were generated using a random number generator in a standard spreadsheet package. One advantage of this approach is that it allows us to observe how a control chart reacts where the distribution of the data is known.

Whilst this case study is presented in the health care context, it can be readily seen that the process can be viewed as a number of "inspections" (in this case surgical procedures) and that each inspection "passes" or "fails" (results in a complication). The method used here can be used to analyse similar situations both within and outside health care.

Background

A key metric in surgical procedures is the proportion (or percentage) of surgeries that experience complications. Four hospitals in a certain geographical area were working closely together to improve their service to their respective communities and decided to compare data on their percentage of complications during surgery in an attempt to reduce the level.

The hospitals agreed on a definition of a "complication", and have gathered data on a monthly basis over a 4-year period.

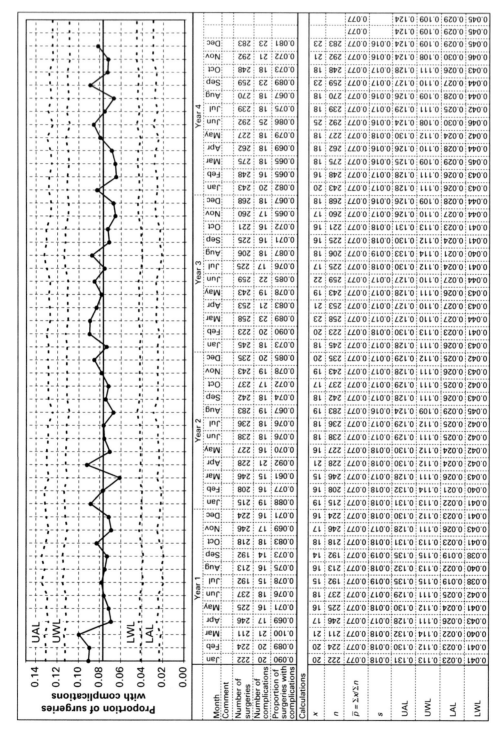

Chart 9.1 p chart: complications during surgery – all hospitals; UWL: upper warning limit; UAL: upper action limit; UWL: upper warning limit; LAL: lower action limit; LWL: lower warning limit

Analysis

Monthly p chart of all data

Chart 9.1 shows the proportion of surgeries with complications for all hospitals. Below the chart is a table which includes the month and year number, comments, number of surgical procedures carried out in the month, number of surgical complications and the proportion of surgeries with complications. The proportion of surgeries with complications is calculated as:

$$\text{proportion of surgeries with complications} = \frac{\text{number of surgeries with complications}}{\text{number of surgical procedures}}.$$

The percentage of surgeries with complications is obtained by multiplying the proportion by 100 and the two terms, proportion and percentage, are used freely in this case study.

The process appears to be stable (in a state of control). There are no obvious runs, patterns or points above the control limits.

The summary statistics for the four hospitals are given in Table 9.1.

The unwary may conclude that as hospital A has the lowest percentage of surgeries with complications, it is the best performing hospital, and that hospitals C and D in particular are performing poorly and need to improve.

Table 9.1 Summary of source data for surgical procedure complications

	Hospitals				
	A	*B*	*C*	*D*	*Total*
Number of surgeries	1426	4065	3630	2391	11,512
Number of surgeries with complications	60	197	338	290	885
Percentage of surgeries with complications	4.2	4.9	9.3	12.1	7.7

Monthly p charts for each hospital

The next step should be to investigate the performance of each hospital separately. To do this, we draw p charts of each hospital. The results are given in Charts 9.2(a)–9.2(d).

The data for these charts were calculated using a random number generator following the rules given in the comments section of each chart, and the calculated values may not be exactly the same as the rules due to sampling from the generator.

For hospital A
As stated in the comment row of the Chart 9.2(a), these data are drawn from a normal distribution with an average of 4% of surgeries having complications. The process appears to be in a state of control with, as we would expect, an average of just over 0.04 (i.e. 4%).

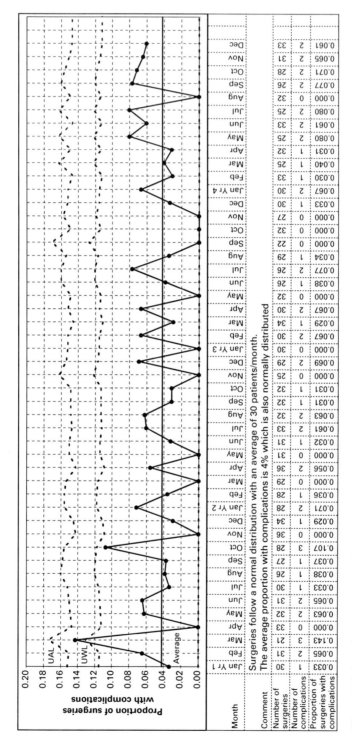

Month	Number of surgeries	Number of complications	Proportion of surgeries with complications
Jan Yr 1	30	1	0.033
Feb	31	2	0.065
Mar	21	3	0.143
Apr	33	0	0.000
May	32	2	0.063
Jun	31	2	0.065
Jul	30	1	0.033
Aug	26	1	0.038
Sep	27	1	0.037
Oct	28	3	0.107
Nov	36	0	0.000
Dec	34	1	0.029
Jan Yr 2	28	2	0.077
Feb	28	1	0.036
Mar	29	0	0.000
Apr	36	2	0.056
May	31	0	0.000
Jun	31	1	0.032
Jul	33	2	0.061
Aug	32	2	0.063
Sep	32	1	0.031
Oct	32	1	0.031
Nov	25	0	0.000
Dec	29	2	0.069
Jan Yr 3	30	0	0.000
Feb	30	2	0.067
Mar	34	1	0.029
Apr	30	2	0.067
May	32	0	0.000
Jun	26	1	0.038
Jul	26	2	0.077
Aug	29	1	0.034
Sep	22	0	0.000
Oct	32	0	0.000
Nov	27	0	0.000
Dec	30	1	0.033
Jan Yr 4	30	2	0.067
Feb	33	1	0.030
Mar	25	1	0.040
Apr	32	1	0.031
May	25	2	0.080
Jun	33	2	0.061
Jul	25	2	0.080
Aug	32	0	0.000
Sep	26	2	0.077
Oct	28	2	0.071
Nov	31	2	0.065
Dec	33	2	0.061

Comment: Surgeries follow a normal distribution with an average of 30 patients/month. The average proportion with complications is 4% which is also normally distributed

Chart 9.2(a) p chart: complications during surgery – hospital A

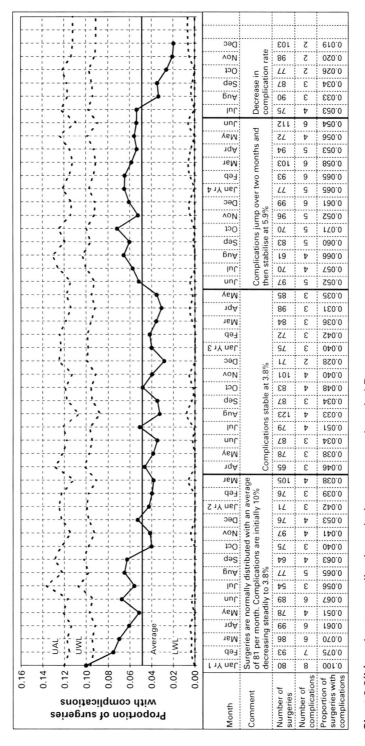

Month	Comment	Number of surgeries	Number of complications	Proportion of surgeries with complications
Jan Yr 1	Surgeries are normally distributed with an average of 81 per month. Complications are initially 10% decreasing steadily to 3.8%	80	8	0.100
Feb		93	7	0.075
Mar		86	6	0.070
Apr		99	6	0.061
May		78	4	0.051
Jun		89	6	0.067
Jul		54	3	0.056
Aug		77	5	0.065
Sep		64	4	0.063
Oct		75	3	0.040
Nov		97	4	0.041
Dec		76	4	0.053
Jan Yr 2	Complications stable at 3.8%	71	3	0.042
Feb		76	3	0.039
Mar		105	4	0.038
Apr		65	3	0.046
May		78	3	0.038
Jun		87	3	0.034
Jul		79	4	0.051
Aug		123	4	0.033
Sep		87	3	0.034
Oct		83	4	0.048
Nov		101	4	0.040
Dec		71	2	0.028
Jan Yr 3		75	3	0.040
Feb		72	3	0.042
Mar		84	3	0.036
Apr		98	3	0.031
May		85	3	0.035
Jun	Complications jump over two months and then stabilise at 5.9%	97	5	0.052
Jul		70	4	0.057
Aug		61	4	0.066
Sep		83	5	0.060
Oct		70	5	0.071
Nov		96	5	0.052
Dec		99	6	0.061
Jan Yr 4		77	5	0.065
Feb		93	6	0.065
Mar		103	6	0.058
Apr		94	5	0.053
May		72	4	0.056
Jun		112	6	0.054
Jul	Decrease in complication rate	75	4	0.053
Aug		90	3	0.033
Sep		87	3	0.034
Oct		77	2	0.026
Nov		98	2	0.020
Dec		103	2	0.019

Chart 9.2(b) p chart: complications during surgery – hospital B

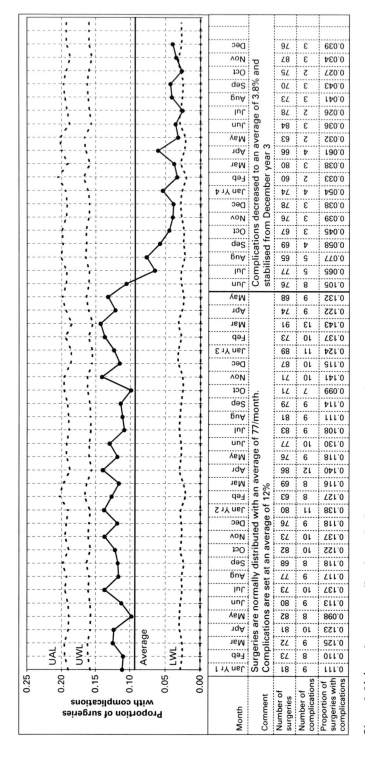

Month	Comment	Number of surgeries	Number of complications	Proportion of surgeries with complications
Jan Yr 1	Surgeries are normally distributed with an average of 77/month. Complications are set at an average of 12%	81	9	0.111
Feb		73	8	0.110
Mar		72	9	0.125
Apr		81	10	0.123
May		82	8	0.098
Jun		80	9	0.113
Jul		73	10	0.137
Aug		77	9	0.117
Sep		68	8	0.118
Oct		82	10	0.122
Nov		73	10	0.137
Dec		76	9	0.118
Jan Yr 2		80	11	0.138
Feb		63	8	0.127
Mar		69	8	0.116
Apr		86	12	0.140
May		76	9	0.118
Jun		77	10	0.130
Jul		83	9	0.108
Aug		81	9	0.111
Sep		79	9	0.114
Oct		71	7	0.099
Nov		71	10	0.141
Dec		87	10	0.115
Jan Yr 3		89	11	0.124
Feb		73	10	0.137
Mar		91	13	0.143
Apr		74	9	0.122
May		68	9	0.132
Jun	Complications decreased to an average of 3.8% and stabilised from December year 3	76	8	0.105
Jul		77	5	0.065
Aug		65	5	0.077
Sep		69	4	0.058
Oct		67	3	0.045
Nov		76	3	0.039
Dec		78	3	0.038
Jan Yr 4		74	4	0.054
Feb		60	2	0.033
Mar		80	3	0.038
Apr		66	4	0.061
May		63	2	0.032
Jun		84	3	0.036
Jul		78	2	0.026
Aug		73	3	0.041
Sep		70	3	0.043
Oct		75	2	0.027
Nov		87	3	0.034
Dec		76	3	0.039

Chart 9.2(c) p chart: complications during surgery – hospital C

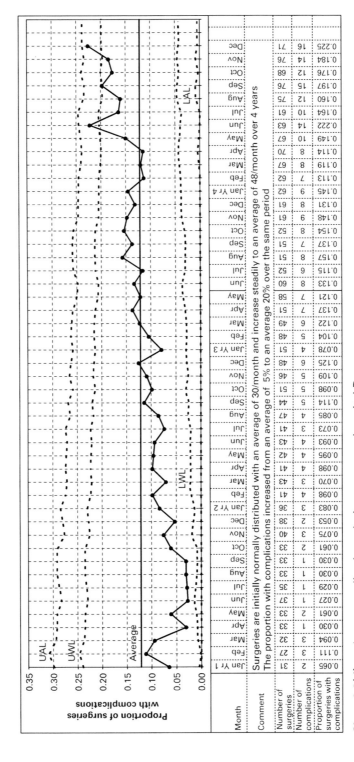

Month	Comment	Number of surgeries	Number of complications	Proportion of surgeries with complications
Jan Yr 1	Surgeries are initially normally distributed with an average of 30/month and increase steadily to an average of 48/month over 4 years. The proportion with complications increased from an average of 5% to an average 20% over the same period	31	2	0.065
Feb		27	3	0.111
Mar		32	3	0.094
Apr		33	1	0.030
May		33	2	0.061
Jun		37	1	0.027
Jul		35	1	0.029
Aug		33	1	0.030
Sep		33	1	0.030
Oct		33	2	0.061
Nov		40	3	0.075
Dec		38	2	0.053
Jan Yr 2		36	3	0.083
Feb		41	4	0.098
Mar		43	3	0.070
Apr		41	4	0.098
May		42	4	0.095
Jun		43	4	0.093
Jul		41	3	0.073
Aug		47	4	0.085
Sep		44	5	0.114
Oct		51	5	0.098
Nov		46	5	0.109
Dec		48	6	0.125
Jan Yr 3		51	4	0.078
Feb		48	5	0.104
Mar		49	6	0.122
Apr		51	7	0.137
May		58	7	0.121
Jun		60	8	0.133
Jul		52	6	0.115
Aug		51	8	0.157
Sep		51	7	0.137
Oct		52	8	0.154
Nov		61	9	0.148
Dec		61	8	0.131
Jan Yr 4		62	9	0.145
Feb		62	7	0.113
Mar		67	8	0.119
Apr		70	8	0.114
May		67	10	0.149
Jun		63	14	0.222
Jul		61	10	0.164
Aug		75	12	0.160
Sep		76	15	0.197
Oct		68	12	0.176
Nov		76	14	0.184
Dec		71	16	0.225

Chart 9.2(d) p chart: complications during surgery – hospital D

Closer inspection reveals that the data appear in "bands". There are a large number of 0 values, then values between 0.03 and 0.04, between 0.06 and 0.08, one value near 0.10 and another near 0.14 and very little elsewhere (e.g. between 0 and 0.03). A histogram would highlight this pattern and you may like to draw one.

When data occur in "bands" like these, one explanation is that because the values have to be integers the range of possible proportions is limited. In this example, the number of surgeries varies between 21 and 36, and the number of complications can only take whole numbers so:

- If there is 1 complication a month of 21 surgeries, the proportion will be $1/21 = 0.048$.
- If there is 1 complication a month of 36 surgeries, the proportion will be $1/36 = 0.028$.

In order to record proportions between 0 and 0.028 we would need more than 36 surgeries.

Similarly if there were 2 complications the proportion would lie between 2/21 and 2/36, that is 0.095 and 0.056, and so values between 0.048 (1 complication in 21 surgeries) and 0.056 will never occur.

To overcome this "graininess" or "chunkiness" there are several techniques we can use including:

- increasing sample sizes (i.e. the number of surgeries in this case);
- collecting data over a longer period (e.g. quarterly rather than monthly);
- monitoring the time between complications incidents.

The case study in Chapter 10 gives an example of collecting data over a longer period, and Chapter 8 monitoring the time between events. Increasing the sample size in this case study would be achieved by collecting data over a longer period, but in many situations, for example sampling, would be achieved by simply taking a larger sample. Since these methods are discussed elsewhere, they are not discussed further in this case study; however, you may like to try combining data into quarters as an exercise.

To continue with the monthly p charts, we conclude from Chart 9.2(a) that the process is in a state of (statistical) control.

For hospital B
Chart 9.2(b) shows that the process is not in a state of control over the 4 years. For the first 15 months the complication rate decreases from 8% to 3.7% and then remains stable until half-way through year 3. The complication rate then jumps over 2 months to 5.8% and remains above the average for 14 months before commencing a decline in month 42.

For hospital C
Chart 9.2(c) shows a steady stable process for the first 2½ years with an average 12%. This is followed by a steady decrease, as might result from a series of improvement measures, until the end of the year after which the complication rate stabilises at 3%.

For hospital D
Chart 9.2(d) is the result of a steady increase in both surgical procedures and in complication rate. This pattern may occur where an organisation is increasing the amount

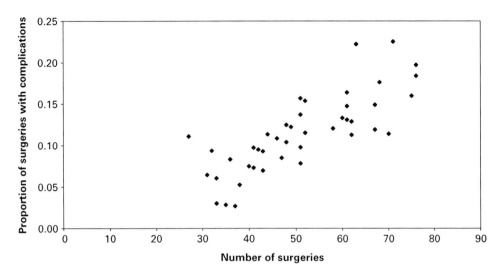

Chart 9.3 Scatter diagram of the proportion of surgeries with complications vs. number of surgeries – hospital D

of work it does, and is unable to maintain the same work standards. Chart 9.3 is a scatter diagram which clearly demonstrates the effect.

In conclusion, despite the fact that the control chart for the combined hospital data (Chart 9.1) suggested that the complication rate was stable, a more detailed analysis has shown that there are four sub-processes: three of which are not in a state of control. In situations like this it is misleading to combine the data. We should always ensure that sub-processes are in a state of statistical control by charting them separately before combining them.

Comparison of hospitals

Since some of the four hospitals show out-of-control signals, is it appropriate to compare hospitals? The answer to that question is a cautious yes. We can compare hospitals which are showing stable conditions, that is, for hospital A, all their data and hospital C from December 2003.

For hospital D we could estimate the rate of increase and comment that the proportion of complications is steadily increasing and currently around 20% (with further analysis we could be more precise, but is unlikely that being any more precise will bring any benefits at this time).

Similarly, for hospital B we could comment that the rate appears to be in steadily improving after a period of stability (June year 3 to June year 4 when it was around 6%) and is currently at around 2%.

However, before making any firm conclusions about hospitals B and, in particular, C (as other hospitals may look to C for learning opportunities to improve) it is necessary to understand what is happening at the hospitals that is driving the changes. For hospital B especially, we should also satisfy ourselves that there is no long-term cycling. (Looking carefully at the data we see that the first 14 months suggest a decrease in complications, followed by a stable year, and then a year of increased complication rate, followed now by a decrease. Will this be followed by an increase?)

What are the next steps?

At this time, suitable actions for the hospitals could be as follows:

- Hospitals A and C have similar performance levels, with C just a little lower than A. They may analyse causes of complications and review each other's practices with a view to finding and implementing improvements.
- Hospitals A and C may also discuss with B the reasons for the perceived improvement. Perhaps there are some good practices that have been/are being implemented.
- Hospital C should aim to continue improving (if they are proactively doing so) until the process stabilises and then investigate possible further improvements.
- Hospital D has an obvious problem and needs to address urgently the relationship between the number of surgeries and the complication rate. If understood, it may be appropriate to reduce the surgery rate and thereby the complication rate until they can implement improvements.

Comments

It would be possible to calculate the average percentage of surgeries with complications by calculating the average of the individual percentages giving $(4.2 + 4.8 + 9.3 + 12.1)/4 = 7.6\%$. This gives a biased result. The correct method is to calculate:

$$\text{average percentage complications} = 100 \times \frac{\text{total number of complications}}{\text{total number of surgeries}} = 7.7\%.$$

The reason that these calculations give different averages is that the first method does not take into account the different number of surgeries at each hospital; for example, hospital A with only 1426 surgeries carries the same weight as hospital B with 4065 surgeries and its low percentage (4.2%) pulls down the average too far. With these data, the difference is not great, but in other situations it may be more significant.

Calculations

The p chart

The results of calculations for Chart 9.1 are given in the table below the chart.

The reject rate for each week, p_i, is calculated as number rejected/number inspected.

$$\text{The average reject rate } (\bar{p}) = \frac{\text{total number of complications}}{\text{total number of surgeries}}$$

$$= \frac{885}{11,512} = 0.077 = 7.7\%.$$

$$\text{Average number inspected } (\bar{n}) = \frac{\text{total number of surgeries}}{\text{total months}}$$

$$= \frac{11,512}{48} = 240 \text{ surgeries per month.}$$

The standard deviation, s, is calculated separately for each month from the formula:

$$s = \sqrt{\frac{\bar{p}(1 - \bar{p})}{n}} = \sqrt{\frac{0.077(1 - 0.077)}{222}} = 0.018, \text{ for January year 1.}$$

The action and the warning limits are calculated in the usual way as $\bar{p} \pm 3s$ and $\bar{p} \pm 2s$, respectively.

Summary

This case study began by analysing a control chart of the combined data from four hospitals. The control chart showed that the process was stable and in a state of statistical control. However, when the data from each hospital were analysed separately we found that only one of the four hospitals had a stable process, the other three had a mix of process changes and trends. This demonstrates the importance of using a control chart to check that sub-processes are in a state of control before combining data.

10

Comparing the frequency of rare medical errors between medical centres

Charts used: p

Introduction

A key metric in health care is the incidence of medical errors. These might be errors in medical procedure, prescribed medication, diagnosis, etc. In each case the analysis method will be similar to methods used with events in other industries, for example, errors in manufacturing procedure, diagnosis of equipment breakdowns, etc. The whole medical industry has, to an extent, a parallel in other industries, where the patient could be replaced by an object needing repair; health care procedures (purchasing, diagnosis, treatment, etc.) are mirrored by maintenance procedures, etc. Of course, health care is a different industry and there are differences in analysis, but there are also common approaches as illustrated in the health care-specific cases in this book.

This case study uses entirely fictional data and is not intended to be indicative of actual error rates in the health care industry. The data were generated using a random number generator in a standard spreadsheet package. One advantage of this approach is that it allows us to observe how a control charts react where the data have been drawn from a specified distributions.

This case study focuses on the difficulties of analysing rare event data. It also illustrates how the method of changing reporting frequency overcomes these difficulties.

Background

A medical centre is aware of their occasional patient care errors, and have decided that they would like to reduce these errors. They realise that one of the most effective methods of improving is to compare error rates with other medical centres and attempt to learn from those with the lowest error rates. They contacted three other medical centres and as a first project agree to focus on medication errors.

Data for the last 4 years are collected and are reproduced in Table 10.1.

Table 10.1 Medical errors for four medical centres A–D

Month	Medical centre A			B			C			D		
	Number of patients	Number of errors	Proportion	Number of patients	Number of errors	Proportion	Number of patients	Number of errors	Proportion	Number of patients	Number of errors	Proportion
Year 1												
Jan	823	5	0.006	671	0	0.000	1619	10	0.006	472	2	0.004
Feb	804	3	0.004	677	1	0.001	1642	10	0.006	473	2	0.004
Mar	821	6	0.007	719	0	0.000	1488	9	0.006	464	2	0.004
Apr	811	3	0.004	769	6	0.008	1591	9	0.006	470	3	0.006
May	801	6	0.007	762	3	0.004	1594	9	0.006	480	3	0.006
Jun	805	0	0.000	740	1	0.001	1563	9	0.006	469	2	0.004
Jul	848	4	0.005	725	0	0.000	1694	9	0.005	459	3	0.007
Aug	819	5	0.006	727	0	0.000	1556	8	0.005	465	3	0.006
Sep	804	7	0.009	674	0	0.000	1546	8	0.005	466	2	0.004
Oct	814	2	0.002	740	3	0.004	1705	10	0.006	490	3	0.006
Nov	810	4	0.005	721	0	0.000	1723	10	0.006	465	3	0.006
Dec	785	4	0.005	729	8	0.011	1710	10	0.006	479	2	0.004
Year 2												
Jan	802	4	0.005	778	0	0.000	1570	11	0.007	498	2	0.004
Feb	788	3	0.004	731	0	0.000	1556	11	0.007	491	2	0.004
Mar	760	2	0.003	718	9	0.013	1632	11	0.007	491	3	0.006
Apr	829	4	0.005	741	0	0.000	1619	9	0.006	479	3	0.006
May	813	2	0.002	713	5	0.007	1495	10	0.007	505	4	0.008
Jun	792	4	0.005	713	1	0.001	1593	9	0.006	471	3	0.006
Jul	849	5	0.006	677	0	0.000	1687	10	0.006	480	2	0.004
Aug	810	2	0.002	749	2	0.003	1690	12	0.007	489	2	0.004

Month												
Sep	829	3	0.004	767	1	0.001	1579	10	0.006	478	3	0.006
Oct	853	3	0.004	746	4	0.005	1616	8	0.005	486	2	0.004
Nov	796	8	0.010	710	6	0.008	1589	10	0.006	481	2	0.004
Dec	847	6	0.007	699	0	0.000	1524	9	0.006	497	3	0.006
Year 3												
Jan	831	5	0.006	684	0	0.000	1536	8	0.005	489	1	0.002
Feb	800	4	0.005	672	0	0.000	1726	11	0.006	490	2	0.004
Mar	798	2	0.003	715	4	0.006	1551	9	0.006	452	2	0.004
Apr	794	5	0.006	786	2	0.003	1608	12	0.007	478	1	0.002
May	833	7	0.008	716	0	0.000	1634	7	0.004	462	2	0.004
Jun	783	6	0.008	717	2	0.003	1520	9	0.006	486	3	0.006
Jul	857	6	0.007	743	0	0.000	1624	10	0.006	507	3	0.006
Aug	805	4	0.005	735	7	0.010	1618	11	0.007	474	3	0.006
Sep	805	4	0.005	767	0	0.000	1659	9	0.005	461	2	0.004
Oct	839	5	0.006	698	0	0.000	1391	8	0.006	481	2	0.004
Nov	807	8	0.010	747	0	0.000	1517	8	0.005	499	3	0.006
Dec	844	5	0.006	702	6	0.009	1522	9	0.006	485	3	0.006
Year 4												
Jan	851	5	0.006	719	6	0.008	1716	11	0.006	481	3	0.006
Feb	846	4	0.005	651	0	0.000	1625	11	0.007	510	2	0.004
Mar	796	4	0.005	683	0	0.000	1670	11	0.007	458	2	0.004
Apr	797	6	0.008	728	0	0.000	1755	11	0.006	515	2	0.004
May	810	3	0.004	754	3	0.004	1594	11	0.007	470	2	0.004
Jun	790	3	0.004	726	2	0.003	1586	9	0.006	497	3	0.006
Jul	826	4	0.005	698	3	0.004	1562	10	0.006	453	1	0.002
Aug	827	3	0.004	736	1	0.001	1683	8	0.005	470	2	0.004
Sep	799	1	0.001	756	11	0.015	1471	10	0.007	465	3	0.006
Oct	829	5	0.006	697	0	0.000	1648	10	0.006	464	3	0.006
Nov	830	7	0.008	719	0	0.000	1583	9	0.006	483	3	0.006
Dec	825	4	0.005	717	7	0.010	1591	10	0.006	471	2	0.004
Average	**815**	**4.3**	**0.0052**	**722**	**2.2**	**0.0030**	**1604**	**9.6**	**0.0060**	**479**	**2.4**	**0.0050**

Analysis

Monthly p chart for each medical centre

Charts 10.1(a)–10.1(d) show the proportion of errors for each of the medical centres, A–D. Below the charts are tables which include the month and year number, comments, the number of patients attended, number of errors and the proportion of patients involved in medical errors. The proportion of errors is calculated as:

$$\text{proportion of errors} = \frac{\text{number of medical errors}}{\text{number of patients}}.$$

The percentage of errors is obtained by multiplying the proportion by 100 and the two terms, proportion and percentage are used interchangeably in this case study.

Medical centre A (Chart 10.1(a))

June year 1 is the only month with zero incidents. This is just below the lower warning limit (LWL), and zero values are occasionally expected. Of more interest is the run of nine values below the average beginning in October of year 1, followed by one value just above the average and another three below. It is also interesting

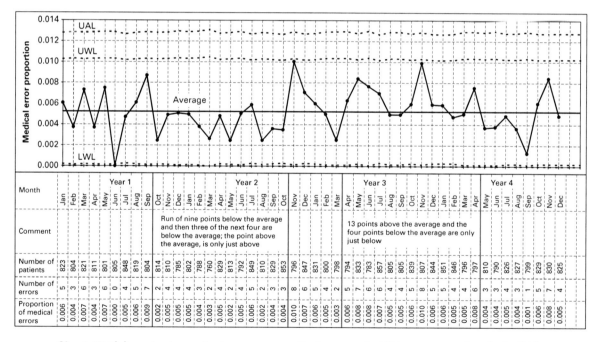

Month	Year	Comment	Number of patients	Number of errors	Proportion of medical errors
Jan	1		823	5	0.006
Feb	1		804	3	0.004
Mar	1		821	6	0.007
Apr	1		811	3	0.004
May	1		801	6	0.007
Jun	1		805	0	0.000
Jul	1		848	4	0.005
Aug	1		819	5	0.006
Sep	1		804	7	0.009
Oct	1	Run of nine points below the average and then three of the next four are below the average; the point above the average, is only just above	814	2	0.002
Nov	1		810	4	0.005
Dec	1		785	4	0.005
Jan	2		802	3	0.004
Feb	2		788	2	0.003
Mar	2		760	4	0.005
Apr	2		829	2	0.002
May	2		813	4	0.005
Jun	2		792	5	0.006
Jul	2		849	2	0.002
Aug	2		810	3	0.004
Sep	2		829	3	0.004
Oct	2		853	2	0.002
Nov	2		796	8	0.010
Dec	2		847	6	0.007
Jan	3		831	5	0.006
Feb	3		800	4	0.005
Mar	3		798	2	0.003
Apr	3	13 points above the average and the four points below the average are only just below	794	5	0.006
May	3		833	7	0.008
Jun	3		783	6	0.008
Jul	3		857	6	0.007
Aug	3		805	4	0.005
Sep	3		805	4	0.005
Oct	3		839	5	0.006
Nov	3		807	8	0.010
Dec	3		844	5	0.006
Jan	4		851	5	0.006
Feb	4		846	4	0.005
Mar	4		796	4	0.005
Apr	4		797	6	0.008
May	4		810	3	0.004
Jun	4		790	3	0.004
Jul	4		826	4	0.005
Aug	4		827	3	0.004
Sep	4		799	1	0.001
Oct	4		829	5	0.006
Nov	4		830	7	0.008
Dec	4		825	4	0.005

Chart 10.1(a) p chart: medical errors – medical centre A; UAL: upper action limit; LWL: lower warning limit; UWL: upper warning limit

to note that between April year 3 and April year 4 the four values below the average are just below. These features suggest that the process is not in a state of control.

Medical centre B (Chart 10.1(b))

The chart has five points above the upper action limit (UAL) suggesting that the process at this centre is not in a state of control. The fact that these five points are scattered throughout the 4 years and there are no indications of a process change (e.g. a run of points above the average), suggests that these were caused by a series of isolated events. On further inspection we see that there are also a large number of zeros, resulting in huge variability. The large number of zeros can be a signal that the grouping (monthly in this case) is too short. Investigating this further, the average proportion of errors is very low at around 0.003, that is, 3 errors per 1000 patients. The average number of patients is 722 per month, giving the actual number of errors as $3 \times (722/1000) =$ 2.2 per month. (We could also average the number of errors to arrive at the same value of 2.2.)

Such low frequencies are known to cause problems when charting, and as the average drops, so the chart becomes less resilient, giving more false "special cause" signals.

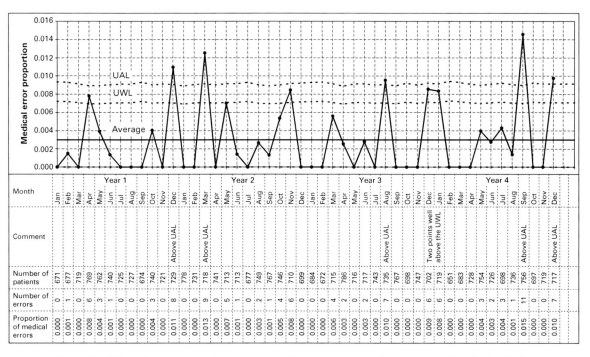

Year	Month	Comment	Number of patients	Number of errors	Proportion of medical errors
Year 1	Jan		671	0	0.000
	Feb		677	1	0.001
	Mar		719	0	0.000
	Apr		769	6	0.008
	May		762	3	0.004
	Jun		740	1	0.001
	Jul		725	0	0.000
	Aug		727	0	0.000
	Sep		674	3	0.004
	Oct		740	0	0.000
	Nov	Above UAL	721	8	0.011
	Dec		729	0	0.000
Year 2	Jan		778	0	0.000
	Feb	Above UAL	731	9	0.013
	Mar		718	0	0.000
	Apr		741	5	0.007
	May		713	1	0.001
	Jun		713	0	0.000
	Jul		677	2	0.003
	Aug		749	1	0.001
	Sep		767	4	0.005
	Oct		746	6	0.008
	Nov		710	0	0.000
	Dec		699	0	0.000
Year 3	Jan		684	0	0.000
	Feb		672	4	0.006
	Mar		715	2	0.003
	Apr		786	0	0.000
	May		716	2	0.003
	Jun		717	0	0.000
	Jul	Above UAL	743	7	0.010
	Aug		735	0	0.000
	Sep		767	0	0.000
	Oct		698	6	0.009
	Nov	Two points well above the UWL	747	6	0.008
	Dec		702	0	0.000
Year 4	Jan		719	0	0.000
	Feb		651	0	0.000
	Mar		683	3	0.004
	Apr		728	2	0.003
	May		754	3	0.004
	Jun		726	1	0.001
	Jul	Above UAL	698	11	0.015
	Aug		736	0	0.000
	Sep		754	0	0.000
	Oct		756	0	0.000
	Nov		697	0	0.000
	Dec	Above UAL	717	7	0.010

Chart 10.1(b) p chart: medical errors – medical centre B; UAL: upper action limit; UWL: upper warning limit

There are a number of guidelines for deciding whether the average is too low. Two common guidelines are that the average is considered too low if either:

- the lower action limit (LAL) is zero,
- the average number of errors <4 per group (i.e. four per month in this case).

We can use these guidelines to determine the minimum recommended group size (see the comments below).

Medical centre C (Chart 10.1(c))

There is a run of nine points just below the average beginning in April of year 2. However, the interesting feature of this chart is the low variability. Whilst the average proportion of errors is high at around 0.006, the variability is remarkably low – all of the 48 points are well within the warning limits. This could indicate that the data are being falsified. Months with high numbers of errors are being under reported and errors "moved" into months with low numbers of errors. We might check to see if there are targets, or limits within which the centre is supposed to operate. Alternatively, it could be that the data are correct, in which case we would want to investigate the reason for such low variability, and the run of nine values below the average (Chart 10.1(c)).

Medical centre D (Chart 10.1(d) see page 109)

Chart 10.1(d) shows the average proportion of errors as virtually the same as Chart 10.1(a), but with a much lower variability.

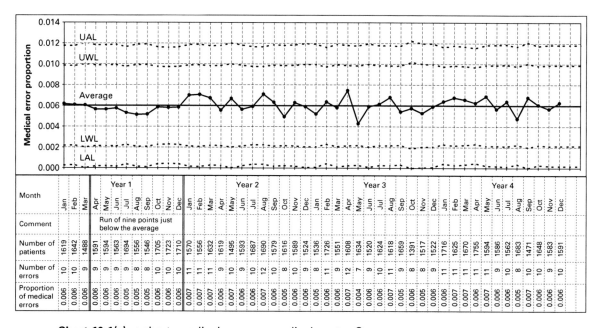

Chart 10.1(c) p chart: medical errors – medical centre C

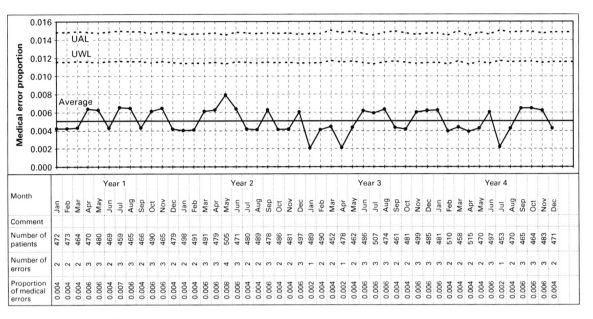

Chart 10.1(d) p chart: medical errors – medical centre D

Table 10.2 provides some more information about the data. For each of the medical centres the table gives the average number of patients seen per month, the average number of errors, the errors as a proportion and as a percentage of the number of patients, the maximum and minimum monthly error percentages and the range. The final two rows are explained below and in the comments.

The average number of errors at the four centres is 4.3, 2.2, 9.6 and 2.4. Using one of the above guidelines, we need at least four errors per time period. Whilst centre C is well above the minimum, A is only just above and B and D are well below.

In this case study, as a solution we have chosen to plot the data quarterly. This will produce error averages well above the minimum of 4 and still provide 16 data points per chart. Charts 10.2(a)–10.2(d) show the result.

Table 10.2 Some summary statistics

	Medical centre			
	A	B	C	D
Average number of patients (monthly), n	815	722	1604	479
Average number of errors (monthly), x	4.3	2.2	9.6	2.4
Average error proportion, $x/n = \bar{p}$	0.0052	0.0030	0.0060	0.0050
Average error percentage $= 100\,\bar{p}$	0.52	0.30	0.60	0.50
Maximum error percentage (monthly)	1.01	1.46	0.75	0.79
Minimum error percentage (monthly)	0.00	0.00	0.43	0.20
Range (maximum–minimum)	1.01	1.46	0.32	0.59
Recommended minimum sample size $= 4/\bar{p}$	765	1349	665	794
Average number of months to reach sample size	0.9	1.9	0.4	1.7

Medical centre A (Chart 10.2(a))

Chart 10.2(a) shows that the process is in a state of control. However, if we compare Chart 10.2(a) with 10.1(a), we noticed above that Chart 10.1(a) had runs of points below and then above the average. The reason that these runs are less marked in Chart 10.2(a) is that we have combined groups of three months into a single point. The run below the average from October year 1 has been condensed into the four points starting year 1 quarter 4, and the high values beginning April year 3 have been condensed into the three values starting in year 3 quarter 2. The fact that there is a difference between these two charts demonstrates that where the choice of chart is not obvious we could consider drawing and comparing different charts. In this particular case we would carry out an investigation to determine if any changes did occur to cause (especially) the run of points below the average in Chart 10.1(a); but should also be aware we do not have a high error rate and the signal may be a false one.

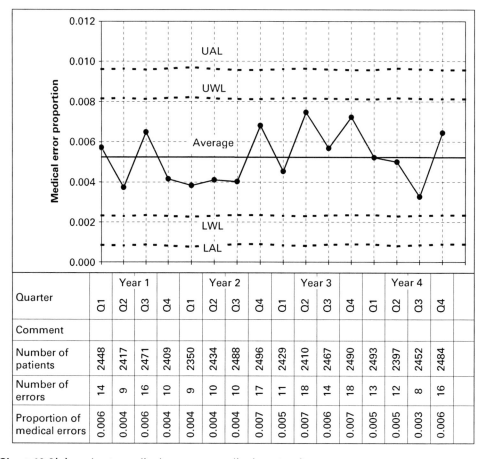

		Year 1				Year 2				Year 3				Year 4		
Quarter	Q1	Q2	Q3	Q4	Q1	Q2	Q3	Q4	Q1	Q2	Q3	Q4	Q1	Q2	Q3	Q4
Comment																
Number of patients	2448	2417	2471	2409	2350	2434	2488	2496	2429	2410	2467	2490	2493	2397	2452	2484
Number of errors	14	9	16	10	9	10	10	17	11	18	14	18	13	12	8	16
Proportion of medical errors	0.006	0.004	0.006	0.004	0.004	0.004	0.004	0.007	0.005	0.007	0.006	0.007	0.005	0.005	0.003	0.006

Chart 10.2(a) p chart: medical errors – medical centre A

Medical centre B (Chart 10.2(b))

Chart 10.2(b) reflects the very high variation seen in Chart 10.1(b), but most of the individual points above the control limit have been smoothed out by combining data together. There is now only one point above the UAL. We would investigate the individual months as indicated in Chart 10.1(b), we could also look into the possibility that points above the UAL occur at the same time of year by drawing an $\bar{X}$/R chart to compare the 12 months of the year (the method of doing this is shown in several other case studies and it is not repeated here).

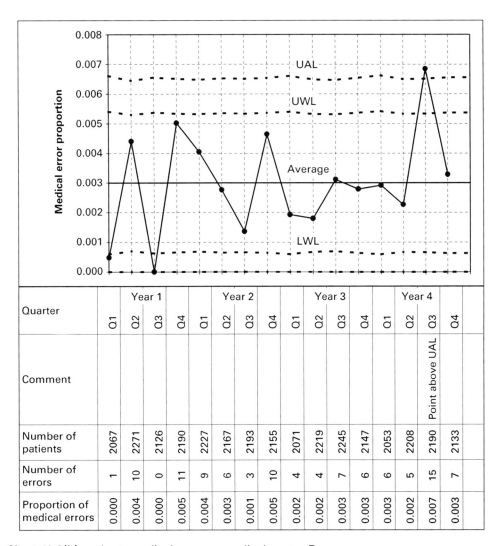

Quarter		Year 1				Year 2				Year 3				Year 4			
		Q1	Q2	Q3	Q4	Q1	Q2	Q3	Q4	Q1	Q2	Q3	Q4	Q1	Q2	Q3	Q4
Comment																Point above UAL	
Number of patients		2067	2271	2126	2190	2227	2167	2193	2155	2071	2219	2245	2147	2053	2208	2190	2133
Number of errors		1	10	0	11	9	6	3	10	4	4	7	6	6	5	15	7
Proportion of medical errors		0.000	0.004	0.000	0.005	0.004	0.003	0.001	0.005	0.002	0.002	0.003	0.003	0.003	0.002	0.007	0.003

Chart 10.2(b) p chart: medical errors – medical centre B

Medical centre C (Chart 10.2(c))

Chart 10.2(c) reflects Chart 10.1(c) is being in control with surprisingly low variation.

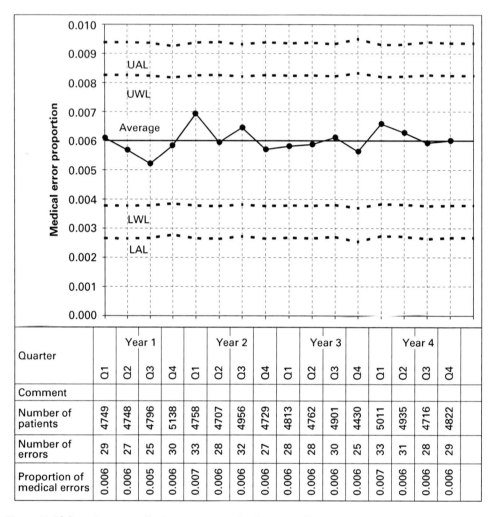

		Year 1				Year 2				Year 3				Year 4		
Quarter		Q1	Q2	Q3	Q4	Q1	Q2	Q3	Q4	Q1	Q2	Q3	Q4	Q1	Q2	Q3	Q4
Comment																	
Number of patients		4749	4748	4796	5138	4758	4707	4956	4729	4813	4762	4901	4430	5011	4935	4716	4822
Number of errors		29	27	25	30	33	28	32	27	28	28	30	25	33	31	28	29
Proportion of medical errors		0.006	0.006	0.005	0.006	0.007	0.006	0.006	0.006	0.006	0.006	0.006	0.006	0.007	0.006	0.006	0.006

Chart 10.2(c) p chart: medical errors – medical centre C

Medical centre D (Chart 10.2(d))

Chart 10.2(d) reflects Chart 10.1(c) is being in control with low variation.

In passing, we also note that the control limits for each of the Charts 10.2(a)–10.2(d) are narrower than those for the corresponding Charts 10.1(a)–10.1(d). This is to be expected because we are combining data and hence smoothing individual high and low values.

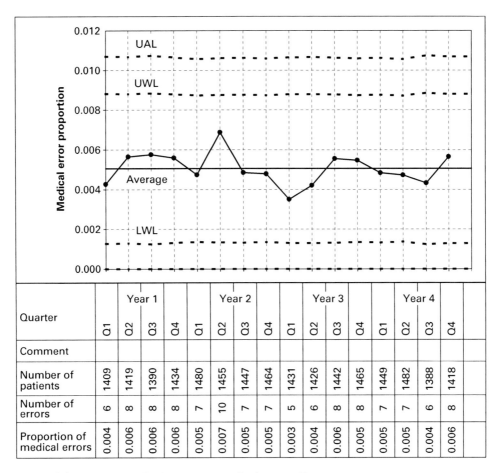

Quarter	Year 1				Year 2				Year 3				Year 4			
	Q1	Q2	Q3	Q4	Q1	Q2	Q3	Q4	Q1	Q2	Q3	Q4	Q1	Q2	Q3	Q4
Comment																
Number of patients	1409	1419	1390	1434	1480	1455	1447	1464	1431	1426	1442	1465	1449	1482	1388	1418
Number of errors	6	8	8	8	7	10	7	7	5	6	8	8	7	7	6	8
Proportion of medical errors	0.004	0.006	0.006	0.006	0.005	0.007	0.005	0.005	0.003	0.004	0.006	0.005	0.005	0.005	0.004	0.006

Chart 10.2(d) p chart: medical errors – medical centre D

Chart 10.3
When comparing results from different groups, it helps if we can draw all the data on one chart. Chart 10.3 is the result of drawing Charts 10.2(a)–10.2(d) on one chart. At a glance we can compare both averages and variability.

The low variability of centre C compared to other centres is very clear from this chart, as is the fact that it has the highest error rate.

Conversely, facility B has the lowest error average at 0.30% but a huge variation which varies from 0 errors one quarter to 15 in another. High variability is often indicative of a poorly managed process.

Chart 10.3 p chart: all medical centre errors

Chart 10.4

To complete the case study, Chart 10.4 compares the average error rates between the four medical centres. As commented for Chart 10.3, centre C has a high average, but low variability, as shown by the narrowing of the control limits. Facility B has a low average, but the limits are similar to centre A. Centres A and D are similar, with D having a slightly lower average and slightly wider limits.

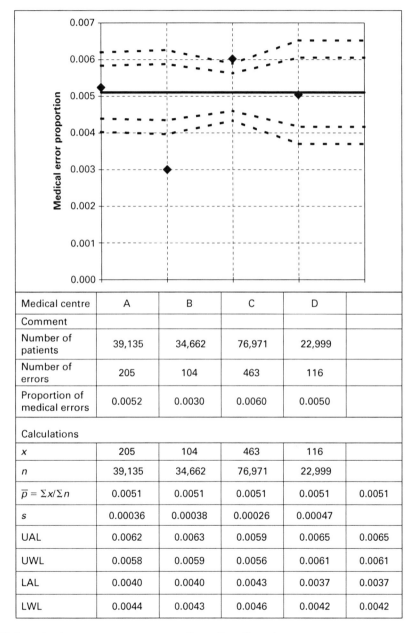

Medical centre	A	B	C	D	
Comment					
Number of patients	39,135	34,662	76,971	22,999	
Number of errors	205	104	463	116	
Proportion of medical errors	0.0052	0.0030	0.0060	0.0050	
Calculations					
x	205	104	463	116	
n	39,135	34,662	76,971	22,999	
$\bar{p} = \Sigma x / \Sigma n$	0.0051	0.0051	0.0051	0.0051	0.0051
s	0.00036	0.00038	0.00026	0.00047	
UAL	0.0062	0.0063	0.0059	0.0065	0.0065
UWL	0.0058	0.0059	0.0056	0.0061	0.0061
LAL	0.0040	0.0040	0.0043	0.0037	0.0037
LWL	0.0044	0.0043	0.0046	0.0042	0.0042

Chart 10.4 p chart: average error proportions for all centres

What Chart 10.4 clarifies is that centre B has a significantly lower error rate and centre C a significantly higher error rate than the other two centres. In this case study we only have four centres to compare and the added value to the analysis of Chart 10.4 is limited. However, as the number of centres increases Chart 10.4 would become more useful for gaining an overall picture of relative performance. Note that the medical centres on the chart are not joined up as a reminder that it is not appropriate to look for trends or sequences.

It would not be appropriate to only draw Chart 10.4 as it assumes that each centre's data is in a state of statistical control (debatable with the data in this case study), and we must check each centre to ascertain the extent to which each is in a state of control.

Comments

Estimating group size

To calculate the minimum group using the first guideline that the LAL > 0.
The LAL is calculated from the formula:

$$\text{LAL} = \bar{p} - 3s = \bar{p} - 3 \times \sqrt{\frac{\bar{p}(1 - \bar{p})}{\bar{n}}} > 0.$$

After some manipulation this leads to:

$$\bar{n} > \frac{9(1 - \bar{p})}{\bar{p}}.$$

Using $\bar{p} = 0.003$, $\bar{n} > 2991$, with an average of 722 patients per month, we need to combine $2991/722 = 4.14$ months.

Rounding down to 4 would be reasonable, or if there were enough data, up to 6 would be preferable.

Using the second guideline $\overline{np} > 4$. Using 1-month grouping gives $\overline{np} = 772 \times 0.003 = 2.2$, so a 2-month grouping would give $\overline{np} = 4.4$, which would be acceptable.

The question arises as to which guideline we should use. The choice will probably be a compromise between the group size and the number of points we need to plot a chart. The larger the group size, the fewer the number of points for plotting, the less chance we have of spotting irregularities.

We chose to group the data into 16 groups of 3 months because 16 data points are enough to draw a control chart, and 4 months results in average numbers of errors well above the minimum recommended. If there had only been 2 years of data, we would probably have chosen to plot 12 groups of 2 months as being the best compromise between having enough data points to plot a chart, and the average number of errors being above the minimum.

Using the first guideline, that the average should be >4, and referring to Table 10.2, the average error rate $\bar{p} = x/n$ and so $n > 4/\bar{p}$. Table 10.2 gives the resulting values for n and the average number of months required to reach this sample size.

Combining data

When we combine data, in this case adding the results over 3 months, the standard deviation of the combined data will always be less than the standard deviation of the averages. See Part 4 selecting the appropriate chart for an explanation.

Effect of sample size on control limits

We can see from Chart 10.3 that the average medical error percentage for centre A (0.52%) and D (0.50%) are virtually the same. The medical error percentage for A ranges from 0.33% to 0.75% whilst centre D has a narrower range, 0.35% to 0.69%. However, the control limits for D are wider than for A. This is because the average number of patients per quarter for D is only 1437 compared to 2446 for centre A. The formula for the standard deviation for the p chart is:

$$s = \sqrt{\frac{\bar{p}(1 - \bar{p})}{\bar{n}}}.$$

Since the calculation for the standard deviation is inversely proportional to the number of patients, n, the standard deviation and hence limits increase as n decreases.

Calculations

The p chart

The results calculations are shown on some of the charts. The calculations are the standard ones for the p chart.

Summary

This case study illustrates how statistical process control (SPC) can be:

- applied in the health care industry,
- used to analyse rare events data,
- used to compare performance between groups of varying sizes carrying out similar tasks.

The methodology explained in this case study is applicable to analysing many types of error or event in many types of industry.

When we are analysing rare events it is important to ensure that we have a large enough sample to make comparisons meaningful. One method of achieving this is to combine groups (in this case, months). If we have some information about event rates it is simple to determine the minimum required sample size for meaningful comparisons.

11 Metrics proposal for a training administration process

Charts used: X/MR, multivariate c, and pare to check sheet

Introduction

Some years ago I used to run one-day introductory courses aimed at explaining the basics of statistical process control (SPC). After one such course, one of the delegates gave me a copy of a document that he had prepared for the training department in which he was working at the time. The document was a proposal to begin measuring the training administration (TA) process and included proposed metrics, analysis methods and benefits. I have kept the proposal as an example of how easily people can assimilate the ideas of SPC and begin to put them into practice. Dave was not a trained statistician or engineer. He just attended the course, took on board what he had learned, and produced the document. If his proposal had related to the examples and case studies in the course or manufacturing (the traditional and well-documented application area of SPC) then his proposal would have been interesting enough. However, his proposal related to a completely different area, the TA process.

This chapter reproduces his proposal in its entirety with no changes except that acronyms have been written in full. Comments and further explanations are inserted in italics underneath each paragraph. Where the same comment applies more than once then it is only given on the first occasion.

Background

The training department processes requests for training (RFT) for a major office employing around 3000 people. The department compiles and updates lists of courses with descriptions, these being a mix of in-house and external courses. In addition employees may ask to attend other courses not on the list. The training department is also available to discuss and advise on training needs. Figure 11.1 is a simplified process flow chart of the training process.

Metric proposal

- RFT: request for training.
- BU: business unit.
- TA: training administration.

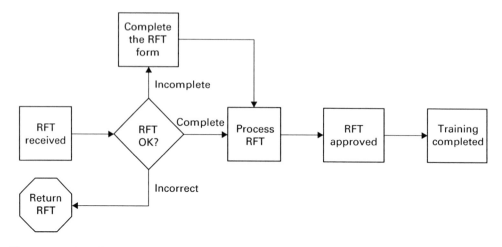

Figure 11.1 RFT flow chart

MEASUREMENT FOR A TA PROCESS *BY DAVID BALDWIN*

Length of time to process RFT in TA

Where in the process will measurement data be collected?
RFT (Figure 11.2) date stamped on receipt and on sending written confirmation to delegate.

> A very useful way of logging the progress of a form though a process is to add the date to the document. This allows for easy analysis and traceability.

Who will collect the data?
TA clerk sending the confirmation.

> Where possible use the people in the process to record data. Only in special situations, such as special measurement skills or if mis-reporting is suspected should someone else be used. For complete independence, an outside consultant can be used.

How will measurement data be collected?
A simple check sheet marked with the number of days from receipt whenever a confirmation is sent.

> It is good practice to mark up an example form (an RFT in this case) to show how the form will be completed. Dave did not do this, but because he was working closely with the training department, they no doubt understood what had to be completed. A check sheet is an extremely simple and useful tool for collecting this type of data. An example is given in Figure 11.3 and an explanation of how and when to use them is given in Part 4 of this book.

```
                    Request for training

RFT no.: 137

Date received: 10 Jan 03

Date confirmed: 17 Jan 03

                        Course requested | Course attended

  Course number

  Course title

  Date

  Others

Originator:
Business unit:
Date:
Name of delegate:


  Comments, including details of discussions with client and provider.
  For example, details of changes, delays, cancellations. Please include dates,
  names of people contacted, etc.
```

Figure 11.2 Some aspects of an RFT form

> Alternatively, record straight onto the control chart! This has the advantage of min-
> imising data transfer/analysis. The only disadvantage is that we would not have a
> tally chart/histogram (see below for further comments on the histogram).

What data will be collected?
Number of days to process an RFT in TA.

> It would, of course, be possible to collect the number of days each RFT spent in dif-
> ferent parts of the process.

Possible additional data:
Reference to the RFT.

> A reference might allow future analyses to determine whether there are differences
> between different types of RFT. Dave gives an example later.

What will we do with it?
Control chart: average number of days to process an RFT in period (week/month).

Check sheet draft: Number of days from receipt of RFT to RFT authorised

Note: Dates should be read off RFT form. EXCLUDE Saturdays, Sundays and public holidays

1									
2									
3	1								
4	1								
5	111								
6	11111	1111							
7	11111	11111	1						
8	11111	11111	11111						
9	11111	11111	11111	111					
10	11111	11111	11111	1					
11	11111	11111	11111	1					
12	11111	11111							
13	11111								
14	11111								
15	111								
16	11								
17									
18									
19	1								
20									

Figure 11.3 Check sheet for days to process RFT

There are a variety of options:
- Average number of days to process an RFT in period would not be my first choice. A month would probably be too long a period for calculating an average as we would need to wait a year or so before we have enough values for a useful control chart. The averages could easily be plotted on an X–/range or X–/s chart depending on the variability of the number of RFTs received each week. A more obvious choice would be an individual/moving range (X/MR) chart of the number of days to process an RFT (Chart 11.1). As each RFT is completed, a new point would be added to the chart. The advantage is that the control limits could be calculated after about 10 RFTs have been completed, and then again after perhaps 15 or 20 when (if) the process has settled down. We would expect little change thereafter.
- If a software package is being used, it would re-calculate the control limits whenever required.
- Training department personnel could be coached in the interpretation of the chart so that in time they could develop and manage their own charts.

Histogram:
Frequency distribution of the days to process an RFT in the period.

One of the advantages of recording data onto a check sheet similar to that in Figure 11.3 is that the paper on which the sheet is drawn can be turned through 90 degrees to give a histogram (try this with Figure 11.3).

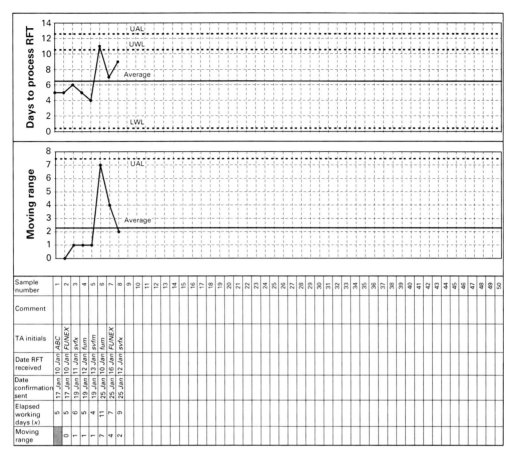

Chart 11.1 X/MR chart of days to process RFT; UAL: upper action limit; UWL: upper warning limit; LWL: lower warning limit

What will it tell us?

The chart enables us to monitor the level of service provided by TA in processing RFTs.

The histogram will indicate the normal level of service that the process is providing, its possible capacity for speedy reaction and the length of time it may take.

Regarding the control chart, correct. It will tell us:
1. Whether the process is in a state of (statistical) control and, if it is:
2. The average time required to process an RFT, along with the maximum and minimum.
   This is useful as we can then make statements such as:
   – "If the process does not change, it WILL take an average of X days to process an RFT, with a minimum of Y and a maximum of Z". This could then be used to form the basis of a service level agreement with the client departments.
   – It will also provide us with a base against which to measure any improvement activities.

Regarding the histogram:
The main benefit is to help confirm, or otherwise, that the data is normally distributed. The problems of non-normality are minimised when using X/MR charts, but if it is still felt that non-normality is a problem, we could resort to an X–/range chart and/or seek advice from a statistician. The first step is, however, to identify whether the data is not normally distributed. Whilst the "histogram will indicate the normal level of service" (and by this I assume Dave means the average), this is also achieved by a control chart, and the histogram will not add any further information. Dave's final comment "its possible capacity for speedy reaction and the length of time it may take" does not seem to convey a clear meaning. Would the client – the training department have understood?

What will we do about it?

The chart and the histogram can be used to measure the effectiveness of any attempts to improve the process of booking the event, for example, by better procedures, training/coaching, etc.

Out-of-control conditions can be recognised and decisions made on whether to investigate.

The additional data, traceable by reference to the RFTs, can be used to show where improvement action might be aimed. Analysis may indicate that the longest times are associated with certain course organisers.

Exactly right. There are only three decisions we can make regarding a process:
1. Do nothing because the process is in a state of control and we can get a better return on investment by improving other processes.
2. Improve the process because the process is in a state of control but we are not happy with the levels of performance.
3. Investigate because the process is not in a state of control and we should find out why and take action to bring it into a state of control.
The last paragraph (the additional data ...) is important. Whenever we collect data for analysis purposes we should be considering what type of further analysis, data collection, etc. may be required. In this example, Dave is already hypothesising as to what the TA department might do next and thinking about what other information may be useful later on.

How many RFTs incomplete/returned?

Where in the process will measurement data be collected?
During the check for RFT completeness, immediately the RFT arrives in the TA department.

Who will collect measurement data?
The TA clerk.

How will measurement data be collected?
Incoming RFTs counted on a simple check sheet as complete, incomplete or returned.

What data will be collected?

(a) Number of incoming RFTs in period (e.g. week).
(b) Number of incomplete RFTs in period.
(c) Number of returned RFTs in period.

Note: There are three possible states of an RFT:
1. Complete.
2. Returned, for example, because it is incomplete or incorrect.
3. Incomplete, but completable by the training staff.

Possible additional data:
Copy of "why returned" tick sheet retained, with sufficient references for traceability to original.

What will we do with it?
Control chart of:

1. % of incoming RFTs incomplete.
2. % of incoming RFTs returned with note of quantities returned.
3. Management information.

I hadn't told Dave about multivariate charts at this stage, so he was not aware of them. There are a variety of options including:
1. To monitor the proportion (or percentage) of incomplete RFTs, we could plot the number of incomplete forms per 10 (or 20 or 50, etc.) received using an np chart. Whether we report per 10 or 20 depends largely on the number we receive. If we are receiving, on average, 2 a day, reporting per 50 would add only one point to the chart every 3 weeks. Alternatively, we could chart the proportion incomplete every week using a p chart. Or we could report the number of forms received between incomplete forms, using an X/MR chart. This last option becomes more appropriate as the number of incomplete forms decreases. For example, if we only receive about one incomplete form every 2 weeks, identifying a reduction in incomplete forms will take a long time. A better approach is to count the number of forms between incomplete forms. The same comments apply to monitoring of returned forms.
2. To monitor the problems within the incomplete forms a multivariate chart could be used. This has the advantage of showing quickly and easily what the key problems are. If the number of "incomplete" and "returned" forms is high, we could collect each separately. However, assuming returned forms are just a more severe case of incomplete forms, we could combine the data in one chart. Taking this idea one step further, we could monitor every RFT and chart and any problem that occurs with it; for example, incorrectly filled in, request rejected, training date/course changed, etc. as the example in Chart 11.2 indicates (Chart 11.2 is discussed in detail below).
Dave is obviously wondering what other (management) information may be useful. It is always worthwhile considering what data to collect and how best to collect and analyse it to address a number of issues rather than one. For example, management may be interested to know if the receipt of RFTs is loaded to the beginning/end of the week, if there has been a steady growth/decline in RFTs and if so in which areas. This could be linked to the possibility, for example, of dropping courses if there is no call for them.

What will it tell us?
The chart will show (i) the level of quality of input to the TA process and (ii) the level of re-work service being provided by TA to try and counter it.

126

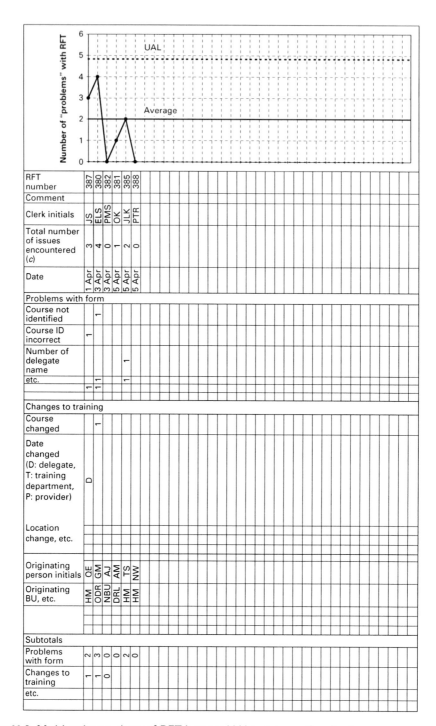

Chart 11.2 Multivariate c chart of RFT issues; UAL: upper action limit

Depending on the final decision as to which data are collected we could determine, amongst other items:

1. The percentage of complete, incomplete and/or returned forms.
2. The key aspects of the RFT causing problems (from the multivariate chart).
3. Whether there are any specific BUs, courses and clerks with more/less incomplete and/or returned forms (also from the multivariate chart).
4. Delays due to RFT returned forms (i.e. by plotting time from returned date to acceptance).
5. Whether the training need is not met because of returned forms (e.g. returned forms that are not re-submitted).

What will we do about it?

Once the natural pattern of variation endemic to the process has been established itself, the chart can be used to measure the effectiveness of any attempts to improve the process of RFT completion (e.g. by training, coaching, education supplying more/ better information on course availability, etc.).

Out-of-control conditions can be recognised and decisions made on whether to investigate.

The additional data provided by copies of the "why returned" tick sheets can be used to show where improvement action might be aimed. Analysis may indicate that most errors originate in one BU or that the problem is equally spread or that one aspect of the RFT gives problems.

A tick sheet is the same as a check sheet. As described above, the multivariate chart can help with the type of analyses proposed. Note that if, for example, significantly more errors originate in one BU the solution is likely to lie in working with that specific BU. If the errors seem to originate randomly from all BUs the solution will lie in changing the process. This is an important distinction.

Chart 11.2:

This chart is a multivariate c chart of issues arising from the whole TA process. Beginning with the table below the chart:

● The RFT number is recorded so that the original RFT can be reviewed if necessary during any analysis.
● Brief comments can be added.
● Initial of the clerk who processed the RFT.
● Total number of issues (e.g. problems, errors, changes) encountered during the processing of the RFT.
● Date of approval of the RFT. It is important to enter the data after the RFT is approved as it is only then that the total number of issues can be entered on the chart.

There is other information that could be entered.

The next section is split into sub-sections detailing the issues arising. For the sake of example, we have included two such sections, the first detailing problems with the form and the second changes to the requested training. The final section provides more information about the form, for example the originating person and the BU. Below the main part of the table subtotals are provided. It is the total of these sub-totals that is plotted on the c chart.

> As an example, details of several forms have been completed. In the first RFT the course ID was incorrect; there was another problem with the form and the delegate changed the date of the course giving a total of three issues. There were no problems with the third or sixth form.

Use of advice desk

Where in the process will the measurement data be collected?
Throughout the (*training*) process, whenever the advice desk is consulted.

Who will collect measurement data?
Anyone providing advice desk service in the training department.

How will measurement data be collected?
All queries entered on an advice desk log sheet at the time of the enquiry.

What data will be collected?
Advisor, customer, date, time, category of query (e.g. RFT completion, status of request, suitable courses, available events, etc.), immediate reply/call back, successful/unsuccessful, advice/action/comment.

What will we do with it?
Control chart:

1. Number of enquiries successful/unsuccessful per week/month, other possible charts:
   - number of enquiries by category,
   - number of enquiries by BU.

> Whenever we are categorising data, for example, the nature of the enquiry, it is likely that a multivariate control chart will be useful to help with further analysis.

What will it tell us?
The chart will tell us (i) the level of advisory service (possible waste) being provided by the training department in support of the TA process and (ii) indicate where the process may be capable of improvement.

What will we do about it?
Once the natural pattern of variation endemic to the process has been established, the chart can be used to measure the effectiveness of any attempts to improve the process (e.g. by better procedures, training/coaching, supplying more/better information on course availability, etc.).

Out-of-control conditions can be recognised and decisions made on whether to investigate.

The additional data provided by customer and enquiry categories can be used to show where improvement action might be aimed. Analysis may indicate that most enquiries come from one BU or that the problem is equally spread, or that one aspect of the process gives problems.

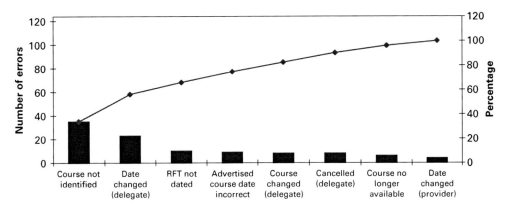

Chart 11.3 Pareto chart of RFT issues arising

To add a little explanation, typical analyses are likely to be:
The control chart of number of enquiries per week/month will tell us whether the process is in control. We could also chart time to close out a query. Pareto charts of, for example, category of enquiry or BU, would help us identify the "vital few" from the "useful many" (see section on Pareto charts for an explanation in Part 4), and so lead us into action that would reduce the number of calls. A sample Pareto chart of issues arising is given in Chart 11.3.

Dave now explains how information on two common problems within the training process may be dealt with.

Wrong course identified

Where in the process will the measurement data be collected?
At any time from receipt of authorised RFT, that the course is identified as "wrong" for the delegate.

Who will collect measurement data?
The TA supervisor.

How will measurement data be collected?
Entry on a log showing customer, course, where rejected, why and action taken.
By "where rejected" Dave means at what point in the process.

What data will be collected?
Number of "wrong course" rejections in period (i.e. week or month).

Possible additional data:
Copy of "wrong course" tick sheet retained with sufficient references for traceability to original (or copy of RFT).

What will we do with it?
Control chart:

● Number of "wrong course" rejections.

Analyse:

- Per annum number of "wrong course" rejections by BU.
- Per annum number of "wrong course" rejections by course originator.
- Per annum number of "wrong course" rejections by rejection category.

> It would be possible to calculate a "per annum" rate over any time span. (For example, if the number of wrong course rejections is 10 over a 2-month period, the per annum rate would be 60.) However, it is probably better to plot reject rates; in other words percentage of wrong course rejections.

What will it tell us?
The chart will show the level of "nuisance" caused by RFTs for wrong courses. Analyses will enable us to pinpoint the most affected courses and the most common originators and reasons for rejection.

What will we do about it?
The chart can be used to measure the effectiveness of any attempts to improve the process of RFT completion with respect to the selection of the right course for the right delegate, with the right prerequisite qualifications (e.g. by better procedures, training, supplying more/better information on course prerequisites), etc.

Out-of-control conditions can be recognised and decisions made on whether to investigate.

The additional data provided by copies of "wrong course" tick sheets can be used to show where improvement action might be aimed. Analysis may indicate that most errors originate in one BU, or the problem is equally spread, or one aspect of the selection procedure gives problems.

Cancellations and changes

Where in the measurement process will the measurement data be collected?
At any time the authorised RFT is cancelled or changed.

Who will collect measurement data?
The TA clerk.

How will measurement data be collected?
Entry on a log showing customer, clerk, course, cancelled or changed, reason for change/cancellation category, what changed, action taken.

What data will be collected?

- Number of cancellations and reason for cancellation category.
- Number of changes and reason for change category.

Possible additional data:
Copy of RFT and change/cancellation sheet retained.

What will we do with it?
Control chart:

1. Number of cancellations per period (week/month).
2. Number of changes per period (week/month).

Analyse:

- Number of changes by reason for change category.
- Number of cancellations by reason for cancellation category.

What will it tell us?
The chart will enable us to monitor the level of waste caused by changes or cancellations.

Analyses will enable us to pinpoint the most affected courses, the most common originators and reasons for changes and cancellations.

What will we do about it?
The chart can be used to measure the effectiveness of any attempts to improve the process of RFT completion with respect to the selection of the right course for the right delegate at the right time (e.g. by better procedures, training, supplying more/better information on course content/applicability, etc.).

Out-of-control conditions can be recognised and decisions made on whether to investigate.

The additional data provided by copies of RFTs and cancellation/change sheets can be used to show where improvement action might be aimed. Analysis may indicate where improvement action might be aimed. Analysis may indicate that the most errors originate in one BU, or that the problem is equally spread, or that one aspect of the selection procedure gives problems.

Final comments

- *A similar approach could also be applied to course appraisal. For example, most course appraisal forms ask the delegate to rate a number of items. An $\bar{X}$/range chart could be used to plot the average rating and range for each question. It would also be possible to collate and analyse course comments and take appropriate action to improve the course.*
- *This is a fairly typical approach to monitoring many business processes. In this situation we are interested in two aspects:*
  1. *Time taken to complete either parts of the process and/or the whole process – monitored by a control chart.*
  2. *Information on problems/delays encountered. The rate of problems/delays is monitored by a control chart and the relative importance of problems by a Pareto (or ranked bar) chart.*

One less usual application is the "No" report. I first encountered the "No" report in a hotel reception setting. Whenever a member of the hotel staff is unable to fulfil a request from a guest (i.e. they say "no") the request is recorded on a "No" report along with the reason for the request being refused. "No" reports can be monitored and analysed using control charts and Pareto analyses. This is an excellent way of identifying unmet needs among hotel guests and taking appropriate action to improve service.

The same idea is, of course, applicable to many other interfaces between an organisation and its customers (both internal and external); for example, airlines, health care, services, retail and help/advice desks.

Summary

This case study demonstrates that it is possible to begin applying SPC with a minimum of experience. It also gives an example of how SPC may be applied to a service, in this case a TA process.

The same ideas as discussed here apply to a large number of other request/application administration processes: for example for loans, membership, purchase order request, holiday booking, college entrants, etc. The results can be used to identify the problem areas, take appropriate improvement action and monitor the effects of the action.

A simple "No" report, completed every time a supplier of a product or service is unable to meet a request can provide an organisation with valuable information on where it is failing to meet its customer's need. The example outlined at the end of the case study relates to a hotel, and the idea is applicable to a great many situations in many organisations.

12 Reducing problems during borehole drilling
An example of monitoring two metrics on one chart

Charts used: u and scatter diagram

Introduction

In this simple monitoring case we present a typical application of the control chart: to monitor process performance and changes. The unusual feature in this case is the plotting of two variables on the same chart.

Background

When drilling boreholes, for example in the oil and gas industry, a hole is drilled into the ground which can be several kilometres long. In order to remove the rock that is being drilled a drilling fluid is pumped down the centre of the drill and out at the drill bit, and then forces its way around the outside of the drill up to the surface carrying the rock cuttings with it.

Two of the main types of problems that can occur whilst drilling are:

1. The rock being drilled can crumble from the wall of the borehole into the hole and be washed away by the drilling fluid (called a washout).
2. The torque on the drill pipe can cause it to break (called a twist-off).

Both problems result in drilling delays. Twist-offs also usually result in loss of expensive equipment and/or the time and expense of attempting to "fish" the lost equipment from the hole.

A process improvement team was commissioned to investigate causes of washouts and twist-offs, investigate potential solutions, and implement improvements. As a first step they set up a control chart to monitor performance.

Analysis

Every month the number of feet drilled, the number of twist-offs and the number of washouts are recorded.

Since the opportunity for washouts and twist-offs occurring depends on the number of feet drilled, the key metrics to be analysed are the:

● number of twist-offs per foot drilled,
● number of washouts per foot drilled.

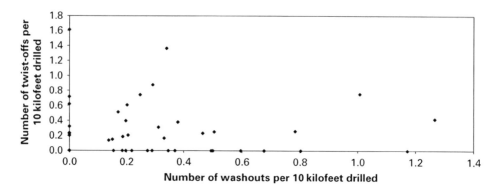

Chart 12.1 Scatter diagram of number of washouts vs. twist-offs showing that there is no relationship between them

The data are reported as the number of incidents per 10,000 (10 kilo) feet, so that the frequency of incidents is 2 per 10 kilofeet.

At the time of the analysis there was no known reason why twist-offs and washouts should be related. However, when working with pairs of variables it is important to check as the existence of a relationship would affect the following analyses.

In order to check the relationship between the occurrence washouts and twist-offs a scatter diagram was drawn (Chart 12.1). For each month a point is plotted on the chart showing the number of washouts per 10 kilofeet drilled and the number of twist-offs per 10 kilofeet drilled. The raw data are provided on Chart 12.2. Analysis of Chart 12.1 suggests:

- There is, as expected, no obvious relationship between the number of twist-offs and washouts.
- The distribution of both washouts and twist-offs is skewed with more low values than high values. A simple way of seeing this on the chart is to look at the number of points lying between equally spaced intervals on the axis. For example, on the washout axis, there are:
  - 28 points lying between 0.0 and 0.2,
  - 12 points between 0.2 and 0.4,
  - only 10 points between 0.4 and 1.4.

  This is what we expect to see in most situations where we are using count or discrete data. It would be possible to draw histograms to highlight the non-normality of the data.
- There is a gap in the number of washouts between 0 and about 0.15 incidents per 10 kilofeet and a similar gap in the number of washouts. This is because the typical number of feet drilled per month is 10 kilofeet. If there is 1 incident in the month this gives a value of 0.2 incidents per 10 kilofeet, and zero incidents gives a value of 0. There can be no values in between.

Chart 12.2 is a u chart with both washout and twist-off data plotted. It is unusual to plot two variables on the same chart, but the engineers involved in investigation were keen to see both together to help identify any similar patterns in the data and to be able to

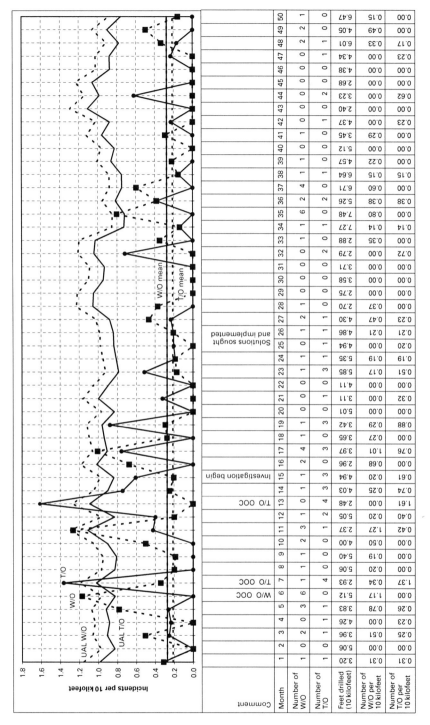

Month	Number of W/O	Number of T/O	Feet drilled (10 kilofeet)	Number of W/O per 10 kilofeet	Number of T/O per 10 kilofeet	Comment
1	1	1	3.20	0.31	0.31	
2	1	0	5.06	0.00	0.00	
3	2	1	3.96	0.51	0.25	
4	0	1	4.26	0.00	0.23	
5	3	1	3.83	0.78	0.26	W/O OOC
6	6	0	5.12	1.17	0.00	W/O OOC
7	1	4	2.93	0.34	1.37	T/O OOC
8	1	0	5.06	0.20	0.00	
9	1	0	5.40	0.19	0.00	
10	2	0	4.00	0.50	0.00	
11	3	1	2.37	1.27	0.42	
12	1	2	5.05	0.20	0.40	
13	0	4	2.48	0.00	1.61	T/O OOC
14	1	3	4.03	0.25	0.74	
15	1	3	4.94	0.20	0.61	Investigation begin
16	2	0	2.96	0.68	0.00	
17	4	3	3.97	1.01	0.76	
18	1	0	3.65	0.27	0.00	
19	1	3	3.42	0.29	0.88	
20	0	0	5.01	0.00	0.00	
21	0	1	3.11	0.00	0.32	
22	0	0	4.11	0.00	0.00	
23	1	3	5.85	0.17	0.51	
24	1	1	5.35	0.19	0.19	
25	0	1	4.94	0.00	0.20	
26	1	1	4.86	0.21	0.21	Solutions sought and implemented
27	2	1	4.30	0.47	0.23	
28	1	0	2.70	0.37	0.00	
29	0	0	2.75	0.00	0.00	
30	0	0	3.58	0.00	0.00	
31	0	0	3.71	0.00	0.00	
32	0	2	2.79	0.00	0.72	
33	1	0	2.88	0.35	0.00	
34	1	1	7.27	0.14	0.14	
35	6	0	7.48	0.80	0.00	
36	2	2	5.26	0.38	0.38	
37	4	0	6.71	0.60	0.00	
38	1	1	6.64	0.15	0.15	
39	1	0	4.57	0.22	0.00	
40	0	0	5.12	0.00	0.00	
41	1	0	3.45	0.29	0.00	
42	0	1	4.37	0.00	0.23	
43	0	0	2.40	0.00	0.00	
44	0	2	3.23	0.00	0.62	
45	0	0	2.68	0.00	0.00	
46	0	0	4.38	0.00	0.00	
47	0	1	4.34	0.00	0.23	
48	2	2	6.01	0.33	0.17	
49	2	0	4.05	0.49	0.00	
50	1	0	6.47	0.15	0.00	

Chart 12.2 u chart: washouts and twist-offs. W/O: washouts; T/O: twist-offs; OOC: out of control

summarise progress on one sheet. The data section of the chart shows the month number, the number of feet drilled, the numbers of washouts and twist-offs, and the resulting washout and twist-off frequency. The comments section of the chart gives information about what was happening in the process and observations on the data, investigations, results and key dates in the project.

The solid line with black dots represents the twist-off data and the dotted line with squares washout data.

There are a number of key features to notice about this chart:

- The control limits for both washouts and twist-offs are parallel. This is because the control limit depends only on the average incident rate and the number of feet drilled. Since the number of feet drilled is the same for both washouts and twist-offs, the upper action limits (UALs) will be parallel and the difference in UAL is accounted for by the difference in incident rates.
- The change in UAL from month to month reflects the changes in the number of feet drilled.
- There are out-of-control conditions at months 6 (washouts), 7 and 13 (twist-offs). There are near out-of-control conditions in months 11, 17 and 35 (washouts), and 14, 17 and 19 (twist-offs). The fact that eight of these high values occur in the first 20 months, and only one afterwards suggests that a process change may have occurred.
- Inspection of the chart suggests that incident frequency decreased from about month 20. The evidence for this is:
  - For the washout data there is a high value of 1.01 in month 17, followed by 2 average months (0.27, 0.29) followed in month 20 by 3 months with zero incidents and four more below the average.
  - After the five high twist-off values between months 13 and 19, the following four values are 0 or near, though above the average, this is followed by eight values below the average.

  Exactly when the change occurred is debatable, and it is likely that the process changed gradually over weeks rather than at the end of a specific month. However, assuming the change to be in month 20 is a reasonable and useful approximation to what was happening.
- The cause of the change is not known for certain, it could be due to the Hawthorn effect. Following the start of the improvement project in month 15 it was also standard practice that obvious and simple improvements would have been carried out and not necessarily logged.

Chart 12.3 shows the same data as Chart 12.2 but treated as two separate processes with the change occurring at month 20. Two separate process averages have been calculated one up to month 20 and one after for both washout and twist-off. The control limits have also been recalculated.

Following the recalculations:

- The average and control limits for the first 20 months have increased, resulting in only 1 month above the UAL.
- The average after month 20 is approximately half of the average before month 20, and the UAL has similarly been reduced.
- The values for the 6 months between months 31 and 37 seem to stand out as being higher than the months either side, with one washout figure being above the UAL.

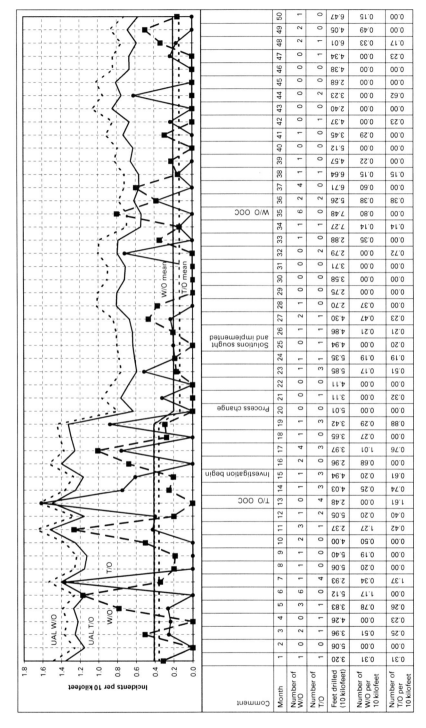

Chart 12.3 u chart: washouts and twist-offs with process change. W/O: washouts; T/O: twist-offs; OOC: out of control

Month	Comment	Number of W/O	Number of T/O	Feet drilled (10 kilofeet)	Number of W/O per 10 kilofeet	Number of T/O per 10 kilofeet
1		1	1	3.20	0.31	0.31
2		0	1	5.06	0.00	0.20
3		2	1	3.96	0.51	0.25
4		0	1	4.26	0.00	0.23
5		3	1	3.83	0.78	0.26
6		6	0	5.12	1.17	0.00
7		1	4	2.93	0.34	1.37
8		1	0	5.06	0.20	0.00
9		1	0	5.40	0.19	0.00
10		2	0	4.00	0.50	0.00
11		3	1	2.37	1.27	0.42
12		1	2	5.05	0.20	0.40
13	T/O OOC	0	4	2.48	0.00	1.61
14		1	3	4.03	0.25	0.74
15	Investigation begin	1	3	4.94	0.20	0.61
16		2	0	2.96	0.68	0.00
17		4	3	3.97	1.01	0.76
18		1	0	3.65	0.27	0.00
19		1	3	3.42	0.29	0.88
20	Process change	0	0	5.01	0.00	0.00
21		0	1	3.11	0.00	0.32
22		0	0	4.11	0.00	0.00
23		1	3	5.85	0.17	0.51
24		1	1	5.35	0.19	0.19
25		0	1	4.94	0.00	0.20
26	Solutions sought and implemented	1	1	4.86	0.21	0.21
27		2	1	4.30	0.47	0.23
28		1	0	2.70	0.37	0.00
29		0	0	2.75	0.00	0.00
30		0	0	3.58	0.00	0.00
31		0	0	3.71	0.00	0.00
32		0	2	2.79	0.00	0.72
33		1	0	2.88	0.35	0.00
34		1	1	7.27	0.14	0.14
35	W/O OOC	6	0	7.48	0.80	0.00
36		2	2	5.26	0.38	0.38
37		4	0	6.71	0.60	0.00
38		1	1	6.64	0.15	0.15
39		1	0	4.57	0.22	0.00
40		0	0	5.12	0.00	0.00
41		1	0	3.45	0.29	0.00
42		0	1	4.37	0.00	0.23
43		0	0	2.40	0.00	0.00
44		0	2	3.23	0.00	0.62
45		0	0	2.68	0.00	0.00
46		0	0	4.38	0.00	0.00
47		0	1	4.34	0.00	0.23
48		2	1	6.01	0.33	0.17
49		2	0	4.05	0.49	0.00
50		1	0	6.47	0.15	0.00

Chart axis label: Incidents per 10 kilofeet (0.0–1.8). Line labels: UAL W/O, UAL T/O, W/O, T/O, W/O mean, T/O mean.

The reason for this apparent increase was not discovered, and as the effect is not strong, it is possible, if unlikely, that it was due to the random variation.

- The number of twist-offs reduced dramatically from month 28 with 15 of the last 23 months returning zero incidents.
- Reducing washout incidents took longer. There was one 3-month performance blip between months 35 and 37 with the washout data this was investigated and fixed. However, from month 40 onwards 7 of the last 11 months had zero incidents. The one concerning feature is the last 3 months which were all non-zero after a run of 6 months with no washouts. The team were investigating this.

Comments

- Traditionally, one set of data is drawn on each chart. However, there is no reason why this convention needs to be followed. Any format that helps meet the needs of the user should be adopted. In this case whilst some people found that viewing two superimposed charts together was confusing, the engineers using the charts elected to keep them through the project and continued to use them for monitoring purposes once the project was completed.
- Process changes sometimes occur when we first begin to focus on the process as people are aware that their activities are being monitored (known as the Hawthorn effect) often these improvements dissipate with time.
- Frequent occurrences of out-of-control or near out-of-control conditions are likely to be due to an unidentified process change rather than a number of special causes. In these situations, the analyst needs to search for a likely point where the process changed and try inserting a change on the chart. If the resulting changes in averages and control limits reduces or eliminates the out-of-control signals, as in this case, then it is likely that a process change did occur at or around that time.
- The team working on reducing washouts and twist-offs proposed and made a number of changes to the drilling process over a period of about a year. Changes included changes to equipment, planning and operations. In some cases they found that procedures were not being followed. Training was provided where necessary, procedures re-written and where appropriate contractors were involved. Throughout this time the process was continually changing and the number of washout and twist-off events gradually decreased. It would be possible to try fitting a regression to investigate the existence of a trend, but this was considered not worthwhile as nothing would have been done with the information.
- From month 28 for twist-offs and month 40 for washouts there are frequently zero incidents in a month. With so few incidents reported the average and control limits drop to such a level that every month with an event would signal an out-of-control condition. In these situations, known as rare events, rather than report the number of incidents per month, we monitor the time or in this case the number of feet drilled between incidents. There are other case studies in the book illustrating charting rare events.
- The chart was used regularly for monitoring and provided a focus for appraising performance, and explaining to management and other interested parties what had been happening in the process.

Calculations

The calculations for washouts and twist-offs are similar and we only give the calculations for washouts.

For Chart 12.3, months 1–19, the average twist-off incident rate is calculated as:

$$\text{average} = \bar{u} = \frac{\text{total number of washouts}}{\text{total feet drilled}} = \frac{31}{75.69}$$

$$= 0.41 \text{ incidents per 10,000 feet drilled.}$$

The standard deviation calculated for each month:

$$s = \sqrt{\frac{\bar{u}}{\text{feet drilled}}}.$$

For month 1:

$$s = \sqrt{\frac{0.41}{3.2}} = 0.36.$$

And the UAL is given by:

$$\text{UAL} = \bar{u} + 3s = 1.48.$$

The results of the calculations are shown in the chart.

Summary

- One of the key applications for control charts is in the monitoring of process performance.
- Control chart design is a matter of using what is useful. In this case study the engineers found that overlaying two variables on one chart was helpful to them.
- Sometimes process changes occur over long periods of time. Control charts can be used in these situations if the results will be of use.
- Studies by Hawthorn demonstrated that process performance may improve temporarily when process operators, aware that their process performance is being measured, exercise more care in the execution of the process than normal.
- Frequent points outside the same control limit often indicates that the process has changed.

13 Applying control charts to benchmarking in the drilling industry

Charts used: $\bar{X}$/range, X/MR and bar charts

Introduction

One of the many uses of control charts is to compare performances of different groups or organisations carrying out similar tasks. This case study illustrates how the concept of control charting can been successfully applied to benchmarking and comes from benchmarking drilling operations in the oil and gas industry. Control charts can be used in a similar way to benchmark groups and activities both within and between organisations. Applications include not only costs and manpower but many other areas where process data can be gathered such as equipment availability, absenteeism and customer satisfaction. This case demonstrates:

- How the use of bar charts to compare performance of different groups may lead to misleading conclusions.
- How control charts can be used to analyse and comment on both levels of performance and consistency of performance.
- How a control chart can be used to compare averages where the numbers of observations in each average differs.
- That control charts can be used with non-time sequenced data.
- The importance and process of screening data for outliers before comparison with other groups of data.

Background

The *Drilling Performance Review* is an annual benchmarking study which collects and compares drilling performance data from operators around the globe. Each participating operator submits data on all the wells they drill. Wells are grouped according to a number of traits to ensure comparison between like wells.

One of the key metrics in the study is the cost per foot drilled (referred to simply as cost per foot) for the entire well. This is calculated by dividing the cost of drilling the well (known as the dry-hole cost) by the total length drilled (drilled interval).

Table 13.1 Cost per foot for all participants

Operator	Number of wells	Average cost per foot	Maximum cost per foot	Minimum cost per foot
B	2	620	665	601
U	1	478	478	478
G	4	409	512	231
S	40	399	852	162
O	2	389	410	368
T	3	388	628	304
G	7	366	512	180
P	4	343	440	122
R	1	332	332	332
A	8	331	608	223
Q	12	325	744	151
H	4	307	354	265
M	11	304	527	148
K	3	278	370	146
D	36	273	955	110
F	2	244	445	213
W	10	239	714	135
I	11	222	346	115
N	7	215	430	83
L	7	212	319	135
E	11	209	295	90
C	7	206	309	159
V	2	140	153	96

Analysis and data

Data for 23 operators are summarised in Table 13.1. This table includes the operator name and the following data for each operator:

- The average cost per foot for all wells.
- The maximum cost per foot of all the wells drilled by the operator.
- The minimum cost per foot of all the wells drilled by the operator.

Gleaning information from tables is difficult, and a common way to present this type of data is as a ranked bar chart, as given in Figure 13.1. In this figure the operators are ranked in decreasing order of average cost per foot.

You might like to review Figure 13.1 and note your conclusions. In particular, which operator would you consult in order to improve your cost performance (assuming that you are not the least cost operator).

Many people would conclude that operator V is the best (i.e. cheapest) operator. In many benchmarking clubs, the data analysis is followed by information exchange where the "better" operators explain how they manage to achieve their results. In this case study, other operators would therefore seek to learn from operator V and may even instill some of operator V's practices into their own organisation incurring any associated costs, for example training, equipment or software.

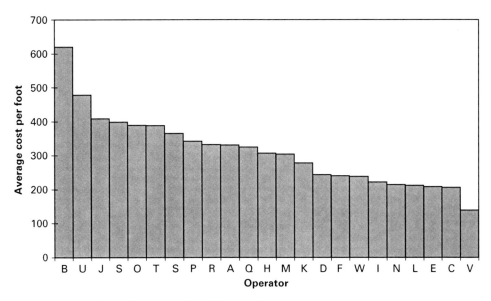

Figure 13.1 Bar chart of operator's drilling performance

Chart 13.1 is an $\bar{X}$/R chart. The top chart shows the average cost per foot drilled for each operator and the lower chart shows the range for each operator. The range is simply calculated as the difference between the maximum and minimum cost per foot for the operator. For example, referring to Table 13.1, operator B, the range is $(665 - 601) = 64$. Control limits are added in the usual way. There are a number of key features about this chart:

- The industry average (£306/foot) estimates the true average cost per foot for all wells drilled. The reason it is not the true industry average is that some operators are not members of the benchmarking club.
- For operators with few wells there is a large uncertainty about the mean. This is reflected in the increasing gap between the upper and lower action limits (UAL and LAL, respectively). For example, the UAL−LAL for operators with seven wells is $(441 - 171) = 270$, whereas for operator S with 40 wells the difference is only $(386 - 226) = 160$. As the number of wells drilled drops to three or less the control limits become very wide and are of diminishing use when interpreting the data. These data are included in the charts for completeness, and because their data does contribute to the industry averages and action limits.
- Operators with high variability in performance, perhaps due to individual "rogue" or "lucky" wells, will be identified on the range chart by their range being above the UAL. Only operator D is above the UAL, however, S, Q and W are quite close to the limit. In situations like this where there are one or more points above or just below a control limit and then a gap to others being much nearer the average, it could be that we have different groups of wells each of which have significantly different variability. For example, with this data we could hypothesise that operators D, S, Q and W have a significantly higher variability than all the other operators. Similarly operators E, I, L, C, H, V, B and Q have a significantly lower variability than the other operators, though none of them are below the LAL. In summary, it may well be that

Operator	S	D	Q	M	E	I	W	A	L	J	N	C	P	G	H	T	K	V	F	B	O	R	U	Total	Average
Number of wells average	40	36	12	11	11	11	10	8	7	7	7	7	4	4	4	3	3	2	2	2	2	1	1	195	9.19
cost per foot ($\bar{X}$)	399	273	325	304	209	222	239	331	212	366	215	206	343	409	307	388	278	140	244	620	389	332	478		306
Range	691	845	592	378	205	231	579	384	184	331	347	150	318	281	89	324	224	57	233	64	42	0	0	6549	
d_2 (from tables)	4.393	4.293	3.258	3.173	3.173	3.173	3.078	2.847	2.704	2.704	2.704	2.704	2.059	2.059	2.059	1.693	1.693	1.128	1.128	1.128	1.128			$d_{2,9}$	2.970
Expected range if $n = 9$	467	585	540	354	192	216	558	401	202	364	381	165	459	405	129	568	393	150	612	168	111				353
Mean range = sum of ranges/21 for average $n = 9$	353	353	353	353	353	353	353	353	353	353	353	353	353	353	353	353	353	353	353	353	353	353	353		
Expected mean range for each operator's $n = \bar{R}$	523	511	388	377	377	377	366	339	322	322	322	322	245	245	245	201	201	134	134	134	134				
Average cost per foot x number of wells ($\bar{n}x$)	15966	9828	3898	3345	2296	2446	2390	2646	1487	2561	1503	1445	1371	1636	1228	1165	835	279	488	1240	779	332	478	59643	
Industry average = $\bar{X}$ 59643/195	306	306	306	306	306	306	306	306	306	306	306	306	306	306	306	306	306	306	306	306	306	306	306		
A_2 (from tables)	0.153	0.153	0.266	0.285	0.285	0.285	0.308	0.373	0.419	0.419	0.419	0.419	0.729	0.729	0.729	1.023	1.023	1.880	1.880	1.880	1.880				
UAL = $\bar{X} + A_2 \times \bar{R}$	386	384	409	413	413	413	419	432	441	441	441	441	484	484	484	512	512	558	558	558	558				
LAL = $\bar{X} - A_2 \times \bar{R}$	226	228	203	198	198	198	193	180	171	171	171	171	127	127	127	100	100	54	54	54	54				
For the range chart: D_4 (from tables)	1.541	1.541	1.717	1.744	1.744	1.744	1.777	1.864	1.924	1.924	1.924	1.924	2.282	2.282	2.282	2.574	2.574	3.270	3.270	3.270	3.270				
UAL = $D_4 \times \bar{R}$	805	787	665	658	658	658	651	631	619	619	619	619	559	559	559	518	518	439	439	439	439				
D_3 (from tables)	0.459	0.459	0.283	0.256	0.256	0.256	0.223	0.136	0.076	0.076	0.076	0.076	0.000	0.000	0.000	0.000	0.000	0.000	0.000	0.000	0.000				
LAL = $D_3 \times \bar{R}$	240	234	110	97	97	97	82	46	24	24	24	24	0	0	0	0	0	0	0	0	0	0	0		

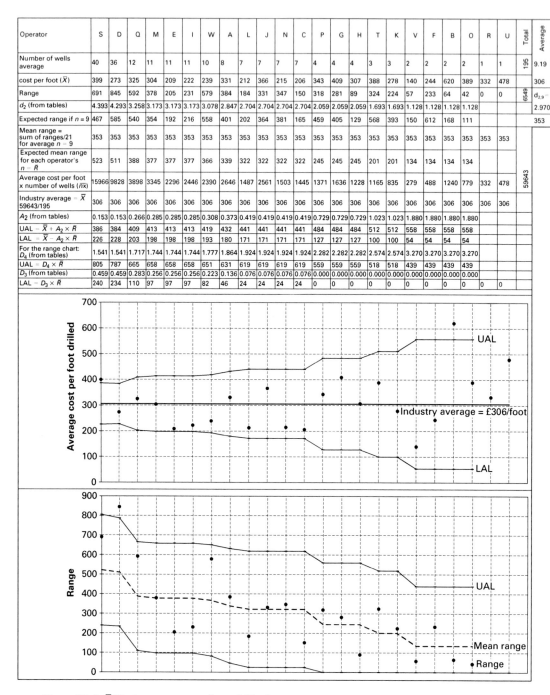

Chart 13.1 $\bar{X}$/R chart: cost per foot drilled

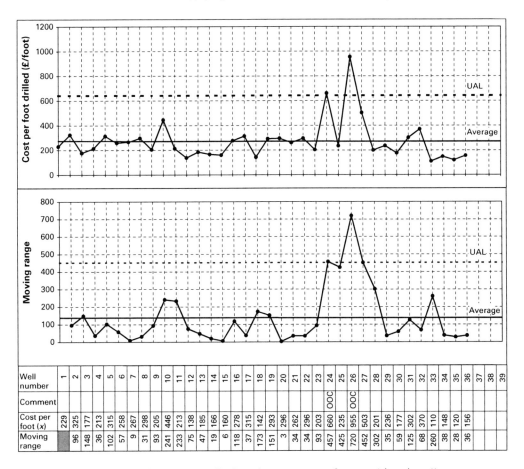

Chart 13.2 X/MR chart for operator D showing two out-of-control (ooc) wells

there are three distinct types of operator: those with high variability, those with medium variability and those with low variability.

● The order in which the points are plotted is irrelevant to the analysis, and so we will not be looking for trends. Therefore the line joining the data points has been omitted. The line joining the control limits is only included for printing clarity.

As an example of the analysis that would be carried out to investigate the high-variability operators D, W, Q and S, operator D is selected. Chart 13.2 is an X/MR chart for operator D. The data are plotted in the order in which drilling commenced on the wells. The chart shows two wells, numbers 24 and 26, above the UAL. The first step in extracting a set of in (statistical) control data for the benchmarking comparison is to remove these two wells from the analysis. In practice the operator would investigate the causes of the unexpected high cost per foot.

Chart 13.3 is a repeat of Chart 13.2 with the two out-of-control points omitted from the calculations. However, they are still drawn on the chart to remind us that they occurred. Chart 13.3 shows one new well, number 27 above the UAL, so this too should be

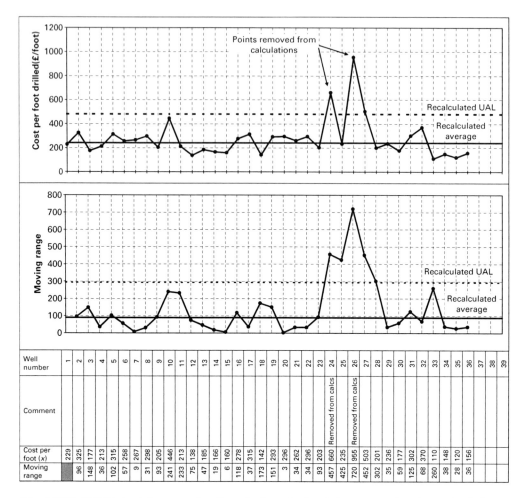

Chart 13.3 X/MR chart for operator D: out-of-control points removed; calcs: calculations

investigated and removed from the analysis. However, it is the purpose of this case study to demonstrate the method of analysis rather then produce final conclusions and so we proceed with omitting only wells 24 and 26.

Chart 13.4 is Chart 13.1 redrawn with the two out-of-control points from operator D removed. The average cost per foot for operator D has dropped from £273/foot to £242/foot. The industry average has dropped, as would be expected with two high-value wells being removed. The action limits have moved closer together, as we would expect having removed two outlier wells, reflecting the reduction in variation.

A similar analysis of every operator's data should be carried out to ensure that any out-of-control wells are removed from the analysis. As our purpose in this case study is to demonstrate methodology, and application, we have omitted these analyses.

To continue the analysis of Chart 13.4:

● Operator V is shown as the lowest cost at £140/foot and with a range of only £57/foot. However, with only two wells drilled, there is not enough data to draw too

Operator	S	D	Q	M	E	I	W	A	L	J	N	C	P	J	H	T	K	V	F	B	O	R	U	Total	Average
Number of wells	40	34	12	11	11	11	10	8	7	7	7	7	4	4	4	3	3	2	2	2	2	1	1	193	9.10
Average cost per foot ($\bar{X}$)	399	242	325	304	209	222	239	331	212	366	215	206	343	409	307	388	278	140	244	620	389	332	478		301
Range	691	393	592	378	205	231	579	384	184	331	347	150	318	281	89	324	224	57	233	64	42	0	0	6097	$d_{2,9}=$
d_2 (from tables)	4.393	4.308	3.258	3.173	3.173	3.173	3.078	2.847	2.704	2.704	2.704	2.704	2.059	2.059	2.059	1.693	1.693	1.128	1.128	1.128	1.128	1.128			2.970
Expected range if $n=9$	467	271	540	354	192	216	558	401	202	364	381	165	459	405	129	568	393	150	612	168	111	0			338
Mean range = sum of ranges/21 for average $n=9$	338	338	338	338	338	338	338	338	338	338	338	338	338	338	338	338	338	338	338	338	338	338	338		
Expected mean range for each operator's $n = \bar{R}$	500	491	371	362	362	362	351	324	308	308	308	308	235	235	235	193	193	129	129	129	129				
Average cost per foot x number of wells ($\bar{n}\bar{x}$)	15966	8228	3898	3345	2296	2446	2390	2646	1487	2561	1503	1445	1371	1636	1228	1165	835	279	488	1240	779	332	478	58043	
Industry average = $\bar{\bar{X}}$ 59643/195	301	301	301	301	301	301	301	301	301	301	301	301	301	301	301	301	301	301	301	301	301	301	301		
A_2 (from tables)	0.153	0.154	0.266	0.285	0.285	0.285	0.308	0.373	0.419	0.419	0.419	0.419	0.729	0.729	0.729	1.023	1.023	1.880	1.880	1.880	1.880				
UAL = $\bar{\bar{X}} + A_2 \times \bar{R}$	377	376	399	404	404	404	409	422	430	430	430	430	472	472	472	498	498	542	542	542	542				
LAL = $\bar{\bar{X}} - A_2 \times \bar{R}$	224	225	202	198	198	198	193	180	172	172	172	172	127	130	130	103	103	59	59	59	59				
For the range chart:																									
D_4 (from tables)	1.541	1.542	1.717	1.744	1.744	1.744	1.777	1.864	1.924	1.924	1.924	1.924	2.282	2.282	2.282	2.574	2.574	3.270	3.270	3.270	3.270				
UAL = $D_4 \times \bar{R}$	771	757	637	630	630	630	623	605	593	593	593	593	535	535	535	496	496	420	420	420	420				
D_3 (from tables)	0.459	0.458	0.283	0.256	0.256	0.256	0.223	0.136	0.076	0.076	0.076	0.076	0.000	0.000	0.000	0.000	0.000	0.000	0.000	0.000	0.000				
LAL = $D_3 \times \bar{R}$	230	225	105	93	93	93	78	44	23	23	23	23	0	0	0	0	0	0	0	0	0				

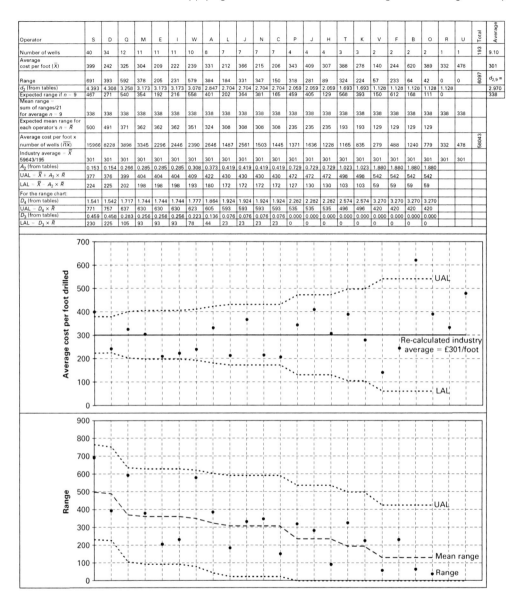

Chart 13.4 $\bar{X}$/R chart: cost per foot drilled; two operator D wells removed

firm a conclusion. It is also interesting to note that there is a large gap between V and the next lowest group of operators at just over £200/foot. Whilst V may be a lower cost per foot operator, another possible explanation is that operator V has mis-reported costs and this possibility should be checked with the operator.

- There is another objection to selecting operator V as the "best" operator. The LAL is the point at which we conclude that an operator is significantly lower than the industry average and the nearer we get to the LAL the more confident we become that the operator is lower than the average. The Operator V is further away from its LAL than operators D, E, I, W, L, N and C, so if any operators are significantly cheaper than the industry at drilling wells, it will be one of these and not operator V. This is perhaps counter-intuitive since V has a lower cost per foot, but it takes into account the consistency with which operators drill. Drilling two cheap wells could be luck, drilling 34, as operator D did or 11 as operators E and I did, is very unlikely to be luck.

- When identifying the performance of operators we need to consider two aspects: the average cost per foot and the consistency of achieving low cost per foot. Operator W, for example, has a low cost per foot, but a very high variability, whereas operator M consistently reports a low cost per foot as evidenced by the low range.

- On the average chart, operators S and B are the only operators whose averages are above the UAL. All the other operators are within the limits and so this analysis suggests that S and B are significantly higher cost operators than the other operators. As with operator V, B only has two wells and the cautionary comments made about V also apply to B.

Further analysis

The elements of the analysis have been explained above, but there is much more that can be done. The steps would be:

- Review the operators with high variability in their results (i.e. S, Q and W) in the same manner as operator D's data were analysed.
- Once all the "out-of-control" wells have been removed, update the $\bar{X}/R$ chart. If some operators are outside the control limits, identify these as being significantly higher (or lower) cost operators, remove them from the calculations, and recalculate the averages and limits to investigate the comparative performance of other operators.

Once the analysis is complete we would be able to:

- Identify individual wells for each operator that warrant investigation due to abnormally high cost per foot.
- Calculate the average cost per foot drilled for each operator, and predict limits within which the cost per foot of future wells will lie if the drilling process does not change. These predictions will less robust the fewer wells they are based on.
- Identify which (if any) operators are significantly cheaper/more expensive with respect to cost per foot, and hence recommend which operators should be approached to help with learning best practices.

Returning again to the bar chart (Figure 13.1), it would not be possible to deduce this information from this type of chart.

Comments

- The fact that it was not possible to calculate $\bar{\bar{X}}$ in the correct way (see calculations in the next page), and that there is a huge variation in the number of wells drilled, leads to concern about the accuracy of control limits and averages. For this reason, we may choose to treat operators with values near as well as outside the control limits as if they were outside the limits.

- The order of plotting operators is not relevant, and so looking for runs, cycling or patterns in these charts is not appropriate. The data were ordered in decreasing numbers of wells drilled. This allows us, at a glance, to see if there is obvious trend with number of wells drilled and cost per foot. In this case, it does not appear to be true that operators drilling more wells have higher/lower cost per foot than those drilling fewer wells. What is clear is that as the number of wells decreases there is a greater variability in the average. For example, the average cost per foot for all operators drilling seven or more wells lies between £200 and £400/foot. Those drilling less than four wells vary between £140 and £620/foot. This is probably due to occasional lucky or rogue wells having a large effect where only small numbers of wells are drilled.

- The methodology of the charting of averages when the number of points making up each average varies does not appear to be well established. It may be that the formula used here need to be modified. However, useful information can still be gleaned from the chart even if the mathematics is not rigorous. The purpose of charting is insight, not numbers! A more rigorous analysis could be carried out by a statistician if required.

- When individual points are found to be above or below the action limits, it may be tempting to delete them from the chart altogether and ignore them. A better approach is to keep the points plotted, exclude them from the calculations and add a comment to the effect that they being treated as "out-of-control" points and have been omitted from calculations.

- Having identified two wells exhibiting out-of-control conditions, the reason(s) for their exceptionally high values should be investigated. If no cause for their high value was found, we may need to consider including them in the calculations.

- A standard deviation chart could have been used in place of a range chart. Standard deviation chart are often preferred as the sample size increase, but the calculations are more complex. Another alternative would have been a median-range or median-standard deviation chart.

- In this case study, we tacitly made the assumption that each operator's data was in a state of control before comparing the averages. It is advisable to first check that each operator's data are in a state of control before making the comparison of averages. It is meaningless to compare results of operators (or of any processes) if the processes being compared are not in a state of statistical control over the period being compared. The checking of each operator's data prior to drawing the average range chart was omitted due to lack of space. Any out-of-control situations and data exclusions or modifications would normally be identified in the benchmarking report.

Calculations

The calculations are for Chart 13.1, the $\bar{X}$/range chart. Those for Chart 13.4 are similar.

The calculations are complex because there are a varying number of wells, ranging from 40 for operator S to 1 for operator U. For those interested in following through the calculations, an explanation of the methodology is given in the procedure for drawing charts in Part 4 of the book. A simpler method of analysis is to use the s chart, of which there are examples elsewhere in the book. Whichever chart is used, the above analysis method is the same.

The table above the chart gives many of the calculations.

The rows provide for each operator:

- Operator name.
- Number of wells drilled.
  To the right is the total number drilled = 195 and the average number drilled per operator = 9.19 (excluding the operator drilling only 1 well) (taken as 9 in the calculations),

- average cost per foot = $\dfrac{\text{total cost for all the operators' wells}}{\text{total number of feet drilled}}$

- Range = the cost per foot for the well with the highest cost per foot – the cost per foot for the well with the lowest cost per foot.
  To the right is the total of the ranges, 6549.

In order to calculate the average range, we could use the sum of the ranges divided by 21, the number of operators drilling more than one well. This would give 6549/21 = 312. However, since the range increases with the sample size, this is not a good method of estimation. First we need to adjust the range for the number of wells drilled. Since the average number of wells drilled is nine, we estimate what the range would be for each operator if the number of wells drilled was nine.

- $d_{2,n}$ are the d_2 constants for sample size n taken from tables and used for estimating a standard deviation from a range. For example, $d_{2,9} = 2.970$.
- The expected range if $n = 9$ is the range we would have expected if nine wells had been drilled. Operator S drilled 40 wells, and the range in cost per foot is £691/foot. To estimate what the range would have been if only nine wells were drilled we calculate:

$$\text{expected range} = \frac{\text{range} \times d_{2,9}}{d_{2,40}} = \frac{691 \times 2.970}{4.3926} = 467.$$

The expected range will increase if the number of wells drilled is <9.
The expected range will decrease if the number of wells drilled is >9.

- Having adjusted the ranges we can now calculate the mean range, $\bar{R}$ = average of the expected ranges if $n = 9$ (as calculated above). In Chart 13.1 this is shown to be 353. It could be argued that we should weight the averages according to the number of wells drilled. If we do this we get 493, which is significantly higher than the 353 calculated above. The fact that they are different suggests that we may need to look for outliers in each operators data.
- Expected mean range for each operator. This is the range we would expect for each operator based on the ranges from all operators. For operators drilling nine wells the mean range is 353. However, if number of wells is not equal to nine we need to adjust the expected mean range for the number of wells actually drilled. To do this we calculate:

$$\text{expected mean range} = \bar{R} = \frac{\text{mean range} \times d_{2,n}}{d_{2,9}} = \frac{353 \times 4.3926}{2.970} = 522$$

for operator S with $n = 40$ wells.

Comparing this value with the actual range of 691, we see that operator S has a much larger range than we would expect based on all operators in the analysis.
The expected mean range will decrease if the number of wells drilled is <9.
The expected mean range will increase if the number of wells drilled is >9.
- The average cost per foot drilled for all operators should be calculated as:

$$\overline{\overline{X}} = \text{industry average} = \frac{\text{total cost for all wells}}{\text{total feet drilled}}.$$

Unfortunately we do not have data for all the wells drilled, and so we have had to estimate the value by:

$$\overline{\overline{X}} = \frac{\text{average cost per foot} \times \text{number of wells}}{\text{total number of wells drilled}} = \frac{59,643}{195} = 306.$$

Average cost per foot × number of wells, $n\overline{x}$, and $\overline{\overline{X}}$ are given in the next two rows.
- A_2 is a constant from tables for calculating chart limits
- The UAL for the average chart for $= \overline{\overline{X}} + A_2\overline{R} = 306 + 0.373 \times 339 = 432$ for operator A.
- The LAL for the average chart $= \overline{\overline{X}} - A_2\overline{R} = 306 - 0.373 \times 339 = 180$ for operator A.
- D_4 is a constant from tables for calculating chart limits for the range chart.
- The UAL for the range chart $= D_4 \times \overline{R} = 1.864 \times 339 = 631$.
- D_3 is a constant from tables for calculating chart limits for the range chart.
- The LAL for the range chart $= D_3 \times \overline{R} = 46$.

Summary

- Control charts are a useful, if seldom used, tool for comparing performances of different groups.
- Using bar charts to compare performance gave only a cursory (and incorrect) view of comparative performance. Use of control charts led us to identifying individual "out-of-control" wells, commenting on both level of performance and consistency of performance as well as commenting on the relationship between sample size and performance level.
- Where data are not in time or any other natural sequence we can choose the order in which data are plotted. In this case study we choose to plot the data in order of sample size and were able to comment on the possible relationship between sample size and performance.
- Using range charts where sample size varies requires intricate calculations and an s chart may be a better option.
- When comparing performance levels from different groups it is necessary to check that each group's data is in a state of statistical control.

14

Comparing the results of using different charts to analyse a set of data
An application to a batch production process

Charts used: c, u, p, X/MR and scatter diagram

Introduction

In many situations it may not be obvious which chart type should be used to analyse a set of data. In this case study we investigate the question: "What happens if I use the wrong type of chart to analyse a set of data" by comparing conclusions when applying X/MR, c, u and p charts to monitor the same situation. We also discuss the appropriateness of each chart type.

Background

Celto is produced in batches and divided into smaller variable size quantities for storage, testing and distribution. All batches are tested. There is some concern over the level of rejected batches at testing and data for the last 25 months has been made available for analysis.

Table 14.1 presents the data over a 25-month period. For each of the last 25 months we have the following data:

- Number of off-specification batches.
- Number of off-specification tons.
- Number of tons tested.

Production batch sizes vary and are approximately 10 tons. In January year 1 approximately 185 batches produced a total of, we are told, 1849 tons. The storage batch sizes used for testing are smaller and in January year 1, 8 batches totalling 26 tons were found to be off-specification.

Table 14.1 Raw data for Celto and comparison of off-specification tons and batches

Month number	Month	Number of off-specification batches	Number of off-specification tons	Tons tested	Average size of batch = Number of tons/number of batches
1	Jan year 1	8	26.0	1849	3.3
2	Feb	14	52.0	2169	3.7
3	Mar	17	57.5	4718	3.4
4	Apr	12	44.1	2748	3.7
5	May	12	38.0	2847	3.2
6	Jun	15	67.5	4297	4.5
7	Jul	13	57.6	2834	4.4
8	Aug	16	44.0	4235	2.8
9	Sep	18	42.0	4018	2.3
10	Oct	12	34.0	3377	2.8
11	Nov	17	154.0	3169	9.1
12	Dec	9	25.0	2578	2.8
13	Jan year 2	9	64.0	4198	2.8
14	Feb	15	63.0	3586	4.2
15	Mar	6	20.0	4615	3.3
16	Apr	7	36.0	2556	5.1
17	May	7	22.0	3536	5.1
18	Jun	7	24.0	2369	3.1
19	Jul	1	5.0	3713	5.0
20	Aug	7	66.0	4377	9.4
21	Sep	8	43.5	4390	5.4
22	Oct	4	25.0	3548	6.3
23	Nov	6	34.0	5459	5.7
24	Dec	4	21.0	4363	5.3
25	Jan year 3	14	143.0	4769	10.2

Analysis

The c chart (Charts 14.1 and 14.2)

The easiest chart to plot is a c chart of the number of off-specification batches per month. The underlying assumption is that the opportunity for failure is the same each month, that is that number of batches tested each month is the same (which, as well will see later is not true). The chart is given in Chart 14.1 and the run of points below the average from March in year 2 strongly suggests a process change.

Chart 14.2 shows the result of applying the process change in March year 2. The corresponding calculations are:

- There are 187 batches over the first 14 months giving an average of $\bar{c} = 187/14 = 13.36$.
- There are 71 batches over the last 11 months giving an average of $\bar{c} = 71/11 = 6.45$. The standard deviations, $s = \sqrt{\bar{c}} = 3.65$ and 2.54 respectively, and the upper action limits (UALs) are: $\bar{c} + 3s = 13.36 + (3 \times 3.65) = 24.3$ and 14.07, respectively.

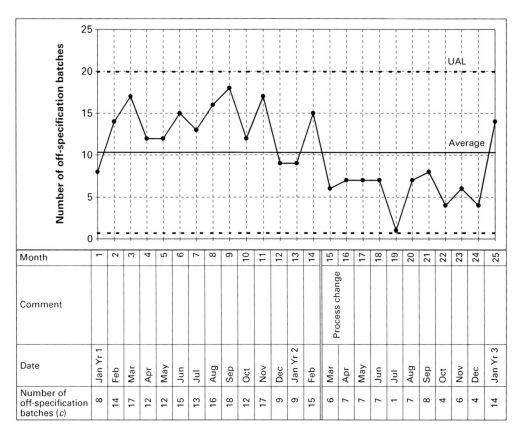

Month	1	2	3	4	5	6	7	8	9	10	11	12	13	14	15	16	17	18	19	20	21	22	23	24	25
Comment															Process change										
Date	Jan Yr 1	Feb	Mar	Apr	May	Jun	Jul	Aug	Sep	Oct	Nov	Dec	Jan Yr 2	Feb	Mar	Apr	May	Jun	Jul	Aug	Sep	Oct	Nov	Dec	Jan Yr 3
Number of off-specification batches (c)	8	14	17	12	12	15	13	16	18	12	17	9	9	15	6	7	7	7	1	7	8	4	6	4	14

Chart 14.1 c chart: number of off-specification batches per month

The lower action limits (LALs) are: $\bar{c} - 3s = 13.36 - (3 \times 3.65) = 2.4$ and $6.45 - (3 \times 2.54) < 0$ so zero is used.

The other interesting feature is the very last point in the data series – is this a single out-of-control point or the onset of another process change? Either way, it should be investigated in an attempt to discover the cause.

The np chart

We could chart the number of off-specification tons in an np chart, the argument being that each ton has been tested and a certain number failed each month. In January 2001, for example, 1849 tons were tested and 26 tons were off-specification. However, the np chart would only be appropriate if the number of tons tested each month were approximately constant (the guideline is that the tonnage does not vary more than 25% from the average). In this case the tonnages vary from 1849 in January year 1 to 5459 tons in November year 2, so an np chart is not appropriate and we would need to use the p chart.

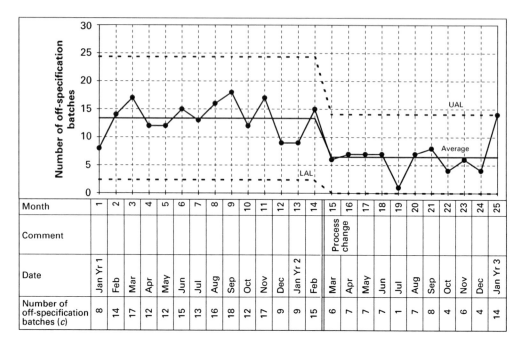

Month	1	2	3	4	5	6	7	8	9	10	11	12	13	14	15	16	17	18	19	20	21	22	23	24	25
Comment																Process change									
Date	Jan Yr 1	Feb	Mar	Apr	May	Jun	Jul	Aug	Sep	Oct	Nov	Dec	Jan Yr 2	Feb	Mar	Apr	May	Jun	Jul	Aug	Sep	Oct	Nov	Dec	Jan Yr 3
Number of off-specification batches (c)	8	14	17	12	12	15	13	16	18	12	17	9	9	15	6	7	7	7	1	7	8	4	6	4	14

Chart 14.2 c chart: number of off-specification batches per month with process change

The p chart (Chart 14.3)

Chart 14.3 is the p chart of the proportion of off-specification tonnages. The proportion of off-specification tons, p, is calculated as:

$$p = \frac{\text{number of off-specification tons}}{\text{number of tons tested}} = \frac{26}{1849} = 0.014 \text{ for the first month.}$$

The data and value for p are shown below the chart.

Like the results shown in the c chart, the p chart also suggests a drop in off-specification tons beginning in month 15. The striking difference between the p and c charts is the very high value in November of year 1. Reviewing the raw data, we see that in that month there were 17 off-specification blends, which is high, but not exceptional (see March, August and September of year 1). However, the number of off-specification tons is 154, a great deal higher than other months. This may be a data error, and should be queried. Assuming that it is correct, the difference in the c and p charts leads us to question the ratio between the number of off-specification tons and the number of batches. Table 14.1 shows the average batch size for each month.

Scatter diagram (Chart 14.4)

In most cases, the ratio given in Table 14.1 lies between 3 and 5. However, the two very high points on the p chart, January year 3 and November year 1, have ton to batch ratios of 10 and 9, that is the sampling batch sizes are large. This leads us to speculate that the batch size may be related to the proportion of off-specification tons, p. We can

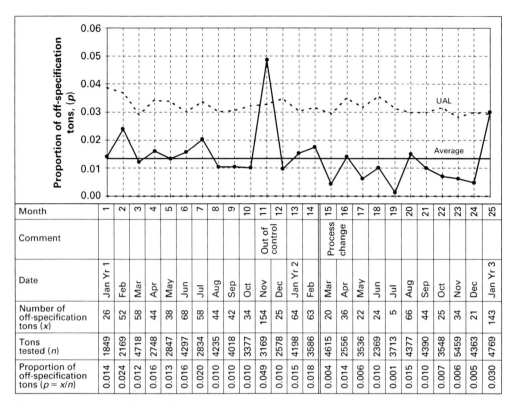

Month	1	2	3	4	5	6	7	8	9	10	11	12	13	14	15	16	17	18	19	20	21	22	23	24	25
Comment											Out of control				Process change										
Date	Jan Yr 1	Feb	Mar	Apr	May	Jun	Jul	Aug	Sep	Oct	Nov	Dec	Jan Yr 2	Feb	Mar	Apr	May	Jun	Jul	Aug	Sep	Oct	Nov	Dec	Jan Yr 3
Number of off-specification tons (x)	26	52	58	44	38	68	58	44	42	34	154	25	64	63	20	36	22	24	5	66	44	25	34	21	143
Tons tested (n)	1849	2169	4718	2748	2847	4297	2834	4235	4018	3377	3169	2578	4198	3586	4615	2556	3536	2369	3713	4377	4390	3548	5459	4363	4769
Proportion of off-specification tons (p = x/n)	0.014	0.024	0.012	0.016	0.013	0.016	0.020	0.010	0.010	0.010	0.049	0.010	0.015	0.018	0.004	0.014	0.006	0.010	0.001	0.015	0.010	0.007	0.006	0.005	0.030

Chart 14.3 p chart: off-specification tons

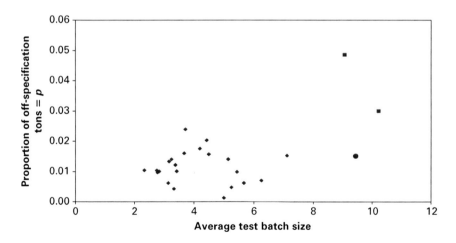

Chart 14.4 Scatter diagram of average batch size vs. proportion of off-specification tons

check this further by drawing a scatter diagram of average batch size against proportion of off-specification tons as shown in Chart 14.4. For example, the last test in January year 3 has an average batch size of 10.2 (shown in Table 14.1) and a proportion of off-specification batches of 0.030 as shown in Chart 14.3.

Chart 14.4 is interesting, but not conclusive. The highest two p values, drawn as squares on the chart, have large average batch sizes. However, August of year 2, plotted as a circle, which has an average batch size of 9.4 tons has only an average proportion of off-specification tons, 0.15. This finding warrants further investigation to determine whether there is something unusual about these three data points.

The p chart (Chart 14.5)

To continue with the analysis, we re-plot the p chart removing the out-of-control point in month 11 from the calculations. The process change does not seem as marked as for the c chart. The January year 3 point is further outside the UAL than on the c chart, and the chart should again be re-drawn omitting the last point from the calculations.

The calculations for the first half of chart are as follows (excluding November year 1):

$$\bar{p} = \begin{bmatrix} \text{average proportion of} \\ \text{off-specification tons} \end{bmatrix} = \frac{614 \text{ off-specification tons}}{43454 \text{ tons tested}} = 0.014$$

$$p = \text{proportion of off-specification tons}$$

$$\text{For January year 1, } p = \frac{26 \text{ off-specification tons}}{1849 \text{ tons tested}} = 0.014.$$

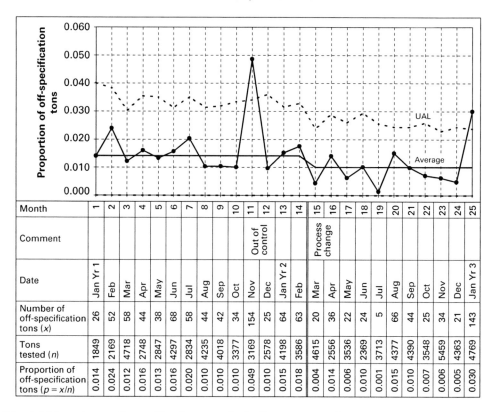

Chart 14.5 p chart: off-specification tons including process change

When calculating s, the standard deviation, we need to remember that the batch size is 10 tons. The s must be calculated separately for each value. For the first month the calculation is:

$$s = \sqrt{\frac{p(1 - p)}{tons/10}} = \sqrt{\frac{0.014(1 - 0.014)}{1849/10}} = 0.0086.$$

UAL is therefore = $0.014 + (3 \times 0.0086) = 0.040$ (0.014 being the value of $\bar{p}$).

The u chart (Chart 14.6)

Like the np chart, the c chart assumes that the number of tests carried out were similar each month. In addition, the c chart assumes that the number of off-specification events is small compared to the tons tested. However, the former assumption is certainly not valid, as explained for the np chart. Therefore, a more accurate interpretation would be obtained by taking into account the differing amounts tested. To do this we could chart the number of off-specification batches per ton tested each month.

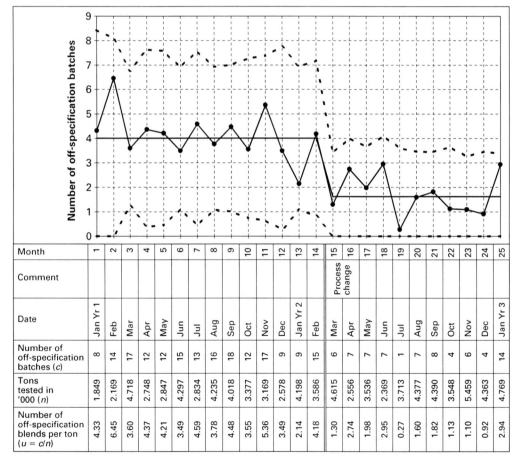

Month	1	2	3	4	5	6	7	8	9	10	11	12	13	14	15	16	17	18	19	20	21	22	23	24	25
Comment															Process change										
Date	Jan Yr 1	Feb	Mar	Apr	May	Jun	Jul	Aug	Sep	Oct	Nov	Dec	Jan Yr 2	Feb	Mar	Apr	May	Jun	Jul	Aug	Sep	Oct	Nov	Dec	Jan Yr 3
Number of off-specification batches (c)	8	14	17	12	12	15	13	16	18	12	17	9	9	15	6	7	7	7	1	7	8	4	6	4	14
Tons tested in '000 (n)	1.849	2.169	4.718	2.748	2.847	4.297	2.834	4.235	4.018	3.377	3.169	2.578	4.198	3.586	4.615	2.556	3.536	2.369	3.713	4.377	4.390	3.548	5.459	4.363	4.769
Number of off-specification blends per ton (u = c/n)	4.33	6.45	3.60	4.37	4.21	3.49	4.59	3.78	4.48	3.55	5.36	3.49	2.14	4.18	1.30	2.74	1.98	2.95	0.27	1.60	1.82	1.13	1.10	0.92	2.94

Chart 14.6 u chart: number of off-specification batches

Chart 14.6 is the resulting chart, which has already had the very obvious process change incorporated. Having taken the process change into account, Chart 14.6 is now in a state of control. In particular, note that November year 2 is well within the control limits and January year 3 is within, though close to the limit. Closer inspection reveals that there are a number of large differences between the c and u charts, and quite often the shape of the chart varies (e.g. a high value in one chart is a low value in the other chart). This highlights the importance of using a u chart when the opportunity for recording an event varies. It is comforting to note that despite the differences, both c and u chart identify a process change and both identify a concern with the very last point, even though the u chart plots it below the UAL, whilst the c chart plots it on the UAL.

Like the p chart, the limits for each point must be calculated separately. Note that for convenience we have changed the units to thousands of tons. The number of off-specification batches, the tons tested and the resulting number of off-specification batches per ton are shown on the chart:

$\bar{u}$ = average number of off-specification batches for the first 14 months

$$= \frac{\text{total number of off-specification batches}}{\text{tons tested}} = \frac{187}{46.62} = 4.01 \text{ for the first point}$$

$$u = \frac{\text{number of off-specification batches}}{\text{tons tested}} = \frac{8}{1.849} = 4.33.$$

$$s = \sqrt{\frac{\bar{u}}{n}} = \sqrt{\frac{4.01}{1.849}} = 1.47$$

UAL = $4.01 + (3 \times 1.47) = 8.42$.

X/MR chart: off-specification batches (Chart 14.7)

The charts we have been considering are all types of attribute chart, and as such they all make assumptions about the underlying distribution of the data. If we feel this is inappropriate, we may conclude that the X/MR chart is appropriate.

We could use the X/MR chart to plot either the number of off-specification batches or the number of off-specification tons and both are discussed.

Chart 14.7 is the resulting X/MR chart for the number of off-specification batches, which has already had the very obvious process change incorporated. Again we see that the last point is suspiciously high, and so could be removed from the calculations.

For the first part of the chart, $\bar{x}$, the average number of batches is calculated as:

$$\bar{x} = \frac{187 \text{ off-specification batches}}{14 \text{ months}} = \frac{13.36 \text{ batches}}{\text{month}}$$

$\overline{mr}$ is the average moving range = $49/13 = 3.77$.

$$S_{mr} = \frac{3.77}{1.128} = 3.34$$

$$\text{UAL} = \bar{x} + 3S_{mr} = 13.36 + (3 \times 3.34) = 23.4.$$

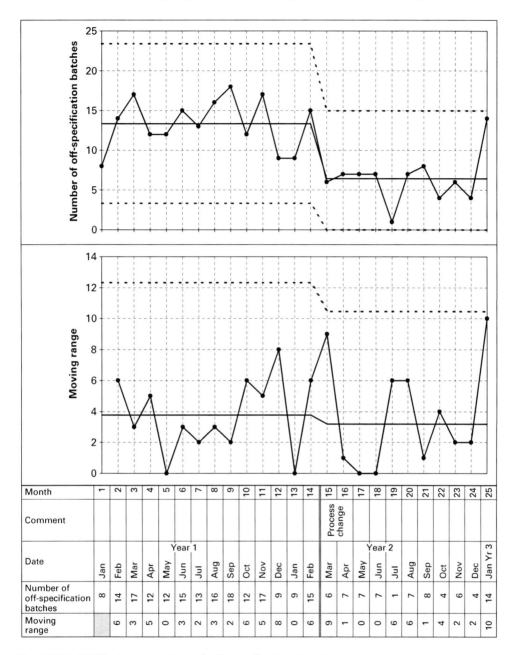

The table at the bottom of the chart:

Month	1	2	3	4	5	6	7	8	9	10	11	12	13	14	15	16	17	18	19	20	21	22	23	24	25
Comment															Process change										
Date	Jan	Feb	Mar	Apr	May	Jun	Jul	Aug	Sep	Oct	Nov	Dec	Jan	Feb	Mar	Apr	May	Jun	Jul	Aug	Sep	Oct	Nov	Dec	Jan Yr 3
Number of off-specification batches	8	14	17	12	12	15	13	16	18	12	17	9	9	15	6	7	7	7	1	7	8	4	6	4	14
Moving range		6	3	5	0	3	2	3	2	6	5	8	0	6	9	1	0	0	6	6	1	4	2	2	10

The Date row spans: Year 1 (Jan–Feb), Year 2 (Mar–Dec), Jan Yr 3.

Chart 14.7 X/MR chart: number of off-specification batches

X/MR chart: tons (Chart 14.8)

Finally, we could draw an X/MR chart of the number of tons off-specification each month. The resulting chart, with the obvious process change incorporated, is given in Chart 14.8. The interpretation is comfortingly similar to the other charts with the obvious process change, the last point above the UAL, and as identified in the p chart of tons, November year 2 above the UAL (and removed from the calculations).

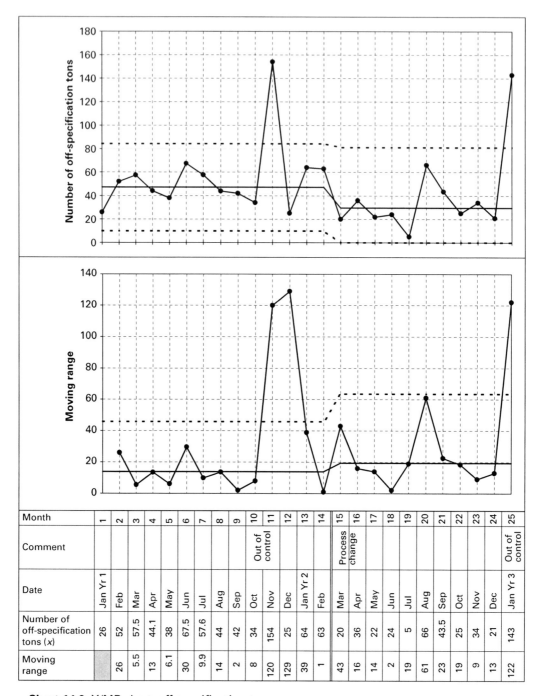

Chart 14.8 X/MR chart: off-specification tons

Month	1	2	3	4	5	6	7	8	9	10	11	12	13	14	15	16	17	18	19	20	21	22	23	24	25
Comment										Out of control				Process change										Out of control
Date	Jan Yr 1	Feb	Mar	Apr	May	Jun	Jul	Aug	Sep	Oct	Nov	Dec	Jan Yr 2	Feb	Mar	Apr	May	Jun	Jul	Aug	Sep	Oct	Nov	Dec	Jan Yr 3
Number of off-specification tons (x)	26	52	57.5	44.1	38	67.5	57.6	44	42	34	154	25	64	63	20	36	22	24	5	66	43.5	25	34	21	143
Moving range		26	5.5	13	6.1	30	9.9	14	2	8	120	129	39	1	43	16	14	2	19	61	23	19	9	13	122

Comments and conclusions

- When beginning to use control chart it is often confusing as to which chart should be used. Perhaps the safest guideline is to draw those charts that you think may be appropriate and compare the results. If the interpretation varies, investigating the differences may yield important clues about the data being analysed.
- Despite the very different approaches and assumptions inherent in the different control charts, it is comforting to realise that they all identified the process change and all charts identified a concern about the last point in the data set, albeit that in some cases the point was within the control limits, and in others outside the limits. The only major difference is the interpretation of November year 2, and this depends on whether we analyse off-specification batches or tons. The difference in interpretation appears to be due to the larger batch size of the tests, and this potential relationship should be investigated.
- The fact that all the charts led to similar conclusions should not be an excuse for complacency – in many cases there will be no harm done, but in others maybe the interpretation and resulting conclusions will be different.
- The c and u charts are appropriate when we are counting occurrences of incidents such as accidents, spills or flaws in a pane of glass. There is no logical limit to the number of these incidents that can occur in each time period. In this case study we have pass or fail criteria which suggest a p or an np chart. If the number of rejects is low compared to the potential number of rejects, then we are approaching the conditions where a c or u chart could be used in place of an np or p chart. This explains why in this case study the results for the c and u charts are in line with the other charts.
- The X/MR change chart is frequently a "safe" option as explained by Don Wheeler (*Making Sense of Data*), and could perhaps be viewed as the fall back if we are not sure which chart to use. Note especially the closeness of the UAL for the corresponding c and X/MR charts (Charts 14.2 and 14.7).
- The data presented in this chapter came from reasonably long production runs. If production runs are too short to draw a control chart for each one, difference and/or Z charts can be used to chart the performance over several different runs.

Calculations

Many of the calculations are included above. Table 14.2 provides more details.

Summary

The data as presented appeared to be quite straightforward to analyse. However, we have seen that there are a variety of charts that could have been used to monitor off-specifications blends.

For those new to statistical process control it can be confusing deciding which chart(s) to use in certain situations.

Whilst it is important to consider carefully what chart is appropriate in any situation, this case study has shown that using the "incorrect" chart may still provide accurate information about process performance. In addition, charting different variables and monitoring the same events, off-specification batches and tons in this case, provide different clues as to what is happening in the process.

Table 14.2 Calculation details for charts

Charts	Number of off-specification batches (c)	Number of off-specification tons	Number of tons tested	c chart (UAL)	Number of off-specification batches per thousand tons	u chart (UAL)	p = off-specification tons/tested tons	p chart (UAL)	X/MR batches (UAL)	X/MR tons (UAL)
	14.2 and 14.7	14.8		14.2	14.6	14.6	14.5	14.5	14.7	14.8
Jan year 1	8	26	1849	24	4.33	8.43	0.010	0.040	23	85
Feb	14	52	2169	24	6.45	8.09	0.024	0.038	23	85
Mar	17	57.5	4718	24	3.60	6.78	0.012	0.030	23	85
Apr	12	44.1	2748	24	4.37	7.64	0.016	0.036	23	85
May	12	38	2847	24	4.21	7.57	0.013	0.035	23	85
Jun	15	67.5	4297	24	3.49	6.91	0.016	0.031	23	85
Jul	13	57.6	2834	24	4.59	7.58	0.020	0.035	23	85
Aug	16	44	4235	24	3.78	6.93	0.010	0.031	23	85
Sep	18	42	4018	24	4.48	7.01	0.010	0.032	23	85
Oct	12	34	3377	24	3.55	7.28	0.010	0.033	23	85
Nov	17	154	3169	24	5.36	7.39	0.049	0.034	23	OOC
Dec	9	25	2578	24	3.49	7.75	0.010	0.036	23	85
Jan year 2	9	64	4198	24	2.14	6.94	0.015	0.031	23	85
Feb	15	63	3586	24	4.18	7.18	0.018	0.033	23	85
Total	**187**	**768.7**	**46,623**		**4.01**					**154**	**175**
					Process change						
Mar	6	20	4615	14	1.30	3.41	0.004	0.024	15	81
Apr	7	36	2556	14	2.74	4.02	0.014	0.029	15	81
May	7	22	3536	14	1.98	3.66	0.006	0.026	15	81
Jun	7	24	2369	14	2.95	4.11	0.010	0.030	15	81
Jul	1	5	3713	14	0.27	3.61	0.001	0.026	15	81
Aug	7	66	4377	14	1.60	3.45	0.015	0.024	15	81
Sep	8	43.5	4390	14	1.82	3.45	0.010	0.024	15	81
Oct	4	25	3548	14	1.13	3.66	0.010	0.026	15	81
Nov	6	34	5459	14	1.10	3.26	0.007	0.023	15	81
Dec	4	21	4363	14	0.92	3.46	0.006	0.024	15	81
Jan year 3	14	143	4769	14	2.94	3.38	0.030	0.020	15	OOC
Total	**71**	**439.5**	**43,695**		**1.62**						**175**

There are 14 data values from Jan year 1 to Feb year 2. There are 11 data values from Mar year 2 to Jan year 3, OOC: out of control

15 Using control charts to analyse data with a trend
An application to cost management

Charts used: X, X with regression and moving average

Introduction

There is a common belief that control charts cannot be used when there is a trend in the data. The main purpose of this case study is to:

- Demonstrate that control charts can be used to help identify and incorporate trends.

We also use the data to:

- Demonstrate how the moving average behaves with data trends and that it compares very unfavourably with the control chart.
- Demonstrate the effect of changing scales on charts.

Background

One of the key metrics used in the drilling industry to measure the performance of the drilling process is the cost per foot drilled. Apart from being used to appraise drilling performance, cost per foot is also used when budgeting for the drilling of future wells.

The cost per foot drilled is calculated as:

$$\text{cost per foot} = \frac{\text{cost of drilling}}{\text{number of feet drilled}}.$$

Drilling cost data are available for a series of 45 wells.

We focus only on the X chart, but as usual the MR chart would normally be included in the analysis.

Analysis

Chart 15.1 is the X chart of cost per foot for the 45 wells. A well number is used as an identifier rather than the well name in order to maintain anonymity. The cost per foot is calculated for each well and plotted in order of spud date (i.e. the date on which drilling first started in the well). The final row is the moving range (used for calculating the control limits). The calculations are shown below the chart.

The process is in control for the first 10 observations.

Well 11 records a sharp drop to £229/foot. Then follow 13 wells below the average followed by eight wells around the average and finally four above the average. This appears to be a steady increasing trend. The next well, 36, could be part of the series even though there is a big jump. However, since it is followed by an extremely high well at £1351/foot, the likelihood is that the process began to change during the drilling of well 36. It is not clear what is happening with the following two wells, 38 and 39. The final six wells appear to show the beginning of a stable performance level.

The next step would be to investigate the cause(s) of the sudden drop at well 11, the trend from wells 11 to 35 and to gain an understanding of what was happening during the next four wells, 36 to 39.

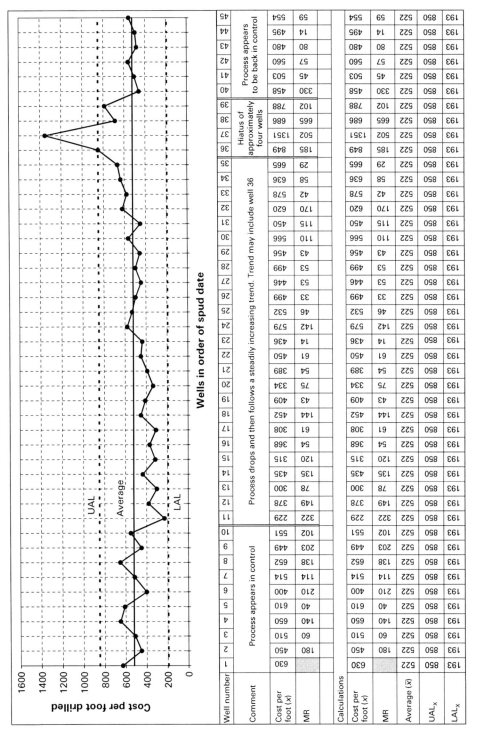

Well number	Comment	Cost per foot (x)	MR	Average (x̄)	UAL_x	LAL_x
1		630		522	850	193
2		450	180	522	850	193
3		510	60	522	850	193
4		650	140	522	850	193
5	Process appears in control	610	40	522	850	193
6		400	210	522	850	193
7		514	114	522	850	193
8		652	138	522	850	193
9		449	203	522	850	193
10		551	102	522	850	193
11		229	322	522	850	193
12		378	149	522	850	193
13		300	78	522	850	193
14		435	135	522	850	193
15		315	120	522	850	193
16		368	54	522	850	193
17		308	61	522	850	193
18		452	144	522	850	193
19	Process drops and then follows a steadily increasing trend. Trend may include well 36	409	43	522	850	193
20		334	75	522	850	193
21		389	54	522	850	193
22		450	61	522	850	193
23		436	14	522	850	193
24		579	142	522	850	193
25		532	46	522	850	193
26		499	33	522	850	193
27		446	53	522	850	193
28		499	53	522	850	193
29		456	43	522	850	193
30		566	110	522	850	193
31		450	115	522	850	193
32		620	170	522	850	193
33		578	42	522	850	193
34		636	58	522	850	193
35		665	29	522	850	193
36		849	185	522	850	193
37	Hiatus of approximately four wells	1351	502	522	850	193
38		686	665	522	850	193
39		788	102	522	850	193
40		458	330	522	850	193
41		503	45	522	850	193
42	Process appears to be back in control	560	57	522	850	193
43		480	80	522	850	193
44		495	14	522	850	193
45		554	59	522	850	193

Chart 15.1 X chart: cost per foot drilled for successive wells ordered by spud date; UAL and LAL: upper and lower action limits

Chart 15.2 incorporates the trend by fitting a regression line between wells 11 and 35. The next four wells do not seem to be part of either the earlier trend or the stable period that is appearing during the last six wells and so neither average nor control limits are drawn for them.

The chart shows no further out of control conditions, and so the interpretation may be confirmed as follows:

- The first 10 wells form a stable process with an average £542/foot drilled.
- The process changes suddenly at well 11 and commences a steady upward trend until well 35. During the period the average increase in cost per foot per well can be calculated by fitting a regression equation. As described in Chart 15.2 this gives an increase of £13/foot per well.

   Alternatively, a rough estimate of the average increase can be made from the formula:

$$\text{average increase} = \frac{\text{last cost per foot value} - \text{first cost per foot value}}{\text{number of wells drilled in this period} - 1}$$

$$= \frac{£665 - £229}{24} = £18/\text{foot per well.}$$

This is different to the estimate from the regression equation because the first well in the sequence, well 11, is below the regression line and the last well is above the regression line.

Further calculation details are given below.

- The following four wells (36–39) appear to show no stable conditions and so we are unable to determine process averages or limits for them.
- The last six wells suggest that a new period of stability may have begun with an average of £508/foot. The average and control limits should be re-calculated regularly, perhaps initially with every new value, until the process has settled down or until another change is identified. Based on these six wells, we can predict that the cost per foot for the next well will lie between the control limits, £373/foot and £664/foot, and the best point estimate is the average, that is £508/foot.

Moving average

In Part 2, we explained the theory of why moving averages should not be used. We now compare the information given by the moving average with that of the control chart when there is a trend in the data.

Chart 15.3 shows the moving average plotted on the control chart.

The table below the chart gives the well number, cost per foot drilled, the control limits and finally the moving average.

Generally, moving averages are calculated over a weekly, monthly or annual basis, the argument being that data are cyclical within this period (e.g. there is a significant difference between the days of the week and so it is not appropriate to compare, for example, a Monday with a Tuesday). An interesting issue arises in the situation of this case study. In some years we may drill 10 wells and in others 20 wells. In this situation the annual moving average would be more correctly calculated by averaging the number of wells drilled in the last 12 months, but we then have to cope with the fact that

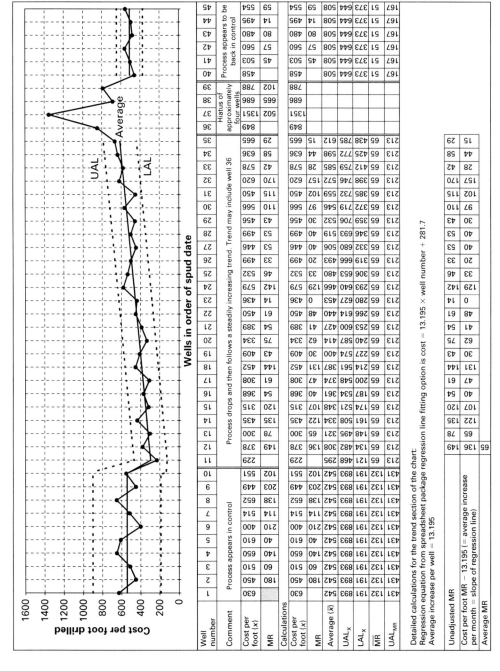

Chart 15.2 X chart: cost per foot drilled with process changes regression line; UAL and LAL: upper and lower action limits

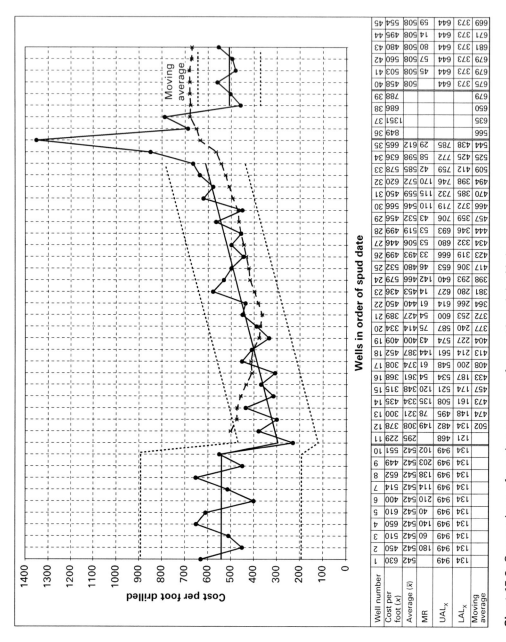

Chart 15.3 Comparison of moving average of span 12 and the X chart

each moving average will consist of different numbers of wells being averaged. Somewhat arbitrarily, the moving average has been taken over 12 values on the rough approximation that around 12 wells are being drilled annually.

As the first moving average point is plotted at well 12, the control chart has already identified that the last two wells are very low, albeit with the control limits bases on only 10 wells.

The moving average then shows a steady decrease in cost per foot – precisely the opposite to the increasing trend that is actually occurring. Eventually, at around well 22 the moving average has picked up the rise in cost per foot – a trend already established in the control chart. Up to well 22, neither the (moving average) trend nor the actual moving average value reflects the actual data except when by chance the moving average line and the control chart trend cross at well 19.

From well 22 the moving average now correctly identifies that the costs are steadily increasing. However, whilst the moving average is parallel to the control chart from wells 22–35, it is below the control chart and hence giving a falsely low estimate of the true cost per foot. If the moving average were used to predict future costs, it would provide an underestimate of approximately £50/foot (the vertical difference in the moving average and control chart regression line).

Generally, the moving average lags the corresponding control chart. However, it does respond correctly to the large jump in cost at well 36.

Unfortunately, thereafter, the moving average levels off at about £680/foot, about £170/foot above the true estimated average.

In summary, for the 34 wells for which a moving average has been calculated:

- Only on one occasion does the moving average reflect the correct cost per foot (well 19).
- For 10 of the 34 wells, the moving average suggests that the cost per foot is decreasing when it is actually increasing.
- Only for 14 of the 24 wells where there is an increasing trend does the moving average correctly reflect the upward trend. However, throughout these 14 occasions, the actual estimate of cost per foot is too low by around £50/foot.
- For the last six wells, the moving average reflects the fact the process may be stable, but even here it over estimates the cost per foot by £170/foot.

Clearly, using the moving average to predict future cost will usually result in significant over or under estimates, and for significant lengths of time will lead management to believe the process is improving when it is actually worsening.

Comments

Chart 15.3 looks startling and seems to show the process changes more dramatically than Chart 15.1 would suggest. There are two reasons for this:

- The line representing the average in Chart 15.3 is a far closer approximation to the true process average than the average line in Chart 15.1 as it includes the process changes.
- The vertical axis scale has been increased and this accentuates differences in the vertical direction.

It is important when selecting chart scales and proportions to ensure that they do not aim to mis-represent the information in the data.

In Charts 15.2 and 15.3 if we were to continue the control limits from well 35 to 36, well 36 would be above the upper action limit (UAL), and so we are probably correct to assume that this is where the process changes.

It would be possible to develop moving averages using spans other than 12. The results will be different, but the same broad conclusions will hold: that the control charts always gives a more accurate reflection of process performance and changes than a moving average.

Looking further ahead, from well 40 it appears that the data may have settled down to a steady average of around £508/foot. However, with only six wells in the series, drawing firm conclusions is risky, and the best we can say is that it looks hopeful.

It could be argued that the data should be plotted not just chronologically, but against a date scale along the horizontal axis. However, changes in process are likely to be more related to the learning gained from previous wells than whether they were drilled a day, a week or a month earlier. It is also debatable as to whether the spud date or the end of drilling should be considered the defining date of a well. The case for using spud date is that once a well has begun learning's from wells spudded later are unlikely to have an influence.

There is another argument that since wells are of different lengths an X chart is not appropriate as it gives the same weighting to all wells. This is a good point. The relationship between cost per foot and feet drilled could be investigated using a scatter diagram and it would probably be found that longer wells are cheaper per foot to drill. The analysis could be modified to take this relationship into account by "adjusting" the cost per foot. This and other aspects of drilling that affect cost per foot are beyond the scope of this book.

Calculations

The limits and average for Chart 15.1 are calculated in the normal way, as explained in Part 4.

Chart 15.2

Regression
The calculations given for the regression line are an application of standard regression analysis. However, Don Wheeler in his book *Making Sense of Data* (2003) argues for a different approach to fitting a regression line. In practice, the analyst is likely to be restricted to whatever methods are available in the software being used. A good practice is to try different methods and if the results are similar, it does not matter which is used. If the results are different there is an opportunity to learn.

Many spreadsheet packages have facilities for fitting regression lines to data. Using one such package for the data from wells 11 to 35 gave the equation:

$$\text{cost per foot} = 13.195 \times \text{well number} + 281.7.$$

So for well 11, the first well of the trend,

$$\text{cost per foot} = 13.195 \times 1 + 281.7 = £294.9/\text{foot}$$

and for the well 35, which is the 25th in the series

$$\text{cost per foot} = 13.195 \times 25 + 281.7 = £612/\text{foot.}$$

The first table under Chart 15.2 gives the usual information previously described. The table underneath, labelled "calculations" gives further information for calculating averages and limits.

The first row of this table gives the cost per foot. The second row gives the moving range. However, note that the moving ranges for the regression section of the chart 6 have been adjusted to take account of the trend by subtracting 13.195 from each value. 13.195 is used because the regression equation tells us that the cost per foot increases by an average of 13.195 from one well to the next.

The $\bar{x}$ value also needs to be adjusted to take account of the trend. This is done by simply substituting the well number into the formula:

$$\text{cost per foot} = 13.195 \times \text{well number} + 281.7.$$

remembering that we start counting from the first well in the trend, that is well 11. For example, well 11, the first in the series the $\bar{x}$ value calculated is $13.195 \times 1 + 281.7 = 295$. For well 12, the second in the series $= 13.195 \times 2 + 281.7 = 308$ and so on until well 35, the 25th in the series $= 13.195 \times 25 + 281.7 = 612$.

The limits are calculated in the usual way. For example, for well 20:

$$\text{UAL} = \bar{x} + 3 \times \left(\frac{\text{average MR}}{1.128} \right) = 414 + 3 \times \frac{65}{1.128} = 587.$$

$$\text{UAL} = \bar{x} - 3 \times \left(\frac{\text{average MR}}{1.128} \right) = 414 - 3 \times \frac{65}{1.128}$$
$$= 241 \ (240 \text{ on the chart, allowing for rounding).}$$

Below is a third part of the table giving further details of the calculation of regression line. For areas of the chart where there is no regression line, no adjustments are necessary. However, where there is a trend we need to adjust the MR by subtracting 13.195. The first moving range is then $149 - 13 = 136$. The average of these adjusted MR's is 65, and it is this adjusted average that is used in the calculations for the control limits.

Estimate of the average

The average cost per foot, and the moving averages have been calculated by simply averaging the separate figures for each data value in the average. This is only an approximation, albeit a reasonable one, to the value that would be obtained by using the correct formula which is:

$$\text{average cost per foot} = \frac{\text{total cost of wells included in the average}}{\text{total feet drilled by wells included in the average}}.$$

The approximation has been used because the individual costs and feet drilled were not available.

Summary

In this case study we have demonstrated that:

- Control charts are adept at identifying and modelling trends in data.
- When there are trends in data, the moving average:
  - does not estimate the true process average;
  - does eventually identify that a trend exists, and estimates the trend correctly, but is slow to do so.
- Care should be taken when selecting chart scaling as inappropriate scaling can repress or accentuate the signals in the chart.

16

Identifying a decrease in the use of hospitality suites

Charts used: X/MR, c, np, cusum, scatter diagram and histogram

Introduction

This case study is an unusual application of statistical process control (SPC) to the use of hospitality suites. Like several other case studies it demonstrates the wide use of SPC outside manufacturing.

The case study includes the effect of using different charts to analyse the same situation, and highlights that sometimes if we do use an inappropriate chart then it is likely that the interpretation will still be valid or that the chart itself will tell us that it is not appropriate for the data being analysed.

Background

One large organisation operated its own hospitality suites that had been purpose-built many years before. After a policy change there was a feeling amongst several people that there had been an adverse effect on the use of the hospitality suites and it was decided to review the use of its three hospitality rooms.

Data was available on a monthly basis over several years of both the number of functions and the number of guests. However, for the first 2 months only the number of guests was recorded and not the number of functions. The data are given in Table 16.1.

Analysis

X/MR chart (Charts 16.1 and 16.2)

As a first step X/MR charts were drawn of both the number of functions (Chart 16.1) and number of guests (Chart 16.2). Note that there is no data for the first 2 months of Chart 16.1.

Table 16.1 Hospitality data

	Year	Jan	Feb	Mar	Apr	May	Jun	Jul	Aug	Sep	Oct	Nov	Dec
No. of functions	1	–	–	50	27	36	36	34	29	44	53	39	36
No. of guests	1	175	294	367	226	277	267	279	210	376	398	304	283
No. of functions	2	35	42	35	42	38	34	36	21	36	45	38	37
No. of guests	2	274	326	235	357	281	253	324	146	321	360	377	337
No. of functions	3	22	49	50	34	25	45	37	29	41	29	45	23
No. of guests	3	202	379	440	266	177	331	272	181	328	243	323	172
No. of functions	4	29	40	36	31	31	32	27	17	28	29	35	33
No. of guests	4	248	290	308	307	196	231	255	120	230	231	321	237

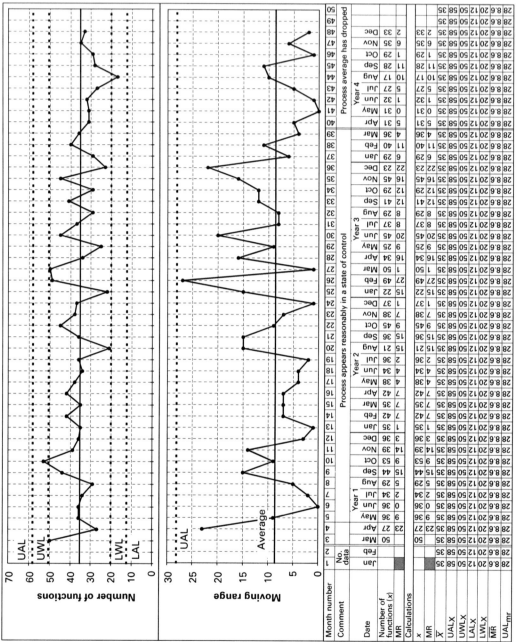

Chart 16.1 X/MR chart: number of functions per month

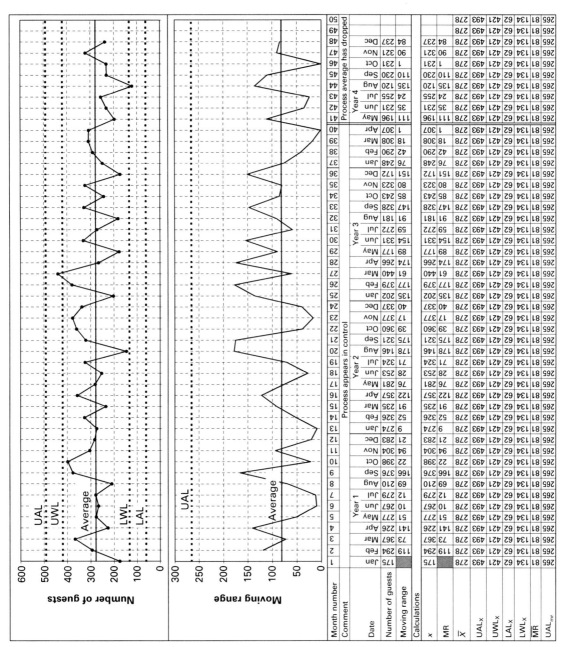

Chart 16.2 X/MR chart: number of guests

Month number	Comment	Year	Date	Number of guests (x)	Moving range (MR)
1		Year 1	Jan	175	
2			Feb	294	119
3			Mar	367	73
4			Apr	226	141
5			May	277	51
6			Jun	267	10
7			Jul	279	12
8			Aug	210	69
9			Sep	376	166
10			Oct	398	22
11			Nov	304	94
12			Dec	283	21
13	Process appears in control	Year 2	Jan	274	9
14			Feb	326	52
15			Mar	235	91
16			Apr	357	122
17			May	281	76
18			Jun	253	28
19			Jul	324	71
20			Aug	146	178
21			Sep	321	175
22			Oct	360	39
23			Nov	377	17
24			Dec	337	40
25		Year 3	Jan	202	135
26			Feb	379	177
27			Mar	440	61
28			Apr	266	174
29			May	177	89
30			Jun	331	154
31			Jul	272	59
32			Aug	181	91
33			Sep	328	147
34			Oct	243	85
35			Nov	323	80
36			Dec	172	151
37		Year 4	Jan	248	76
38			Feb	290	42
39			Mar	308	18
40			Apr	307	1
41	Process average has dropped		May	196	111
42			Jun	231	35
43			Jul	255	24
44			Aug	120	135
45			Sep	230	110
46			Oct	231	1
47			Nov	321	90
48			Dec	237	84
49					
50					

Calculations (constant across all months):

$\bar{x} = 278$, $UAL_x = 493$, $UWL_x = 421$, $LAL_x = 62$, $LWL_x = 134$, $\overline{MR} = 81$, $UAL_{mr} = 265$

The interpretation from both is very similar:

- In Chart 16.1 the run of points from month 9 to 19 are all above the average except for month 18 which is just below. The last nine values are below the average. Other than these two runs the data are in a state of control.
- In Chart 16.2 beginning in month 9, seven of the following nine months are above the average and month 13 is only just below the average. Of the last eight values seven are below the average, with month 44 being below the lower warning limit (LWL).

On closer inspection we note that the two charts appear to be closely related (as we would expect). See for example, the cycling in both charts between months 31 and 36. Also note that between months 22 and 38 the shape of the two charts is the same; that is, from one month to the next either both charts show an increase or both show a decrease.

The moving range chart in Chart 16.1 is much more erratic than in Chart 16.2, though both appear to be in a state of control up to the process change in month 39, thereafter both charts show a reduced moving range, as is common when the process average falls.

Scatter diagram (Chart 16.3)

To confirm this observation a scatter diagram was drawn of the number of guests per month vs. the number of functions per month (Chart 16.3). The chart shows that as the number of functions increases so the number of guests increases. A regression line has been fitted through the points as a visual aid and the fact that there is little scatter of points away from the line shows that the two variables are closely related. This finding allows us to monitor either the number of guests, or the number of functions; it is not necessary to monitor both.

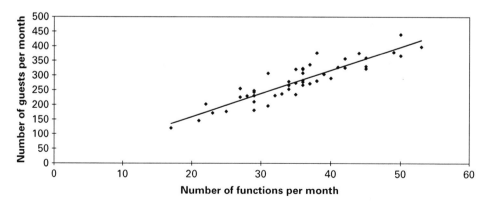

Chart 16.3 Scatter diagram showing the close relationship between number of functions and number of guests

The c chart (Charts 16.4 and 16.5)

We could view the number of functions as counts data and analyse the data using a c chart. Analysis of the c chart (Chart 16.4) for the number of functions results in the same conclusions as Charts 16.1 and 16.2: that there are a group of high values beginning at month 9 and a change which occurred nine observations from the end. Notice, however, that in the c chart there are more points outside the warning and control limits. This is because the limits are closer to the mean for the c chart. (The four limits, the upper action and warning limits (UAL and UWL), and the lower action and warning limits (LAL and LWL), for the X/MR chart are 58, 50, 12 and 20, respectively, and for the c chart the values are 53, 47, 17 and 23, respectively. The means are the same, at 35. For the X/MR chart the standard deviation, s, is based on the moving range and is calculated as $s = \overline{MR} = 8.6/1.128 = 7.6$. For the c chart, s is the square root of the average $= \sqrt{\overline{c}} = 5.9$. Note that with the c chart s increases with the average and takes no account of the actual variability of the data. So, for example, it does not matter whether the actual values vary between 34 and 36 or between 25 and 45, the standard deviation will still be 5.9. The X/MR chart has no such connection between the standard deviation and the average.

One of the assumptions of the c chart is that there is no upper limit on the number of occurrences that could occur. Clearly this is not the case, since in a typical month there are 22 working days and three suites, there is a theoretical upper limit of 66 (though it may be possible to have two functions in the same suite at different times). If the counts were relatively low in relation to this upper limit, the limit would not be important. In this example, however, there are on average 35 functions per month, and occasionally over 60.

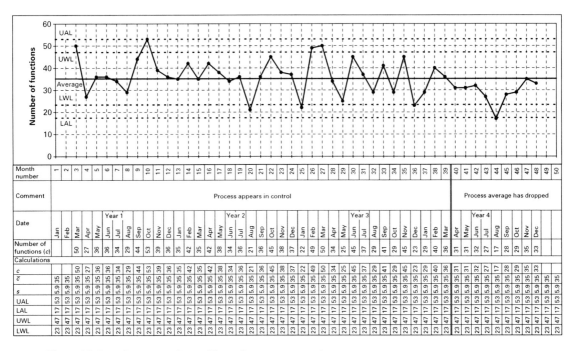

Chart 16.4 c chart: number of functions per month

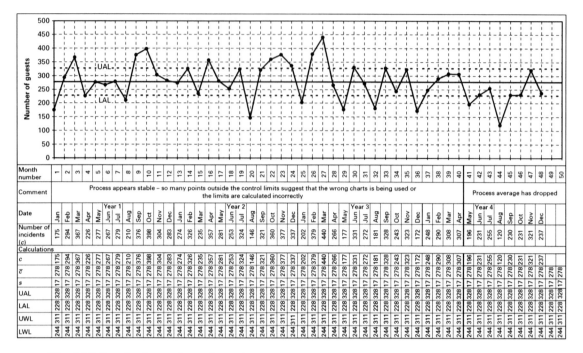

The chart shows "Number of guests" on the y-axis (0 to 500) against Month number (1 to 50), with UAL and LAL control limits marked.

Comment: Process appears stable – so many points outside the control limits suggest that the wrong charts is being used or the limits are calculated incorrectly.   Process average has dropped

Month number	Date	Number of incidents (c)	c	c̄	s	UAL	LAL	UWL	LWL
1	Jan (Year 1)	175	175	278	17	328	228	311	244
2	Feb	294	294	278	17	328	228	311	244
3	Mar	367	367	278	17	328	228	311	244
4	Apr	226	226	278	17	328	228	311	244
5	May	277	277	278	17	328	228	311	244
6	Jun	267	267	278	17	328	228	311	244
7	Jul	210	210	278	17	328	228	311	244
8	Aug	376	376	278	17	328	228	311	244
9	Sep	398	398	278	17	328	228	311	244
10	Oct	304	304	278	17	328	228	311	244
11	Nov	283	283	278	17	328	228	311	244
12	Dec	274	274	278	17	328	228	311	244
13	Jan (Year 2)	326	326	278	17	328	228	311	244
14	Feb	235	235	278	17	328	228	311	244
15	Mar	357	357	278	17	328	228	311	244
16	Apr	281	281	278	17	328	228	311	244
17	May	253	253	278	17	328	228	311	244
18	Jun	324	324	278	17	328	228	311	244
19	Jul	146	146	278	17	328	228	311	244
20	Aug	321	321	278	17	328	228	311	244
21	Sep	360	360	278	17	328	228	311	244
22	Oct	377	377	278	17	328	228	311	244
23	Nov	337	337	278	17	328	228	311	244
24	Dec	202	202	278	17	328	228	311	244
25	Jan (Year 3)	379	379	278	17	328	228	311	244
26	Feb	440	440	278	17	328	228	311	244
27	Mar	266	266	278	17	328	228	311	244
28	Apr	177	177	278	17	328	228	311	244
29	May	331	331	278	17	328	228	311	244
30	Jun	272	272	278	17	328	228	311	244
31	Jul	181	181	278	17	328	228	311	244
32	Aug	328	328	278	17	328	228	311	244
33	Sep	243	243	278	17	328	228	311	244
34	Oct	323	323	278	17	328	228	311	244
35	Nov	172	172	278	17	328	228	311	244
36	Dec	248	248	278	17	328	228	311	244
37	Jan (Year 4)	290	290	278	17	328	228	311	244
38	Feb	308	308	278	17	328	228	311	244
39	Mar	307	307	278	17	328	228	311	244
40	Apr	196	196	278	17	328	228	311	244
41	May	231	231	278	17	328	228	311	244
42	Jun	255	255	278	17	328	228	311	244
43	Jul	120	120	278	17	328	228	311	244
44	Aug	230	230	278	17	328	228	311	244
45	Sep	231	231	278	17	328	228	311	244
46	Oct	321	321	278	17	328	228	311	244
47	Nov	237	237	278	17	328	228	311	244
48	Dec			278	17	328	228	311	244
49				278	17	328	228	311	244
50				278					244

Chart 16.5 c chart: number of guests

In practical terms that may mean that the suites were fully booked on some occasions, and further bookings had to be refused. Therefore, the c chart, though it gives reasonable results (compared to the X/MR chart), would not be the preferred choice.

Similarly, a c chart could be drawn for the number of guests (Chart 16.5). The resulting chart exhibits many points beyond the control limits (the warning limits have not been plotted on the chart in an attempt to keep it uncluttered). These points are not occasional outliers well separated from the rest of the data, as we would normally expect with an "out-of-control" process. Instead they appear to be part of the process. When data exhibit this feature, it is more likely that the chart being used is not appropriate, rather than that the process is wildly out of control. On this occasion, not only is the c chart inappropriate, but also it would lead to the incorrect conclusion that the process is continually not in a state of control.

The np chart of the number of functions (Chart 16.6)

We noticed earlier that there were a maximum of about 66 functions that could be held per month, made up of three suites being in use for each of the 22 working days of the month. In light of this, we could view the data as giving 66 opportunities for rooms to be booked, and hence chart the data as an np chart. (It may help to consider this as 66 "inspections" and if the room is booked we record a "failure".) The resulting np chart (Chart 16.6) has several points beyond the control limits. The question is, are there so many out-of-control signals because the process is not in a state of control or is it because the np chart is not appropriate for this data?

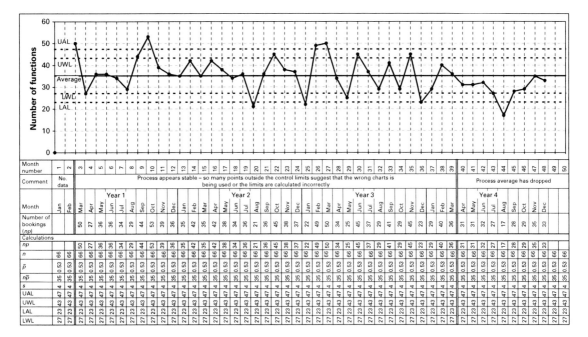

Chart 16.6 np chart: Number of functions per month

The arguments against using this chart are:

- In some cases, one booking takes up more than one function room, so the assumption that there is a maximum of 66 potential bookings a month is not strictly correct. In addition, we should consider the effect of months having different numbers of days available for booking, especially February, and holiday periods.
- Bookings do not occur randomly. Some are regular weekly or monthly meetings, and if the first choice of room and date are not available, the client will probably try several others.

With experience, it becomes easier to interpret control charts. Looking at Chart 16.6, regardless of where the warning and control limits are, the process looks to be reasonably well in a state of control (bearing in mind the process change in the latter months and nine high values between months 11 and 20).

Use of the histogram to investigate data distribution (Charts 16.7 and 16.8)

We have seen how different charts behave when charting the same data. The key difference between them is the location of the limits relative to the average. When deciding which chart is appropriate for our data it is useful to know how the data are distributed. Attributes data plotted on c, u, np and p charts are skewed. For the X/MR chart, we would like the data to be normally distributed, but the chart is robust against non-normality.

To investigate the distribution of the data we draw a histogram. Chart 16.7 is the histogram for the number of functions and is double peaked. The lower peak could be due

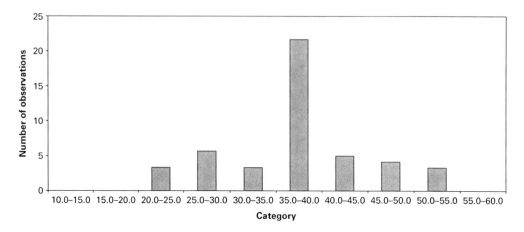

Chart 16.7 Histogram of the number of functions

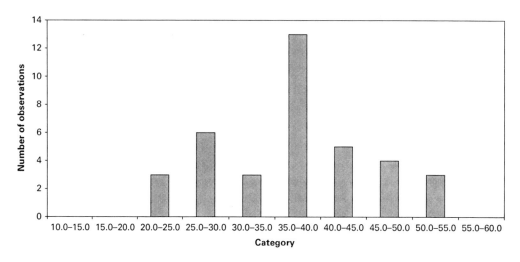

Chart 16.8 Histogram of the number of functions after removing the nine values

to the last nine values signalling the process change. Chart 16.8 is the resulting histogram when these nine values are removed. Still we see the double peak. In addition, it includes the high values from months 11 to 20. This does not look like the distribution for either a c or an np chart, and the X/MR chart appears to be the most appropriate.

The fact that the histograms have two peaks is disconcerting and leads us to suspect that there may be two processes with the output of the two being mingled or that there has been a change in process.

$\bar{X}$/range chart (Chart 16.9)

One possible investigation is to see whether the data vary over the year. To achieve this an $\bar{X}$/range chart (Chart 16.9) was drawn to compare the average number of guests per

Month number	Jan	Feb	Mar	Apr	May	Jun	Jul	Aug	Sep	Oct	Nov	Dec
Comment	Low?							Out of control				Low?
Mean	225	322	338	289	233	271	283	164	314	308	331	257
Range	99	89	205	131	104	100	69	90	146	167	73	165

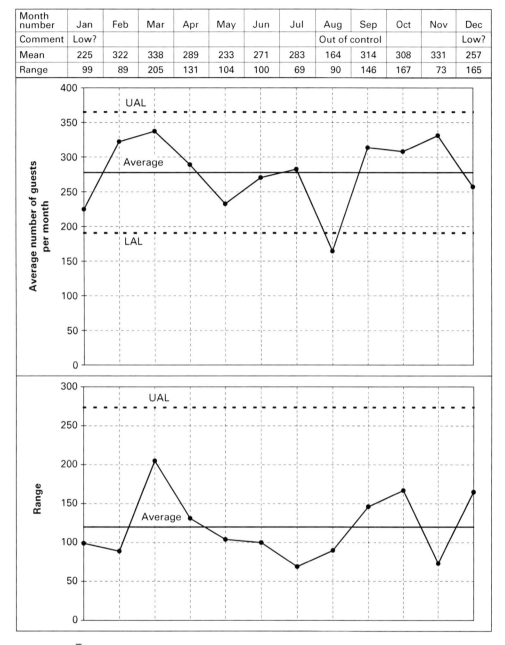

Chart 16.9 $\bar{X}$/range chart of the number of guests per month

month (equally an $\bar{X}$/range chart of the number of functions could have been drawn, and the decision to use guests is somewhat arbitrary).

The chart shows clearly that the number of bookings unusually low during August which is the holiday period, and is quite low over the Christmas period, December and January. It is likely that bookings drop during the second half of December and increase again in mid-January. It might be possible to adjust the December and January data by estimating what they would be if there were no holiday.

The August drop coincides with the summer holiday period and could either be removed from the study or the data could be adjusted for seasonality.

Having made appropriate adjustments, we should then return to the histogram to check that the data follows something more akin to the normal distribution.

Cusum chart (Chart 16.10)

There is one final chart worth drawing – a cumulative sum chart. Chart 16.10 is the resulting chart of the number of functions per month. The chart plots the cumulative difference between the number of functions each month and the average number of functions each month, 35.

There are conventions on the scaling on the chart. On this occasion we are only looking for confirmation of the process change over the last 10 months or so and scaling is not important.

The table below the chart includes, among other items, a comment section, the number of functions and the cumulative sum. If you want to follow through the calculations they are given below the chart. The calculation section includes the number of functions, the mean, the difference between the monthly number of functions and the mean, and finally the cusum (the cumulative differences).

The chart is interesting. Remember that to interpret a cusum chart we analyse the slope. The target value for the chart has been taken as the average of all data, 35. Months 3–8 show a slight downward trend with an average of 32, but this is not enough to indicate a process change. Between months 8 and 35 the trend is upwards, indicating that the average of these values is a little above the average of 35. However, it is arguable that the slope has become horizontal from observation 27, but the signals are not large enough to be significant. The two large drops in months 20 and 25 correspond to the low values seen in the corresponding X/MR, c and np charts. From month 36 there is a steady and steep decline. The average is 30, well below the average of 35 for the whole data set and is the most significant change on the whole data set.

Conclusion

This analysis, somewhat expanded to compare the results of using different charts on the same data, demonstrated to management that their actions had resulted in a drop in both bookings, guests and income. Further changes were made in an effort to increase usage of the hospitality suites.

Comments

Related variables

As demonstrated in Chart 16.3, two variables may be closely related. In these situations we can choose to monitor only one of the two, as the state of the process is expected to be reflected in both metrics in a similar way.

Chart 16.10 Cumulative sum chart: number of functions per month

Month number	Date	Comment	Number of functions (x)	Cusum	Calc: x	Calc: Mean	Calc: x-mean	Calc: Cusum
1	Jan	No data						
2	Feb	No data						
3	Mar		50	14.8	50	35	15	14.8
4	Apr	Downward slope	27	6.6	27	35	-8	6.6
5	May		36	7.3	36	35	1	7.3
6	Jun	Year 1	36	8.1	36	35	1	8.1
7	Jul		34	6.9	34	35	-1	6.9
8	Aug		29	0.7	29	35	-6	0.7
9	Sep		44	9.5	44	35	9	9.5
10	Oct		53	27.3	53	35	18	27.3
11	Nov		39	31.0	39	35	4	31.0
12	Dec		36	31.8	36	35	1	31.8
13	Jan		35	31.6	35	35	0	31.6
14	Feb		42	38.4	42	35	7	38.4
15	Mar		35	38.2	35	35	0	38.2
16	Apr		42	45.0	42	35	7	45.0
17	May	Upward slope / Two large drops at months 20 and 25 / Two large increases at months 26 and 27	38	47.7	38	35	3	47.7
18	Jun		34	46.5	34	35	-1	46.5
19	Jul	Year 2	36	47.3	36	35	1	47.3
20	Aug		21	33.1	21	35	-14	33.1
21	Sep		36	33.9	36	35	1	33.9
22	Oct		45	43.7	45	35	10	43.7
23	Nov		38	46.4	38	35	3	46.4
24	Dec		37	48.2	37	35	2	48.2
25	Jan		22	35.0	22	35	-13	35.0
26	Feb		49	48.8	49	35	14	48.8
27	Mar		50	63.6	50	35	15	63.6
28	Apr		34	62.3	34	35	-1	62.3
29	May	Possible change in slope to be horizontal	25	52.1	25	35	-10	52.1
30	Jun		45	61.9	45	35	10	61.9
31	Jul	Year 3	37	63.7	37	35	2	63.7
32	Aug		29	57.5	29	35	-6	57.5
33	Sep		41	63.3	41	35	6	63.3
34	Oct		29	57.0	29	35	-6	57.0
35	Nov		45	66.8	45	35	10	66.8
36	Dec		23	54.6	23	35	-12	54.6
37	Jan		29	48.4	29	35	-6	48.4
38	Feb		40	53.2	40	35	5	53.2
39	Mar		36	54.0	36	35	1	54.0
40	Apr		31	49.7	31	35	-4	49.7
41	May	Dramatic decrease in slope suggests process change may have started from month 36	31	45.5	31	35	-4	45.5
42	Jun		32	42.3	32	35	-3	42.3
43	Jul	Year 4	27	34.1	27	35	-8	34.1
44	Aug		17	15.9	17	35	-18	15.9
45	Sep		28	8.7	28	35	-7	8.7
46	Oct		29	2.4	29	35	-6	2.4
47	Nov		35	2.2	35	35	0	2.2
48	Dec		33	0.0	33	35	-2	0.0

The fact that we can monitor either of two highly correlated variables can be usefully generalised. Firstly, where number of variables are closely related to each other, we can select any one to reflect the state of the others. For example, in this case study we would expect the number of hours usage and the income generated from hiring out the function suite to be highly correlated with both the number of functions and the number of guests, and we need only to monitor one of all these variables.

This concept is particularly useful if it is not possible, or is very expensive, to monitor the variable in which we are interested, but cheap/easy to monitor another variable that is highly correlated to it. Common examples are vibration and temperature monitoring as surrogate measures of the condition of equipment.

Calculations

The calculations for all these charts are straightforward, and the intermediate calculations are given underneath each chart, except for Chart 16.9 where the workings are above the chart.

Summary

- Control charts can be applied to usage/utilisation of equipment, plant and, as in this case study of hospitality suites, facilities.
- Attributes charts generally assume that the data follow specific skewed distributions (see Part 4).
- X/MR charts are the "standby" chart if we are concerned about the appropriateness of using attributes charts.
- Histograms are a useful tool for investigating the distribution of data and help selecting the appropriate control chart for a particular set of data.
- If we do use an inappropriate chart for a particular data set then it is likely that the interpretation will still be valid or that the chart itself will tell us that it is not appropriate for the data being analysed.
- The cumulative sum chart, though more difficult to interpret, is very powerful at identifying process changes.
- When it is not feasible to monitor the variable that we are interested in, we can monitor a surrogate metric which is highly correlated to the metric of interest.

17 Increase in reject rate at manufacture due to inspectors' fear of losing their jobs

Charts used: X/MR, p and cusum

Introduction

This example is taken from the final inspection and test function in a manufacturing plant. In this case study we:

- demonstrate the use of a p and weighted cusum chart for monitoring;
- investigate the effect of using an X/MR chart in place of a p chart;
- show how the working environment can significantly effect reported process outputs.

Background

A regular major client had been carrying out inspections before using goods supplied from the manufacturer. The reject rate was high resulting in expensive delays as well as the usual inefficiencies of having to return faulty items.

As a first step, the client brought forward the time of inspection from the end-user inspection to the warehouse (Figure 17.1). This was to ascertain whether damage was occurring during storage or shipment to the end user or prior to receipt at the warehouse. Analysis of the results suggested that although some damage was indeed occurring during this period, the reject rate at arrival at the warehouse was nearly as high as at the end user. The client then contacted the manufacturer with the aim of working together to reduce, if not eliminate, faults occurring before arrival at the warehouse.

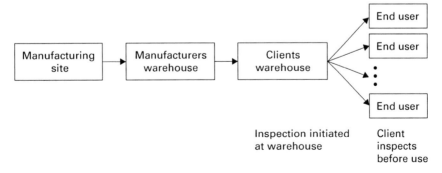

Figure 17.1 Manufacturing, transportation and use process including inspection points

Various causes of damage were identified and steps taken to reduce rejects, and the manufacturing staff were trained in the inspection methods and standards required by the end user. As part of the client's requirement, the supplier was required to keep records and chart reject rates. This is the story of what happened.

Analysis

The p chart

Since all items to be shipped are inspected, and each item either passes or fails the test, a p or np chart should be used. In this case the number of items inspected varies widely and so a p chart is appropriate, and is reproduced in Chart 17.1.

The date is given on the chart as week number/year number, and it is clear from the week number that some weeks are missing (e.g. year 00 weeks 41, 47, 51, 52). Whenever we suspect missing data, the cause should be investigated. Usually, as in this case, there is a good reason. It so happens that whilst goods are ordered and shipped by the week, there are some weeks where none are ordered. It would not be appropriate to record zero rejects on zero inspections, and so the week is omitted.

Inspection of the chart clearly shows that the last three points are above the UAL and should be investigated. However, if we look a little more closely, it appears that a general increase in reject rate began some weeks earlier, perhaps at year 01 week 18. No limits have been breached, and arguably we do not break any of the "rules" for identifying a process change, but it does look suspicious. Going further back to the beginning of the chart, the first 12 observations appear to form a well behaved in control process, with five of the 10 observations at zero. We then have 10 points before the next zero, with five in a row at or above the average (observations 13–17) followed by six below the average (observations 18–23). Does this reflect a process in control, or has something happened? Is there anything we can do to investigate further?

Cusum chart

One very powerful tool for identifying process changes is a weighted cusum chart, which is reproduced in Chart 17.2.

Cusum charts are interpreted by looking at changes in slope. The key features of the chart are the following:

- it is more or less horizontal for the first 17 observations, followed by
- a downward slope to observation 23 and then
- a slope consisting of just two observations, 14 and 25
- a flat section to observation 28 and finally
- a steep upward section from observation 29 on, which includes the last three out control points identified on the p chart.

The main change in slope occurs at observation 23 with the slope from the start of the data to that point being downward and afterwards being up. This suggests that the most likely point at which the process average changed was observation 23. It is possible to construct decision lines on cusum charts to determine whether a process

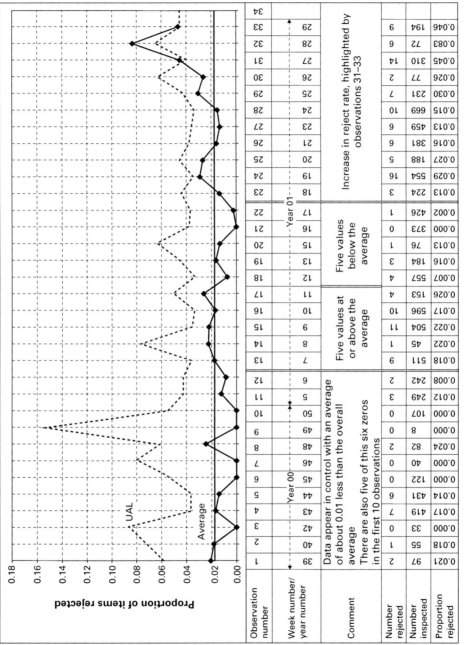

Observation number	Week number/ year number	Comment	Number rejected	Number inspected	Proportion rejected
1	39	Data appear in control with an average of about 0.01 less than the overall average. There are also five of this six zeros in the first 10 observations	2	97	0.021
2	40		1	55	0.018
3	42		0	33	0.000
4	43		7	419	0.017
5	44 — Year 00		6	431	0.014
6	45		0	122	0.000
7	46		0	40	0.000
8	48		2	82	0.024
9	49		0	8	0.000
10	50		0	107	0.000
11	5		3	249	0.012
12	6		2	242	0.008
13	7	Five values at or above the average	9	511	0.018
14	8		1	45	0.022
15	9		11	504	0.022
16	10		10	596	0.017
17	11		4	153	0.026
18	12	Five values below the average	4	557	0.007
19	13		3	184	0.016
20	15		1	76	0.013
21	16		0	373	0.000
22	17 — Year 01		1	426	0.002
23	18	Increase in reject rate, highlighted by observations 31–33	3	224	0.013
24	19		16	554	0.029
25	20		5	188	0.027
26	21		6	381	0.016
27	23		6	459	0.013
28	24		10	669	0.015
29	25		7	231	0.030
30	26		2	77	0.026
31	27		14	310	0.045
32	28		6	72	0.083
33	29		9	194	0.046
34					

Chart 17.1 p chart: rejects at manufacture; UAL: upper action limit

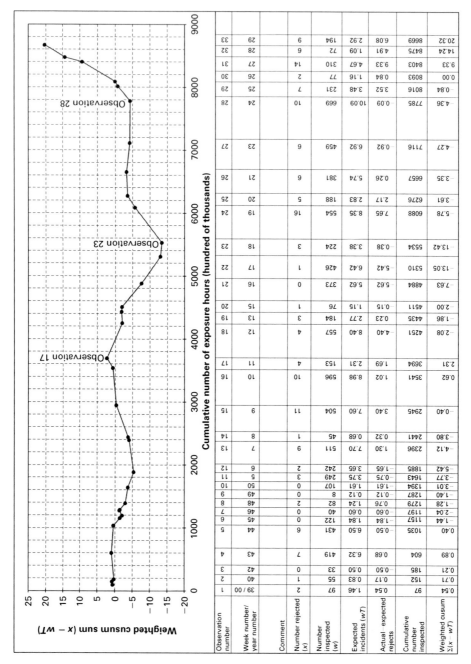

Observation number	Week number/ year number	Comment	Number rejected (x)	Number inspected (w)	Expected incidents (wT)	Actual − expected rejects	Cumulative number inspected	Weighted cusum Σ(x − wT)
1	39/00		2	97	1.46	0.54	97	0.54
2	40		1	55	0.83	0.17	152	0.71
3	42		0	33	0.50	−0.50	185	0.21
4	43		7	419	6.32	0.68	604	0.89
5	44		6	431	6.50	−0.50	1035	0.40
6	45		0	122	1.84	−1.84	1157	−1.44
7	46		0	40	0.60	−0.60	1197	−2.04
8	48		2	82	1.24	0.76	1279	−1.28
9	49		0	8	0.12	−0.12	1287	−1.40
10	50		0	107	1.61	−1.61	1394	−3.01
11	5		3	249	3.75	−0.75	1643	−3.77
12	6		2	242	3.65	−1.65	1885	−5.42
13	7		9	511	7.70	1.30	2396	−4.12
14	8		1	45	0.68	0.32	2441	−3.80
15	9		11	504	7.60	3.40	2945	−0.40
16	10		10	596	8.98	1.02	3541	0.62
17	11		4	153	2.31	1.69	3694	2.31
18	12		4	557	8.40	−4.40	4251	−2.08
19	13		3	184	2.77	0.23	4435	−1.86
20	15		1	76	1.15	−0.15	4511	−2.00
21	16		0	373	5.62	−5.62	4884	−7.63
22	17		1	426	6.42	−5.42	5310	−13.05
23	18		3	224	3.38	−0.38	5534	−13.42
24	19		16	554	8.35	7.65	6088	5.78
25	20		5	188	2.83	2.17	6276	3.61
26	21		6	381	5.74	0.26	6657	3.35
27	23		6	459	6.92	−0.92	7116	4.27
28	24		10	669	10.09	−0.09	7785	4.36
29	25		7	231	3.48	3.52	8016	0.84
30	26		2	77	1.16	0.84	8093	0.00
31	27		14	310	4.67	9.33	8403	9.33
32	28		6	72	1.09	4.91	8475	14.24
33	29		9	194	2.92	6.08	8669	20.32

Chart 17.2 Weighted cusum chart of rejects at manufacture

change has occurred, and the method of doing so is explained in Part 4. However, for the purposes of this case study we were looking for confirmation, or otherwise of the onset of a change, and so have not included them.

Armed with this information some research was carried out as to why the reject rates were increasing. The answer, when it came, was intriguing. Management had deduced that reject rates had begun to drop after observation 17. The inspectors, who were contractors, had heard a rumour that they would be made redundant because the reject rates had been low for some time and the process seemed to be under control. In order to justify their continued employment, they became exceptionally meticulous about inspection, and rejected any item they possibly could. As they became better at finding rejects, so the reject rate climbed.

Comments

Increasing the sample size narrows the control limits (p chart)

The effect of the varying number of items inspected each week is that the limits on the p chart varies in the opposite direction; that is, as the number of items inspected increases, the upper limits are reduced. The mathematical reason for this can be seen by considering the formula for the standard deviation in which the sample size is in the denominator. This also makes logical sense as we would expect that the more we inspect, the nearer our sample reject rate will be to the true reject rate. To help visualise the effect consider spinning a fair coin. We know that the probability of the outcome being a head is 0.5. If we spin the coin five times, we might reasonably expect to see between one and four heads; that is, between 20% and 80% heads and we would still not consider the coin to be biased. However, if we spun the coin 100 times, we would be very suspicious to return either <20% or >80% heads. Though explained for the p chart, the same argument applies to the u chart.

Shortcuts for manual calculations

Calculating the control limits on p charts for every observation by hand can be tedious. It is possible to reduce the number of calculations to only those occasions where we suspect the action limit (or warning limit if it is used) may be breached. In this case, since the lower limits are zero, we need never calculate the limits for reject rates less than the average. However, with a little thought we can save more effort. In year 01 week 7 we see that for a sample size of 511 the control limit is at just over 0.3. Therefore we can conclude that whenever the sample size is less than 511 the control limit will always be above 0.3. Using this information, and the fact that we would not need to calculate limits for observations below or just above the average, we would not have needed to calculate the limits until year 01 week 19 where we have a higher sample size and a high reject rate, and thereafter for the last 3 weeks.

* * *

The importance of being aware of the moving limits in situations where the p and u charts are appropriate becomes clear with observation 31. The reject rate is 0.045 and

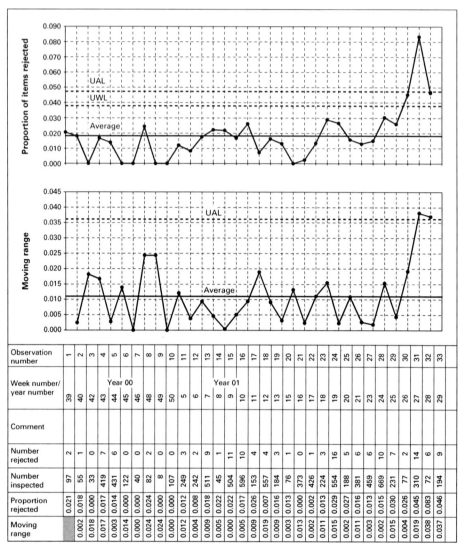

Observation number	1	2	3	4	5	6	7	8	9	10	11	12	13	14	15	16	17	18	19	20	21	22	23	24	25	26	27	28	29	30	31	32	33
Week number/ year number	39	40	42	43	44	45	46	48	49	50	5	6	7	8	9	10	11	12	13	15	16	17	18	19	20	21	23	24	25	26	27	28	29
				Year 00										Year 01																				
Comment																																	
Number rejected	2	1	0	7	6	0	0	2	0	0	3	2	9	1	11	10	4	4	3	1	0	1	3	16	5	6	6	10	7	2	14	6	9
Number inspected	97	55	33	419	431	122	40	82	8	107	249	242	511	45	504	596	153	557	184	76	373	426	224	554	188	381	459	669	231	77	310	72	194
Proportion rejected	0.021	0.018	0.000	0.017	0.014	0.000	0.000	0.024	0.000	0.000	0.012	0.008	0.018	0.022	0.022	0.017	0.026	0.007	0.016	0.013	0.000	0.002	0.013	0.029	0.027	0.016	0.013	0.015	0.030	0.026	0.045	0.083	0.046
Moving range		0.002	0.018	0.017	0.003	0.014	0.000	0.024	0.024	0.000	0.012	0.004	0.009	0.005	0.000	0.005	0.009	0.019	0.009	0.003	0.013	0.002	0.011	0.015	0.002	0.011	0.003	0.002	0.015	0.004	0.019	0.038	0.037

Chart 17.3 X/MR chart: rejects at manufacture; UAL and UWL: upper action and warning limits, respectively

is above the control limit. However, the same reject rate would be below the limit in both the preceding and following weeks where the sample size were smaller.

The cause of the process change – the inspectors' fear that they might loose their jobs – had nothing to do with the manufacturing process and everything to do with management.

Frequently when managers turn their attention on a process it improves. Reject rates and failures drop, performance levels rise as people endeavour to assure the manager that things are going well. In this case when the inspectors thought that their activities were being focused on, they deliberately "managed" the process to seemingly degrade it and so save their jobs.

Finally, since the average number of rejects is >1, it is possible to plot the proportion of rejects as an X/MR chart. Chart 17.3 is the corresponding X/MR chart and a brief review suggests that the interpretation is similar to the p chart, with one exception. The third data point from the end is shown as being just below the upper action limit (UAL) (though above the warning limit). We also note that in general points on the X/MR chart are further away from the action limit than on the p chart. In conclusion, the X/MR chart would lead to similar conclusions to the (more appropriate) p chart.

Calculations

The p chart

The reject rate for each week, p_i, is calculated as number rejected/number inspected:

$$\text{average reject rate } (\bar{p}) = \frac{\text{total number rejected}}{\text{total number inspected}} = \frac{151}{8669} = 0.0174.$$

$$\text{average number inspected } (\bar{n}) = \frac{\text{total number inspected}}{\text{total observations}}$$
$$= \frac{8669}{33} = 263 \text{ inspected per week.}$$

$$s = \sqrt{\frac{\bar{p}(1 - \bar{p})}{\bar{n}}} = \sqrt{\frac{0.0174(1 - 0.0174)}{263}} = 0.0081.$$

We can use this standard deviation for any week when the number inspected is within 25% of the average weekly number inspected, 263 (i.e. 198–329).

Alternatively, and when the number inspected is outside this range, we calculate the standard deviation in the same way but replacing by the $\bar{n}$ number of items inspected that week. For example, for observation 32, 72 items were inspected and the standard deviation will be:

$$s = \sqrt{\frac{0.0174(1 - 0.0174)}{72}} = 0.015.$$

The action limits are calculated in the usual way as $\bar{p} \pm 3s$.

The weighted cusum chart

Since from the p chart we know that the last three values are above the action limits, we omit them from some of the calculations as outlined below. This is not necessary and should not change the interpretation of the chart, but it removes one source of variation from the analysis, and helps us focus on the data of concern (i.e. up to the last three values).

For week i, w_i items are inspected of which x_i are rejected. w_i is known as the weight factor and it reflects the varying opportunity for recording rejects.

For a cusum chart we plot the differences between the observed values and a target, T. The target can be any value we choose, and for reasons of scaling is usually chosen to be a value equal to or near the average (see Part 4 for further information). For convenience we select the target as being the average reject rate up to and excluding the last three values. This is calculated as:

$$T = \frac{\text{number rejected}}{\text{number inspected}} = \frac{122}{8093} = 0.0151.$$

The calculations for the individual weeks are on the chart and are calculated as follows:

In any week the expected number of rejects equals the number inspected $\times$ $0.0151 = w_i T$.

The difference between the actual and expected number of rejects is $(x_i - w_i T)$.

The weighted cusum is the cumulative sum of the weekly differences $=$

$\sum\limits_{i=1}^{m} (x_i - w_i T)$ for the mth observation.

The total inspected to date is calculated as $\sum\limits_{i=1}^{m} w_i$ for the mth observation.

The "x" axis is the cumulative number of items inspected and the "y" axis is the weighted cusum. Since the number inspected each week varies, the horizontal distance between the points also varies. The data relating to the observations have been printed below and as near as possible to the point to which they relate.

It is worth pointing out that:

- if T is selected as being greater than the average then the overall slope of the chart will be down;
- if T is selected as being less than the average then the overall slope of the chart will be up;
- identifying changes in slope is easiest when the overall slope is zero (i.e. horizontal, relating to a target equalling the average).

Summary

In this case study we examined reject rates at the final inspection step of a manufacturing process. A p chart was initially used to analyse the reject rates and hinted at a process change. A cusum chart, which is very powerful at identifying small changes in process average, was used to investigate the suspected change, and the point at which the process changed was identified. On investigating the process it was found that an increase had indeed occurred. The cause of the increase was found to be due to an increase in diligence by the inspectors, who in fear of losing their contract, became more rigorous in their application of the criteria for rejecting items.

The case study highlights one of the uses of the cusum chart, and demonstrates that a change in process results may be due to changes in monitoring rather than the production process. It also demonstrates the importance of soft aspects of managing an organisation.

Clearly, these lessons are applicable to a wide range of applications both in non-manufacturing and manufacturing organisations.

18

Comparison of test results of production process

From a batch production process to identify a key cause of variation and that the process is not capable of producing within specification

Charts used: X/MR and $\bar{X}$/R

Introduction

In this case study we look at continuous batch paper manufacturing process discover that:

- the process was not capable of regularly producing in specification material;
- a key cause of variation in the results is the person carrying out the test.

And we see how to:

- use control charts to generate and test theories as to causes of variation;
- combine data from different batches thus enhancing the data available for analysis;
- use control charts to determine whether a process is capable of producing outputs within specification.

Background

The manufacture of paper using recycled materials is a complex multi-stage process and is shown in simplified form in Figure 18.1.

Paper products are made to order on several machines. One of the feeds is recycled paper whose properties vary widely in real time. Automatic continuous monitoring and process adjustment is carried out in real time, but there is a time delay between the monitoring of product and the effects of the adjustment.

Once the paper has been produced, various tests, including one which we call here the bond test, are carried out on samples and the whole roll of paper is then accepted or rejected. There were concerns over the amount of paper being rejected. Control charting was not commonly in use, and it was decided to investigate the potential of control charts to help improve product quality.

Paper is produced in rolls, and orders for a number of rolls may take up to several days to produce. Tests are typically carried out 10 times per 8 hour shift and the results recorded. Test results from three separate orders A, B and C were analysed.

In each case the nominal, or target, value is 90 and the specification limits are at 80 and 100.

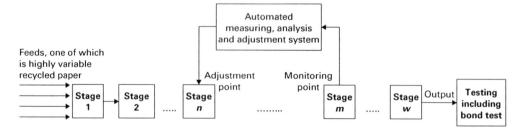

Figure 18.1 Simplified process flow highlighting monitoring, adjustment and testing

Analysis

Data set A

The data for product run A consists of 141 measurements taken over 5 days. Chart 18.1 is the X/MR chart showing the bond values plotted in the order in which they were measured. The action limits (annotated as upper and lower action limits, respectively, UAL and LAL) and specification limits (annotated as upper and lower specification limits, respectively, USL and LSL) are drawn on the chart. The table below the chart includes the sample number, day number and shits, the measured bond value and the moving range. The chart shows the key features:

- The process is not producing within specification. This is clear from the fact that there are many points outside the specification limits. There are 6 points below the LSL and 17 above the USL as identified in the comment section of the chart.
- The process is not capable of producing within specification. Regardless of whether there are points outside the specification limits, it is possible to tell if the process is capable of producing within specification. Remembering that the control limits tell us the limits within which a stable process will operate, if:
  - the LAL is lower than the LSL then the process will sometimes produce results below the LSL;
  - the UAL is higher than the USL then the process will sometimes produce results above the USL.

  This fact is particularly useful in situations where although we do not have any data outside the specification limits, we can predict that we will produce results outside the specification limits unless the process is improved.
- There is evidence of out-of-control conditions including:
  - six consecutive samples above the average (samples 4–9);
  - six of seven samples below the average (samples 17–23). The sample above the average is only just above;
  - cycling of values (samples 22–27, 29–35, 37–45, etc.).

  Further inspection will yield other evidence including points outside the action limits.

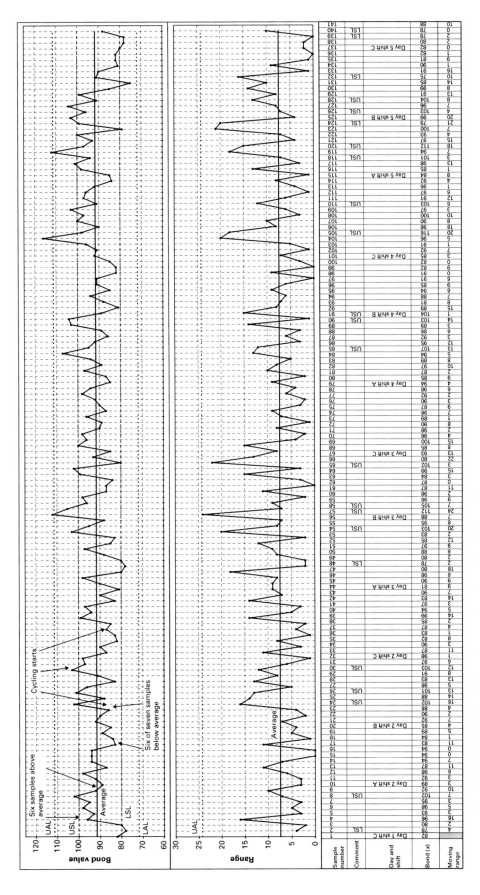

Chart 18.1 X/MR chart: bond sample values for data set A

- Since there are clear indications of out-of-control conditions in the individuals chart, no detailed analysis of the moving range chart is reported here. You may like to draw your own conclusions.

Having identified from Chart 18.1 that the process is not in a state of control, we use the chart to identify possible process changes (Chart 18.2). With experience, and as a starting point, this can be done by eye. All the proposed process changes have been identified solely from the data, and should be investigated to ascertain their authenticity, cause and onset. Unfortunately this was a historic analysis and no such investigation was possible. The position of the changes were selected for the following reasons:

- Change 1 at sample 4 was chosen as a set of low values is followed by a set of high values with a large jump.
- Change 2 at sample 17 coincides with a run of sample values below the average.
- Change 3 at sample 23 coincides with the end of the run of low sample values.
- Change 4 at sample 92 coincides with the apparent beginning of a short run of low-variability samples as identified in the range chart, and with a lower average. Nine of the first 10 points lie below the previous average.
- Change 5 at sample 104 coincides with a very high value followed by a group that has a higher average than the previous average. The first 10 sample values are above the previous average.
- Change 6 at sample 132 coincides with a large reduction in sample value, all 10 points being below the previous average.

When imposing process changes we expect either the process average to change or the variability to change or, as in most situations, both the average and the variability to change. The process changes identified above certainly result in large shifts in average and variability which are unlikely to be due to random variation.

Note that another analyst, especially one with process knowledge and information, and if done at the time the data were collected may identify different changes and may also be able to identify the causes and onset of changes.

Analysis of Chart 18.2 suggests the following:

- There is one long period of reasonable processes stability, at least as far as the average is concerned and this gives a good indication of the potential process capability. Unfortunately the conclusion is that the process is not capable of producing consistently within specification even when in a state of control, as verified by the specification limits being within the action limits.
- Within the long period of stability there are still concerns, for example, the cycling noted earlier.
- Considering the other periods of stability, in most cases the specification limits are within the action limits. Two exceptions are the first and third periods. In both cases the process is off centered (i.e. the process average is not the same as the nominal value of 90). We cannot draw any conclusions from the first period as it only consists of three values. In the third period there are only seven values, and there is a suspicion that these may be showing an upward drift (and hence not be in a state of statistical control). Therefore, whilst it is true that if the process had been centered,

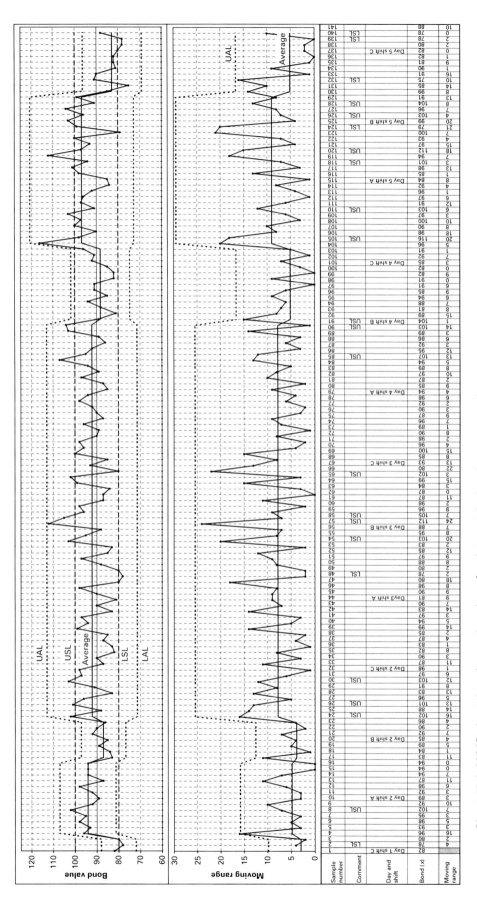

Chart 18.2 X/MR chart: bond sample values for data set A showing process changes

the specification and action limits would nearly coincide, it would be unwise to deduce that therefore the process is capable of regularly producing within specification.

- A common response to producing off-specification material is to adjust or "center" the process. The conclusion from the above observations is that the process is not capable of consistently producing within specification. This is an important observation as it implies that merely changing the process average to coincide with the nominal value will not solve the problem: off-specification material will still be regularly produced. The variability must be reduced.
- Process changes and out-of-control points occur irrespective of the shifts, and shift changes do not appear to induce an out-of-specification value or process change.

Data sets B and C (Charts 18.3 and 18.4)

Charts 18.3 and 18.4 show the data for orders B and C. You might like to spend a little time analysing the charts. The conclusions for both charts are similar to that for order A and we do not give a detailed analysis.

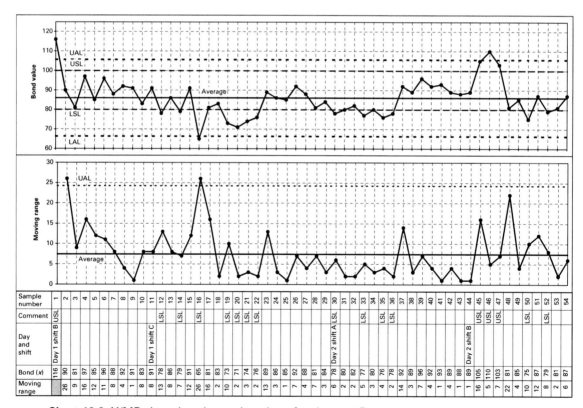

Chart 18.3 X/MR chart: bond sample values for data set B

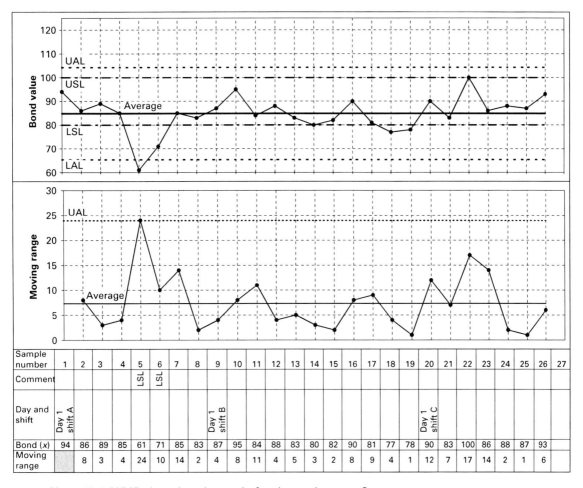

Chart 18.4 X/MR chart: bond sample for data values set C

Analysis by shift (Chart 18.5)

There is a three shift system operating at the plant (morning, afternoon, night). It is known that operators often start their shift by adjusting the process to the values they believe are correct. An obvious theory is that the bond values vary according to shift. This theory is tested by drawing an $\bar{X}/R$ chart of bond by shift for data set A (Chart 18.5). Data set A was chosen as having the most data, for complete analysis similar charts for data sets B and C could be drawn. The table above the chart gives the day number, the average bond value and the range of bond values for each shift. There are three shifts per day. Production began on shift three of day 1 and was completed by the end of shift three on day 5.

The chart shows that the averages are in a state of statistical control. The only item of interest is the last point which has a very low average of 81 and a very low range of 10. Reference to Chart 18.2 shows that there were only five readings during the shift. It is

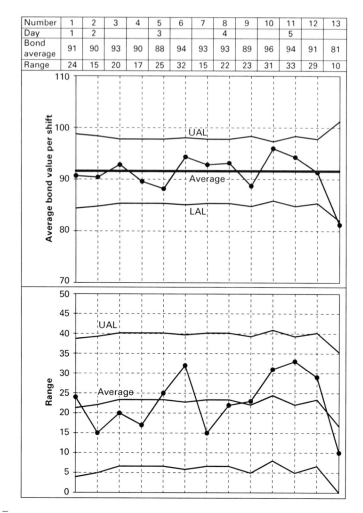

Number	1	2	3	4	5	6	7	8	9	10	11	12	13
Day	1	2			3			4			5		
Bond average	91	90	93	90	88	94	93	93	89	96	94	91	81
Range	24	15	20	17	25	32	15	22	23	31	33	29	10

Chart 18.5 X̄/R chart: bond data set A averages plotted by shift

possible that the operators became aware that the bond value was too low and adjusted the machine settings as the very last value is much higher than the previous few values. However, this is pure conjecture as no log is kept of operator setting changes.

If we compare the first shift of each day (average values of 90, 88, 93, 94) we see that there is a good spread. Similarly comparing the second and third shift averages with the first shift average yields nothing of interest. (Note that had there been more data, it may have been appropriate to draw an average range control chart ordered by shift and then date. This would make the analysis easier.) We conclude that if there are any changes due to shift, they are not evident from this chart.

Analysis by analyst (Chart 18.6)

A further theory was advanced that there may be differences due to analyst. Analysts are tied to particular crew, of which there are five. To test the theory an average range

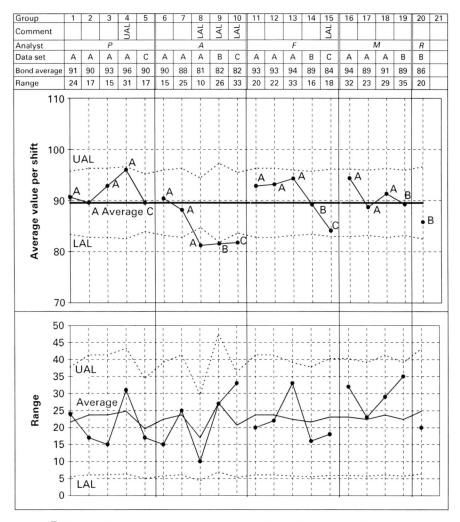

Group	1	2	3	4	5	6	7	8	9	10	11	12	13	14	15	16	17	18	19	20	21
Comment				UAL				LAL	LAL	LAL					LAL						
Analyst			P					A					F					M		R	
Data set	A	A	A	A	C	A	A	A	B	C	A	A	A	B	C	A	A	A	B	B	
Bond average	91	90	93	96	90	90	88	81	82	82	93	93	94	89	84	94	89	91	89	86	
Range	24	17	15	31	17	15	25	10	26	33	20	22	33	16	18	32	23	29	35	20	

Chart 18.6 X̄/R chart: bond value summary by analyst, data sets A, B and C

chart of bond values by analyst was developed for data set A. The results did show some interesting features and it was decided to add data sets B and C. The result is shown in Chart 18.6.

The table in Chart 18.6 identifies the analyst, the data set from which the data are taken, the average and the range. The comments identify where a control limit was breached. Note that the data are ordered first by analyst, then time. (Since the runs ran consecutively, A followed by B and then C, all the A data appear first followed by B and then C.)

The use of a line joining the points highlights which points belong to which analyst. The line is broken to highlight a change in analyst, and that to look for trends or other patterns which cross from one analyst to another is inappropriate.

All five analyst *P* values are at or above the mean and one is on the UAL. Four of the five analyst *A* values are below the mean and three are below the LAL. Analyst *F* has a large variability with three of the five values quite near the UAL, and one near the

LAL. Analyst M's four values exhibit no exceptional features. There is only one value from analyst R and this is half way between the average and the LAL.

In conclusion there appear to be significant differences between the analysts, and an explanation for these differences should be sought.

There is one last point of interest. Two of the three averages from data set C are very low, with the third on the average. Due to lack of data we cannot say for certain if this is coincidence, but further analysis from more than three data sets could investigate the possible differences due to new product runs.

Finally, the range chart is in a state of control. The reason that the mean varies is explained below.

Comments

- The action limits tell us the capability of the process. For Chart 18.2, at no time, even in times of statistical control, is the process capable of consistently producing within specification. In general, the control chart will help us recognise a process which is incapable of consistently producing within specification **before** it strays outside the specification limits, enabling us to take preventive action. In the current example, the manufacturer would be able to estimate the percentage of product that would need to be scrapped, and hence how many rolls of paper are likely to be needed to meet the order. This would allow better scheduling of the machines.
- Suppose a process is not capable of operating within the specification limits as is the case with Chart 18.2. If the process is centered on the nominal (90), it would be tampering to change the process whenever a value outside the specification limits was observed. This tampering is costly on three counts:
  1. It costs money to investigate the cause.
  2. It costs money to adjust the process.
  3. Frequently we make the process worse (because we have now adjusted the process average away from the nominal or, if the process was not centered on the nominal, we may have adjusted it further away). The best we can do is to center the process so that either:
     – the LAL = LSL, so that we will seldom breach the LSL;
     – the UAL = USL, so that we seldom breach the USL;
     – the process average = nominal which will minimise breaches of both LSL and USL.
     In these situations it is important to adjust the process only if we have evidence that the process is not centered where we would like it to be. We must accept out-of-specification results as inevitable until we have reduced variability.
- Specification limits and targets are often arbitrarily imposed on a process. When these values are rounded numbers (80, 90 and 100 in this case) it is always worthwhile investigating the cause and validity of these values. A more realistic view of targets is expounded by Taguchi's Loss Function. Taguchi, a Japanese engineer, proposed that specification limits are a crude approximation to the true situation which can be summarised as follows:
  – There is a nominal value at which we would like our process to operate.
  – Any departure from the nominal value incurs a loss.
  – The loss increases exponentially as we move away from the nominal.

- The loss includes producer's, customer's and society's loss.
- Effort should be expended to centre the process on the nominal and continually reduce variation (and hence loss).

This view is better than the somewhat false assertion that as producing anywhere within specification is equally good, and as soon as the specification is breached the output is in some sense a failure. It also assumes that the measurement system is 100% accurate. For a more complete discussion see, for example, Forth Generation Management by Brian Joiner.

- When carrying out an historical analysis, it is usual to first remove points exhibiting out-of-control conditions and incorporate process changes. This should leave a chart with out of (statistical) control points omitted from the analysis but still plotted and the effects of process changes visible on the chart.
- It is useful to consider control charts as the process memory and record both data and process information, such as operator actions and settings, shift and feedstock changes and any other information that may be of use when investigating process behaviour.
- For individuals data it is standard practice to draw moving range charts to monitor variability.
- Given action and specification limits along with process average and nominal value it is quite straight forward to calculate various capability indices which reflect the ability of a process to meet specification, and estimate the proportion of observations that will fall outside the specification limits. More information can be found in Statistical Process Control by Oakland.
- Note that the control limits of the average chart are closer to the average than the X chart, and generally within the specifications limits. This is because we are averaging individual values and so limiting the effect of high and low values.
- The range charts in both Charts 18.5 and 18.6 shows that the average range changes. The reason for this is that the number of sample values varies for each of the 20 groups. The average number of samples taken is 11, and for each group we have adjusted the range to estimate what the range would have been if the number of samples had been 11. The detailed calculations are explained below.

Calculations

X/MR chart

Chart 18.1: data set A plotted by time

- The process average = sum of the bond values/number of values = 12,913/141 = 91.58.
- The moving range is the difference between successive bond values.
- There is no moving range for the first value.
- The average moving range = sum of the moving ranges/number of moving ranges = 1038/140 = 7.41.
- The standard deviation s = average moving range/1.128 = 7.41/1.128 = 6.57.
- The UAL = average + $3s$ = 91.58 + (3 × 6.57) = 111.
- The LAL = average − $3s$ = 91.58 − (3 × 6.57) = 72.

- The nominal value, USL and LSL as provided by the process owner are 90, 100 and 80.
- The UAL for the moving range chart $= 3.27 \times$ average moving range $= 3.27 \times 7.41 = 24.2$.
- The lower limit is 0.

The calculations for Charts 18.2–18.4 are similar.

$\bar{X}$/R charts

Unfortunately the number of observations in each of the averages is not the same, making the calculations more difficult. Part 4 provides the formula and an example where the number of observations are the same for each average. In this section we discuss five alternative methods for dealing with varying numbers of observations in the averages.

The raw data are given in Table 18.1 and consist of up to 20 sample values (in rows) for each of 20 groups of data (columns). In the table are details of calculations for Charts 18.6 and 18.7 which are explained below.

The rows in the calculation section for Chart 18.6 are as follows:

- $n =$ number of observations for each of the 20 averages. The first average has nine observations and the 20th has 14 observations.
- $\bar{n} = 11$ is the average value of n. By chance, the average is exactly 11, if it were not, we would round to the nearest integer.
- $\bar{x}$ is the average of the column of observations. For column 1 the average is 90.7.
- Range is the (maximum − minimum) of the observations. For column 1 this is $(102–78) = 24$.
- $\bar{\bar{x}}$ is the average of all observations $= 89.3$.
- Moving down two rows, $\bar{R}$ is the average of the ranges in the "Range" row $= 23.1$.
- The two rows above the $\bar{R}$ row are the UAL and LAL for the $\bar{X}$.
- The other rows are explained in detail below.

The rows in the calculation section for Chart 18.7 are the same as described above, in addition the control limits for the $\bar{X}$ chart:

- $\text{UAL}_X = \bar{\bar{x}} + A_2\bar{R} = 89.3 + (0.337 \times 23.1) = 97.1$ for the first average, where the values for A_2 are read off tables (see Appendix A) using the appropriate for n, that is for $n = 9$ $A_2 = 0.337$.
- $\text{LAL}_X = \bar{\bar{x}} - A_2\bar{R} = 89.3 - (0.337 \times 23.1) = 81.5$ for the first average.

The last two rows give the UAL_{MR} and LAL_{MR} for the range chart:

- $\text{UAL} = D_4\bar{R} = 1.816 \times 23.1 = 41.9$ where $D_4 = 1.816$ for $n = 9$ (see Appendix A).
- $\text{LAL} = D_3\bar{R} = 0.184 \times 23.1 = 4.3$ where $D_3 = 0.184$ for $n = 9$ (see Appendix A).

Having outlined most of the calculations in the table we now discuss five options for addressing the problem of differing sample sizes beginning with the simplest and ending with the most complex:

1. *Method 1*: Use only the first n observations, where n is the smallest sample size, five in this example. This is the simplest solution, but throws away valuable data, and is not pursued here.
2. *Method 2*: Keep all the data, ignore the fact that n varies and use the average value of n (i.e. $\bar{n} = 11$) in all calculations. The resulting limits will be the same for all groups can be read off the calculations for Chart 18.6 part of the table on either of the two columns where $n = 11$ (i.e. column 15 or 16). This is a little more complex than method 1 and is a reasonable solution if n does not vary much.
3. *Method 3*: As method 2 but use the individual values of n when looking up values of A_2, D_3 and D_4. This is what we have done in this to produce Chart 18.7.
4. *Method 4*: Adjust each value of $\bar{R}$ to estimate what the range would have been if the sample size was n rather than $\bar{n}$ (=11 in this case). To adjust $\bar{R}$ calculate:

$$\bar{R}_{adjusted} = \frac{\bar{R} \times d_{2old}}{d_{2new}} = \frac{23.1 \times 2.97}{3.17} = 21.6 \text{ for the first column}$$

where, d_{2old}, is the value for d_2 for $n = 9$ and d_{2new} is the value for d_2 for $n = 11$. Note that $\bar{R}_{adjusted} < \bar{R}$, which is what we expect since the smaller n is the smaller the range would be. $\bar{R}_{adjusted}$ is then used in calculations. We could then use $\bar{n} = 11$ when looking up values of A_2, D_3 and D_4, or use the individual values of n when looking up A_2, D_3 and D_4, which results in chart 18.6.
5. *Method 5*: In method 4 we adjusted $\bar{R}$. However, $\bar{R}$ has already been based on the assumption that the number of observations, n, for all columns is the same. In theory a better alternative is to estimate what each range would have been if the sample size had been $n = 11$. The formula are similar to the above for the first average:

$$R_{adjusted} = \frac{R \times d_{2old}}{d_{2new}} = \frac{24 \times 3.17}{2.97} = 25.6.$$

$\bar{R}$ would then be calculated in the normal way using the adjusted values of R to give 23.4 and the limits calculated in the normal way using $\bar{n}$.

It is interesting to compare the results of these different methods of calculating $\bar{R}$ and hence the control limits. Chart 18.8 gives the results for the $\bar{X}$ chart:

- The differences between methods 2 and 5 are very small in all cases. The upper limits are constant values of 96.1 and 96.2, respectively.
- Where n is near $\bar{n}$ the differences between methods using n (3 and 4) and $\bar{n}$ (methods 2 and 5) are small. For $n = 10$, 11 or 12 there is hardly any difference. For $n = 9$ (and presumably for $n = 13$ if there was an occasion with $n = 13$), the difference between the methods using variable n and $\bar{n}$ become more noticeable.
- The differences between the methods 2 and 4 using n are reasonably close for $n = 8$ to $n = 20$. It is only for low values of n, up to about 7 or so that there is a significant gap beginning to open up.
- In terms of interpreting the chart, for group 4, the average is just above the limit for methods using n and just below for methods using $\bar{n}$. Whichever method is being used we would be concerned. For group 8, with $n = 5$ there is not much data and

Table 18.1 Bond value summary by analyst, data sets A, B and C data and calculations

Data

Sample value	Group number																			
	1	2	3	4	5	6	7	8	9	10	11	12	13	14	15	16	17	18	19	20
1	82	98	93	85	90	89	81	82	91	94	85	94	84	90	87	88	104	99	105	80
2	78	87	85	92	83	92	90	80	78	86	92	85	85	81	95	112	89	103	110	82
3	80	90	100	91	100	98	98	78	86	89	90	87	98	97	84	105	81	96	103	77
4	96	82	96	96	86	87	80	78	79	85	86	97	101	85	88	96	88	104	81	80
5	93	83	98	116	88	94	78	88	91	61	102	89	94	96	83	98	94	91	85	76
6	98	87	90	98	87	94	80		65	71	88	94	112	88	80	87	85	99	75	78
7	95	85	89	90	93	94	88		81	85	101	107	97	92	82	87	91	85	87	92
8	102	99	96	100		83	97		83	83	96	95	93	91	90	84	91	75	79	89
9	92	94	87	97		84	85		73		83	92	100	83	81	99	82	91	81	96
10		97	90	103		89	83		71		91	86	79		77	102	82	90	87	92
11		83	92	91			103		74		103	89			78	80		81		93
12		90	98	97			95		76		97	103						82		89
13				96					89											88
14				92					86											89
15									85											
16									92											
17									88											
18									81											
19									84											
20									78											

Calculations: for Chart 18.6

Number of observations (n)	9	12	12	14	7	10	12	5	20	8	12	12	10	9	11	11	10	12	10	14
$\bar{n}$	11	11	11	11	11	11	11	11	11	11	11	11	11	11	11	11	11	11	11	11
$\bar{x}$	90.7	89.6	92.8	96.0	89.6	90.4	88.2	81.2	81.6	81.8	92.8	93.2	94.3	89.2	84.1	94.4	88.7	91.3	89.3	85.8
Range	24	17	15	31	17	15	25	10	27	33	20	22	33	16	18	32	23	29	35	20
$\bar{\bar{x}}$	89.3	89.3	89.3	89.3	89.3	89.3	89.3	89.3	89.3	89.3	89.3	89.3	89.3	89.3	89.3	89.3	89.3	89.3	89.3	89.3
UAL_X	96.6	95.6	95.6	95.1	97.6	96.2	95.6	99.1	94.2	97.0	95.6	95.6	96.2	96.6	95.9	95.9	96.2	95.6	96.2	95.1
LAL_X	82.0	83.0	83.0	83.5	81.1	82.4	83.0	79.5	84.4	81.6	83.0	83.0	82.4	82.0	82.7	82.7	82.4	83.0	82.4	83.5
$\bar{R}$	23.1	23.1	23.1	23.1	23.1	23.1	23.1	23.1	23.1	23.1	23.1	23.1	23.1	23.1	23.1	23.1	23.1	23.1	23.1	23.1
$\bar{R}$ adjusted	21.6	23.7	23.7	24.8	19.7	22.4	23.7	16.9	27.2	20.7	23.7	23.7	22.4	21.6	23.1	23.1	22.4	23.7	22.4	24.8
UAL_{MR}	37.7	41.4	41.4	43.3	34.3	39.1	41.4	29.5	47.4	36.1	41.4	41.4	39.1	37.7	40.3	40.3	39.1	41.4	39.1	43.3
LAL_{MR}	5.5	6.1	6.1	6.3	5.0	5.7	6.1	4.3	7.0	5.3	6.1	6.1	5.7	5.5	5.9	5.9	5.7	6.1	5.7	6.3

Calculations: for Chart 18.7

Number of observations (n)	9	12	12	14	7	10	12	5	20	8	12	12	10	9	11	11	10	12	10	14
$\bar{n}$	11	11	11	11	11	11	11	11	11	11	11	11	11	11	11	11	11	11	11	11
$\bar{x}$	90.7	89.6	92.8	96.0	89.6	90.4	88.2	81.2	81.6	81.8	92.8	93.2	94.3	89.2	84.1	94.4	88.7	91.3	89.3	86
Range	24	17	15	31	17	15	25	10	27	33	20	22	33	16	18	32	23	29	35	20
$\bar{\bar{x}}$	89.3	89.3	89.3	89.3	89.3	89.3	89.3	89.3	89.3	89.3	89.3	89.3	89.3	89.3	89.3	89.3	89.3	89.3	89.3	89.3
UAL_X	97.1	95.4	95.4	94.7	99.0	96.4	95.4	102.6	93.5	97.9	95.4	95.4	96.4	97.1	95.9	95.9	96.4	95.4	96.4	94.7
LAL_X	81.5	83.2	83.2	83.9	79.6	82.2	83.2	76.0	85.1	80.7	83.2	83.2	82.2	81.5	82.7	82.7	82.2	83.2	82.2	83.9
$\bar{R}$	23.1	23.1	23.1	23.1	23.1	23.1	23.1	23.1	23.1	23.1	23.1	23.1	23.1	23.1	23.1	23.1	23.1	23.1	23.1	23.1
UAL_{MR}	41.9	39.7	39.7	38.6	44.4	41.0	39.7	48.8	36.6	43.1	39.7	39.7	41.0	41.9	40.3	40.3	41.0	39.7	41.0	38.6
LAL_{MR}	4.3	6.5	6.5	7.6	1.8	5.2	6.5	0.0	9.6	3.1	6.5	6.5	5.2	4.3	5.9	5.9	5.2	6.5	5.2	7.6

Group	1	2	3	4	5	6	7	8	9	10	11	12	13	14	15	16	17	18	19	20	21
Observation number (n)	9	12	12	14	7	10	12	5	20	8	12	12	10	9	11	11	10	12	10	14	
Comment				UAL					LAL												
Analyst	P	P	P	P	P	A	A	A	A	A	F	F	F	F	F	M	M	M	M	R	
Data set	A	A	A	A	C	A	A	A	B	C	A	A	A	B	C	A	A	A	B	B	
Average	91	90	93	96	90	90	88	81	82	82	93	93	94	89	84	94	89	91	89	86	
Range	24	17	15	31	17	15	25	10	26	33	20	22	33	16	18	32	23	29	35	20	

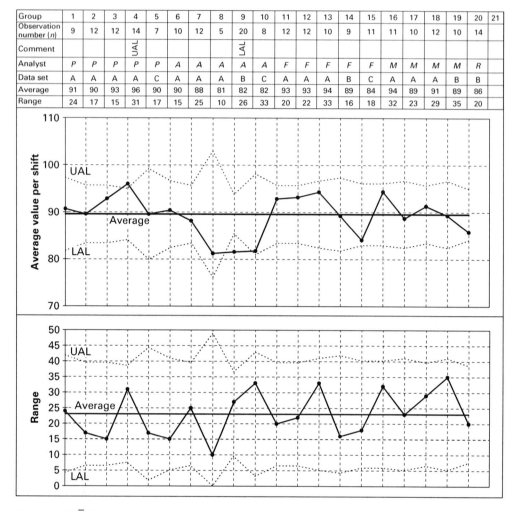

Chart 18.7 $\overline{X}$/R chart: bond summary by analyst, data sets A, B and C using unadjusted range

the limits based on n conclude that the point is just within the control limits, which is not the case for the methods 2 and 5 using $\overline{n}$ whereas the methods used for point: For group 9 all methods show the point below the LAL.

For group 10 the point lies close to or just beyond the limits.

Returning to the purpose of this particular chart, it is to compare different analysts. We are more interested in patterns rather than whether any individual point is within the control limits, and so in this instance it is comforting that the results are reasonably similar to whichever method is used.

However, If this were a plot of means over time and we are more concerned about individual points, my personal recommendation is to use method 4 and to be aware that the limits are approximate, especially when n is small.

Calculations

Calculations																				
n	9	12	12	14	7	10	12	5	20	8	12	12	10	9	11	11	10	12	10	14
$\bar{n}$	11	11	11	11	11	11	11	11	11	11	11	11	11	11	11	11	11	11	11	11
$\bar{X}$	90.7	89.6	92.8	96.0	89.6	90.4	88.2	81.2	81.6	81.8	92.8	93.2	94.3	89.2	84.1	94.4	88.7	91.3	89.3	85.8
Range	24.0	17.0	15.0	31.0	17.0	15.0	25.0	10.0	27.0	33.0	20.0	22.0	33.0	16.0	18.0	32.0	23.0	29.0	35.0	20.0
R adjusted	25.6	16.6	14.6	28.9	19.9	15.5	24.3	13.6	22.9	36.8	19.5	21.4	34.0	17.1	18.0	32.0	23.7	28.2	36.1	18.6
$\bar{\bar{X}}$	89.3	89.3	89.3	89.3	89.3	89.3	89.3	89.3	89.3	89.3	89.3	89.3	89.3	89.3	89.3	89.3	89.3	89.3	89.3	89.3
$\bar{R}$	23.1	23.1	23.1	23.1	23.1	23.1	23.1	23.1	23.1	23.1	23.1	23.1	23.1	23.1	23.1	23.1	23.1	23.1	23.1	23.1
$\bar{R}$ using R adjusted	23.4	23.4	23.4	23.4	23.4	23.4	23.4	23.4	23.4	23.4	23.4	23.4	23.4	23.4	23.4	23.4	23.4	23.4	23.4	23.4
$\bar{R}$ adjusted	21.6	23.7	23.7	24.8	19.7	22.4	23.7	16.93	27.2	20.7	23.7	23.7	22.4	21.6	23.1	23.1	22.4	23.7	22.4	24.8
UAL_X Method 2	95.9	95.9	95.9	95.9	95.9	95.9	95.9	95.9	95.9	95.9	95.9	95.9	95.9	95.9	95.9	95.9	95.9	95.9	95.9	95.9
UAL_X Method 3	97.1	95.4	95.4	99.0	96.4	95.4	102.6	93.5	97.9	95.4	96.4	97.1	95.9	95.9	96.4	95.4	96.4	95.4	96.4	94.7
UAL_X Method 4	96.6	95.6	95.6	95.1	97.6	96.2	95.6	99.08	94.2	97	95.6	95.6	96.2	96.6	95.9	95.9	96.2	95.6	96.2	95.1
UAL_X Method 5	96.0	96.0	96.0	96.0	96.0	96.0	96.0	96.0	96.0	96.0	96.0	96.0	96.0	96.0	96.0	96.0	96.0	96.0	96.0	96.0
LAL_X Method 2	82.7	82.7	82.7	82.7	82.7	82.7	82.7	82.7	82.7	82.7	82.7	82.7	82.7	82.7	82.7	82.7	82.7	82.7	82.7	82.7
LAL_X Method 3	81.5	83.2	83.2	83.9	79.6	82.2	83.2	76.0	85.1	80.7	83.2	83.2	82.2	81.5	82.7	82.7	82.2	83.2	82.2	83.9
LAL_X Method 4	82.0	83.0	83.0	83.5	81.1	82.4	83.0	79.5	84.4	81.6	83.0	83.0	82.4	82.0	82.7	82.7	82.4	83.0	82.4	83.5
LAL_X Method 5	82.6	82.6	82.6	82.6	82.6	82.6	82.6	82.64	82.6	82.6	82.6	82.6	82.6	82.6	82.6	82.6	82.6	82.6	82.6	82.6

Group	1	2	3	4	5	6	7	8	9	10	11	12	13	14	15	16	17	18	19	20
Comment																				
Analyst	P	P	P	P	P	A	A	A	A	A	F	F	F	F	F	M	M	M	M	R
Data set	A	A	A	A	C	A	A	A	A	B	C	A	A	A	B	C	A	A	B	B
Bond average $\bar{X}$	91	90	93	96	90	90	88	81	82	82	93	93	94	89	84	94	89	91	89	86
Range	24	17	15	31	17	15	25	10	26	33	20	22	33	16	18	32	23	29	35	20

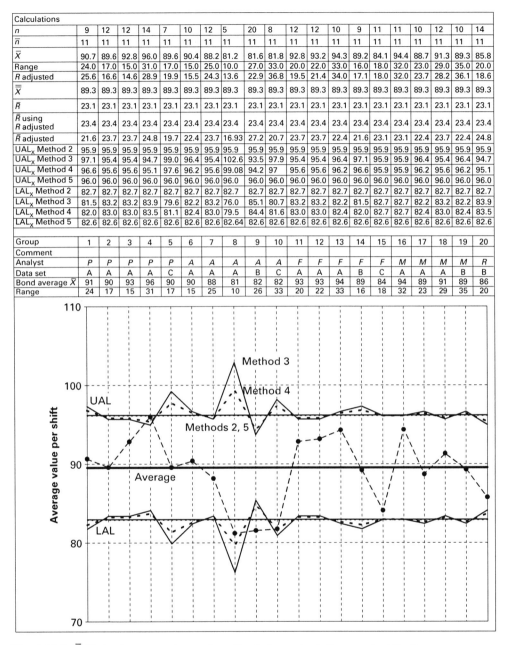

Chart 18.8 $\bar{X}$/R chart: comparison of methods for calculating control limits

Summary

In this case study we say how control charts were applied to a manufacturing unit to process and allow us to:

- determine that the process was not capable of consistently producing outputs within specification;
- investigate alternative potential sources of variation and conclude that a key source of variation in results was due to the measurement process itself.

We also investigated different methods of drawing $\bar{X}$/R charts where the number of observations per group varied and concluded that whilst there were differences in results, these were generally quite small.

19 Categorising, de-seasonalising and analysing incident data using multivariate charts

Charts used: c, $\bar{X}$/R, cusum, Pareto and scatter diagram

Introduction

This case study is taken from the area of health, safety and environment (HSE), but exactly the same method and analysis apply to many other incident/failures/rejects data.

The case study shows:

- How categorising incidents and carrying out further simple analyses can significantly help the engineer/analyst/manager understand the information in the data and help target improvement effort.
- How to check for seasonality and de-seasonalise data.
- Investigate the relationship between variables.

Background

An organisation had been collecting safety data regularly over a number of years. Originally a bar chart showing the number of incidents was updated monthly. When the number of incidents in any particular month was deemed high by the manager, there would be a safety talk. Sometimes notices would appear exhorting the workforce to work safely.

Management eventually realised that there was more they could do with the data. A member of staff was assigned to find out what data was available and draw up a control chart.

He discovered that when an incident occurred that required medical aid, a form was completed which included, amongst other information, the part of the body injured and the immediate cause. He developed a c chart (Chart 19.1) of the injuries and included below the chart a table showing how the data was categorise by part of the body and immediate cause of injury.

Analysis

The c chart: multivariate (Chart 19.1)

Chart 19.1 is the c chart of the number of incidents per month. We can use a c chart because the number of hours worked each month (exposure hours) remained more or

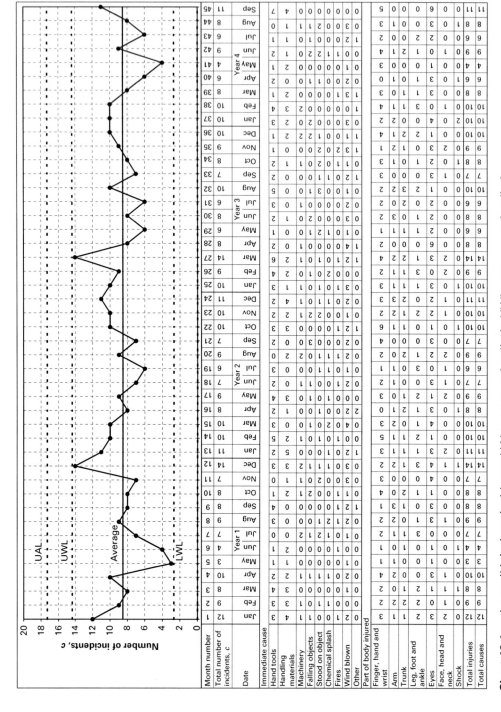

Chart 19.1 c chart: all injurious incidents; UAL: upper action limit; UWL: upper warning limit

less constant. The chart shows that the average number of injuries per month is around 8.5, the upper control limit is just above 17, and the two warning limits are just under 3 and just over 14. The lower control limit is at 0, which is common in u and c charts where the averages tend to be low.

The process appears to be more or less in control. There are no points outside the control or warning limits. The fact that there are none outside the warning limits is in itself suspicious as we would expect two or three in a series of 40 points (the warning limits are set in such a way that about 5% of values should lie outside these limits by chance alone). However, three values are very close.

The chart is called a multivariate chart because it includes a breakdown of the values of c, in this case incidents. They have been categorised in two ways: by immediate cause and by part of the body injured. There are nine identified causes and seven areas of the body, and these are listed below the chart with their corresponding numbers of incidents. Note that the cause does not necessarily indicate the underlying cause of the incident, only the immediate cause of the injury.

Cusum chart (Chart 19.2)

The cusum chart of the number of incidents is given in Chart 19.2. The cusum chart monitors the cumulative difference between a "target" value and the data. The target value (T) is usually chosen to be a value equal to or near the process average. In this case the average number of incidents per month, 8.53, is used. The cusum chart is analysed by looking for changes in slope which indicate a change in process average. There is an obvious change in November of year 1 from a general downward slope to an upward slope. Other changes of slope that may be significant are in May of year 1, March and September of year 2, March and October of year 3, and finally February of year 4. The fact that possible slope changes occur regularly in the Spring and Autumn may be significant, and will be investigated below.

Apart from these potential short-term changes, there seems to be no definitive change in general slope (i.e. no long-term change in average incident rate).

Beneath the chart the table gives:

- The month number.
- The month and year.
- The total number of incidents, c.
- The difference between c and T. The target used is the average number of incidents per month, that is, 8.53.
- The cusum which equals the cumulative sum of the values in the row above.

Investigation of seasonality (Chart 19.3)

The cusum chart hints to us that the data may be seasonal (because the changes in slope occur in Spring and Autumn). To investigate this possibility an $\bar{X}$/Range chart was drawn and is given in Chart 19.3.

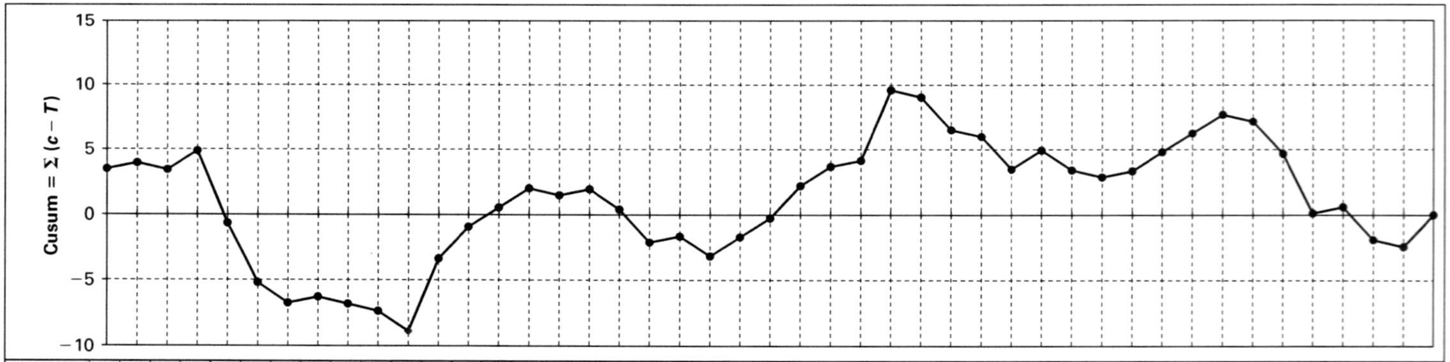

216

Chart 19.2 Cusum chart of all injurious incidents

Month number	Comment	Date	Total number incidents, c	Difference c − T (T = 8.53)	Cusum Σ(c − T)
1		Jan (Year 1)	12	3.5	3.5
2		Feb	9	0.5	3.9
3		Mar	8	−1	3.4
4	Slope change?	Apr	10	1.5	4.9
5		May	3	−6	−0.7
6		Jun	4	−5	−5.2
7		Jul	7	−2	−6.7
8		Aug	9	0.5	−6.3
9		Sep	8	−1	−6.8
10		Oct	8	−1	−7.3
11	Slope change?	Nov	7	−2	−8.9
12		Dec	14	5.5	−3.4
13		Jan	11	2.5	−0.9
14		Feb	10	1.5	0.5
15	Slope change?	Mar	10	1.5	2.0
16		Apr	8	−1	1.5
17		May	9	0.5	1.9
18		Jun (Year 2)	7	−2	0.4
19		Jul	6	−3	−2.1
20		Aug	9	0.5	−1.7
21	Slope change?	Sep	7	−2	−3.2
22		Oct	10	1.5	−1.7
23		Nov	10	1.5	−0.3
24		Dec	11	2.5	2.2
25		Jan	10	1.5	3.7
26		Feb	9	0.5	4.1
27	Slope change?	Mar	14	5.5	9.6
28		Apr	8	−1	9.1
29		May	6	−3	6.5
30		Jun	8	−1	6.0
31		Jul (Year 3)	6	−3	3.5
32		Aug	10	1.5	4.9
33		Sep	7	−2	3.4
34	Slope change?	Oct	8	−1	2.9
35		Nov	9	0.5	3.3
36		Dec	10	1.5	4.8
37		Jan	10	1.5	6.3
38	Slope change?	Feb	10	1.5	7.7
39		Mar	8	−1	7.2
40		Apr	9	−3	4.7
41		May (Year 4)	4	−5	0.1
42		Jun	9	0.5	0.6
43		Jul	6	−3	−1.9
44		Aug	8	−1	−2.5
45		Sep	11	2.5	0.0

Cusum = Σ(c − T)

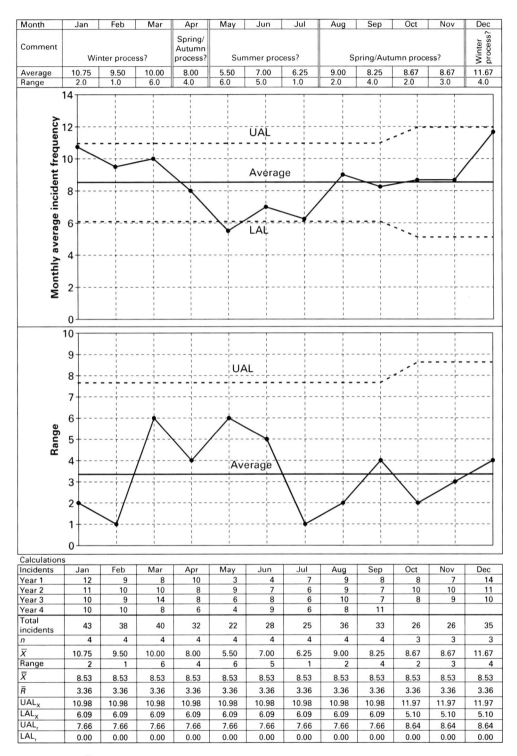

Month	Jan	Feb	Mar	Apr	May	Jun	Jul	Aug	Sep	Oct	Nov	Dec
Comment	Winter process?			Spring/Autumn process?	Summer process?			Spring/Autumn process?				Winter process?
Average	10.75	9.50	10.00	8.00	5.50	7.00	6.25	9.00	8.25	8.67	8.67	11.67
Range	2.0	1.0	6.0	4.0	6.0	5.0	1.0	2.0	4.0	2.0	3.0	4.0

Calculations

Incidents	Jan	Feb	Mar	Apr	May	Jun	Jul	Aug	Sep	Oct	Nov	Dec
Year 1	12	9	8	10	3	4	7	9	8	8	7	14
Year 2	11	10	10	8	9	7	6	9	7	10	10	11
Year 3	10	9	14	8	6	8	6	10	7	8	9	10
Year 4	10	10	8	6	4	9	6	8	11			
Total incidents	43	38	40	32	22	28	25	36	33	26	26	35
n	4	4	4	4	4	4	4	4	4	3	3	3
$\bar{X}$	10.75	9.50	10.00	8.00	5.50	7.00	6.25	9.00	8.25	8.67	8.67	11.67
Range	2	1	6	4	6	5	1	2	4	2	3	4
$\bar{\bar{X}}$	8.53	8.53	8.53	8.53	8.53	8.53	8.53	8.53	8.53	8.53	8.53	8.53
$\bar{R}$	3.36	3.36	3.36	3.36	3.36	3.36	3.36	3.36	3.36	3.36	3.36	3.36
UAL_x	10.98	10.98	10.98	10.98	10.98	10.98	10.98	10.98	10.98	11.97	11.97	11.97
LAL_x	6.09	6.09	6.09	6.09	6.09	6.09	6.09	6.09	6.09	5.10	5.10	5.10
UAL_r	7.66	7.66	7.66	7.66	7.66	7.66	7.66	7.66	7.66	8.64	8.64	8.64
LAL_r	0.00	0.00	0.00	0.00	0.00	0.00	0.00	0.00	0.00	0.00	0.00	0.00

Chart 19.3 $\bar{X}$/R chart: monthly incident analysis; UAL: upper action limit; LAL: lower action limit

For each month of the year the average number of incidents is calculated, along with the range (the maximum number of incidents recorded in that month minus the minimum number of incidents recorded in that month).

The control limits and average are calculated in the usual way, and because we only have 3 years of data for October to December, the control limits are wider than for the other months.

The chart suggests that there may be three "processes":

- A summer process lasting from May to July with a lower incident rate than at other times of year. Although May is the only month showing an out-of-control condition (below the lower action limit (LAL)) June and July are the next two lowest months of the year.
- A winter process lasting from December through to March which has the highest numbers of incidents.
- A Spring/Autumn process for the other months.

This type of seasonal finding is quite common.

The next step in pursuing this theory would be to investigate causes of differences during the year. Weather would be an obvious cause for outside workers, but it could be that different types of work are carried out in summer and winter with a mix in Spring and Autumn, and that it is the different types of work that incur different incident frequency rates. If, on investigation, this is found to be true, the incident frequencies should be charted and analysed by work type.

Investigation of causes of incidents (Pareto Charts 19.4(a) and 19.4(b))

Being aware of seasonality, we can continue to investigate the causes of incidents and aim to improve the process. The first question is where to focus attention? Should it be hand tools, machinery or some other activity? Should it perhaps be a part of the body that is injured?

To help us answer that question we can use a Pareto chart. Chart 19.4(a) is the Pareto chart showing injuries split by immediate cause. We see that hand tools are by far the biggest cause, and fires are the lowest identified cause. Clearly, we have a much larger opportunity to reduce the number of incidents if we try to reduce hand tool injuries rather than fires. Of the 384 incidents, only 16 were due to fires whilst 108 were due to hand tools, so a mere 15% reduction in hand tools would reduce the number of incidents by the same amount as eliminating all the fires.

Similarly, the two categories ((a) fingers, hand and wrist; (b) eyes) are the two injured parts of the body with the greatest number of incidents, both with around 100 incidents or more each compared with less than 50 for the other parts.

Hand tools incidents (Chart 19.5)

Having decided that we wish to start by reducing the number of incidents caused by hand tools, we draw a c chart of the number of hand tool incidents as shown in Chart 19.5. Sometimes, as in this case, interpretation of control charts is not as straightforward as we

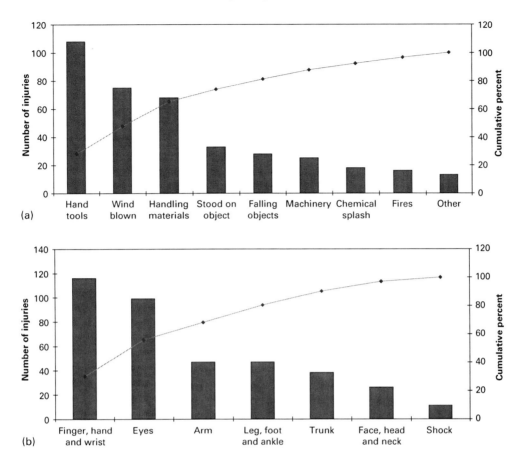

Chart 19.4 Pareto charts of incident frequencies by (a) immediate cause and (b) part of body injured

would wish. There are no clear signals in Chart 19.5, but the first 11 months do appear to be lower than months 12 to 32. (There are 2 zeros and 3 ones in the first 11 months and only 1 zero and no ones in months 12 to 32.) There is also some indication that from month 33 the incident frequency has dropped again. (A run of 4 months below the average, followed by two above and then six below ending in the first zero incident months since month 20.) The very last month with seven incidents, just below the control limit is very worrying, but not a special cause of variation signal (because it is not above the upper action limit (UAL)). We could use a cusum chart to investigate further; however, we have already discovered that there is seasonality in the total number of incidents, hence the useful next step is to investigate the possibility that hand tool incidents are also seasonal.

Chart 19.6 is the resulting $\bar{X}/R$ chart. January to March and November have a relatively high number of incidents whilst the rest of the year have relatively fewer incidents. The seasonality effect does not appear to be the same as for Chart 19.2.

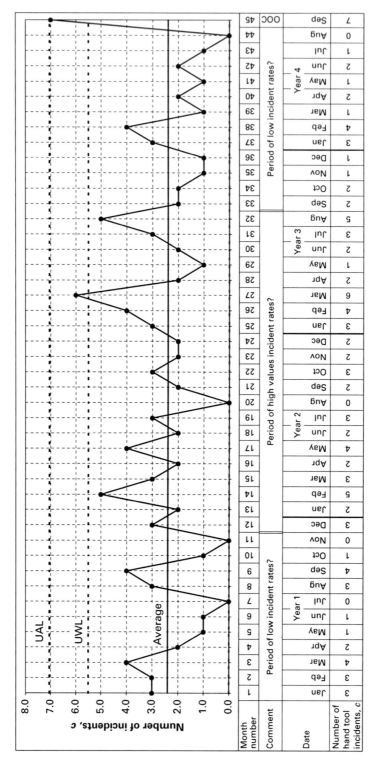

Chart 19.5 c chart of hand tool incidents; UAL: upper action limit; UWL: upper warning limit

Month	Jan	Feb	Mar	Apr	May	Jun	Jul	Aug	Sep	Oct	Nov	Dec
Comment		Chart appears to follow a similar, but not the same, pattern as in Chart 19.3										
Average	2.75	4.00	3.50	2.00	1.75	1.75	1.75	2.00	3.75	2.00	1.00	2.00
Range	1.0	2.0	5.0	.0	3.0	1.0	3.0	5.0	5.0	2.0	2.0	2.0

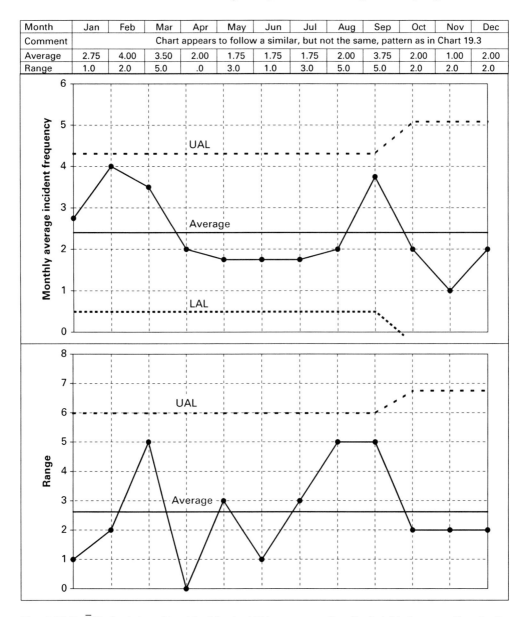

Chart 19.6 $\bar{X}$/R chart: hand tool incidents; UAL: upper action limit; LAL: lower action limit

Hand tools control charts, taking account of seasonality (Charts 19.7–19.10)

There is an indication of seasonality in the hand tools data, and we will use it as an example of how to take seasonality into account using control charts. This will allow us to de-seasonalise the data and check the de-seasonalised data for out-of-control conditions.

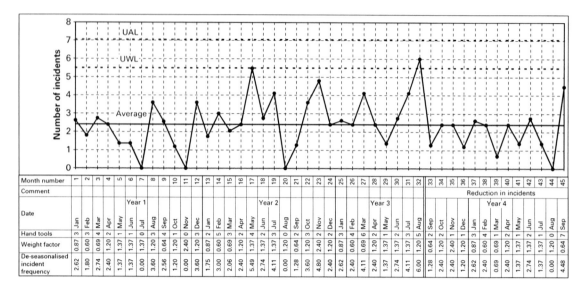

Chart 19.7 c chart: hand tool de-seasonalised incidents; UAL: upper action limit; UWL: upper warning limit

Month number	Date	Hand tools	Weight factor	De-seasonalised incident frequency
1	Year 1 – Jan	3	0.87	2.62
2	Feb	3	0.60	1.80
3	Mar	4	0.69	2.74
4	Apr	2	1.20	2.40
5	May	1	1.37	1.37
6	Jun	1	1.37	1.37
7	Jul	0	1.37	0.00
8	Aug	3	1.20	3.60
9	Sep	4	0.64	2.56
10	Oct	1	1.20	1.20
11	Nov	0	1.20	0.00
12	Dec	3	2.40	3.60
13	Year 2 – Jan	2	0.87	1.75
14	Feb	5	0.60	3.00
15	Mar	3	0.69	2.06
16	Apr	2	1.20	2.40
17	May	4	1.37	5.49
18	Jun	2	1.37	2.74
19	Jul	3	1.37	4.11
20	Aug	0	1.20	0.00
21	Sep	2	0.64	1.28
22	Oct	3	1.20	3.60
23	Nov	4	1.20	4.80
24	Dec	1	2.40	2.40
25	Year 3 – Jan	3	0.87	2.62
26	Feb	4	0.60	2.40
27	Mar	6	0.69	4.11
28	Apr	2	1.20	2.40
29	May	1	1.37	1.37
30	Jun	2	1.37	2.74
31	Jul	3	1.37	4.11
32	Aug	5	1.20	6.00
33	Sep	2	0.64	1.28
34	Oct	2	1.20	2.40
35	Nov	2	1.20	2.40
36	Dec	1	2.40	1.20
37	Year 4 – Jan	3	0.87	2.62
38	Feb	4	0.60	2.40
39	Mar	1	0.69	0.69
40	Apr	2	1.20	2.40
41	May (Reduction in incidents)	1	1.37	1.37
42	Jun	2	1.37	2.74
43	Jul	1	1.37	1.37
44	Aug	0	1.20	0.00
45	Sep	7	0.64	4.48

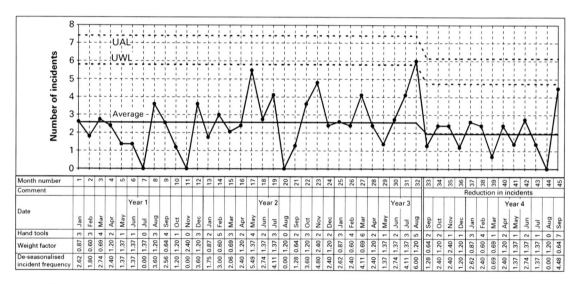

Chart 19.8 c chart: hand tool de-seasonalised incidents with one process change identified; UAL: upper action limit; UWL: upper warning limit

Month number	Date	Hand tools	Weight factor	De-seasonalised incident frequency
1	Year 1 – Jan	3	0.87	2.62
2	Feb	3	0.60	1.80
3	Mar	4	0.69	2.74
4	Apr	2	1.20	2.40
5	May	1	1.37	1.37
6	Jun	1	1.37	1.37
7	Jul	0	1.37	0.00
8	Aug	3	1.20	3.60
9	Sep	4	0.64	2.56
10	Oct	1	1.20	1.20
11	Nov	0	1.20	0.00
12	Dec	3	2.40	3.60
13	Year 2 – Jan	2	0.87	1.75
14	Feb	5	0.60	3.00
15	Mar	3	0.69	2.06
16	Apr	2	1.20	2.40
17	May	4	1.37	5.49
18	Jun	2	1.37	2.74
19	Jul	3	1.37	4.11
20	Aug	0	1.20	0.00
21	Sep	2	0.64	1.28
22	Oct	3	1.20	3.60
23	Nov	4	1.20	4.80
24	Dec	1	2.40	2.40
25	Year 3 – Jan	3	0.87	2.62
26	Feb	4	0.60	2.40
27	Mar	6	0.69	4.11
28	Apr	2	1.20	2.40
29	May	1	1.37	1.37
30	Jun	2	1.37	2.74
31	Jul	3	1.37	4.11
32	Aug	5	1.20	6.00
33	Sep	2	0.64	1.28
34	Oct	2	1.20	2.40
35	Nov	2	1.20	2.40
36	Dec	1	2.40	1.20
37	Year 4 – Jan	3	0.87	2.62
38	Feb	4	0.60	2.40
39	Mar	1	0.69	0.69
40	Apr	2	1.20	2.40
41	May (Reduction in incidents)	1	1.37	1.37
42	Jun	2	1.37	2.74
43	Jul	1	1.37	1.37
44	Aug	0	1.20	0.00
45	Sep	7	0.64	4.48

To take account of seasonality, a simple factor for each month is obtained by dividing the average over all the data (which assumes that any process change has minimal effect) by the average for the month. The calculations are shown in Table 19.1. In January, for example, the average number of incidents is 2.75, and the average over all months is 2.4, giving a seasonality factor of 2.4/2.75 = 0.87. What this factor tells us is

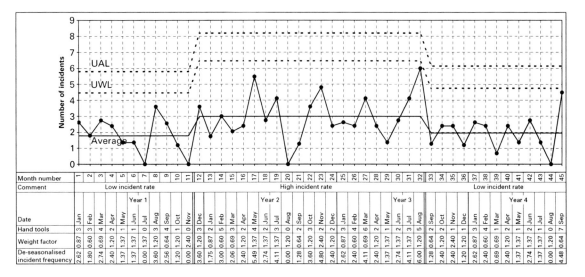

Chart 19.9 c chart: hand tool de-seasonalised incidents with process changes identified; UAL: upper action limit; UWL: upper warning limit

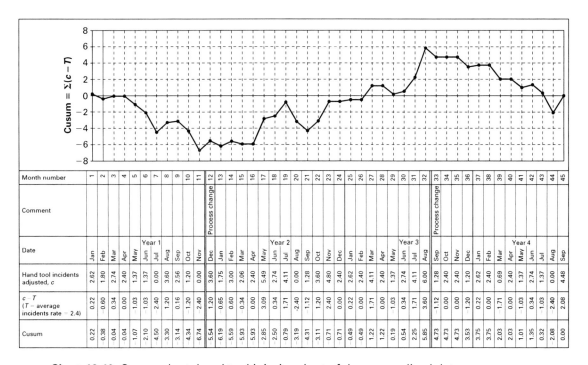

Chart 19.10 Cusum chart: hand tool injuries chart of de-seasonalised data

Table 19.1 Seasonality calculations for hand tools

Month	Jan	Feb	Mar	Apr	May	Jun	Jul	Aug	Sep	Oct	Nov	Dec
Overall average	2.40	2.40	2.40	2.40	2.40	2.40	2.40	2.40	2.40	2.40	2.40	2.40
Monthly average	2.75	4.00	3.50	2.00	1.75	1.75	1.75	2.00	3.75	2.00	1.00	2.00
Weight factor	0.87	0.60	0.69	1.20	1.37	1.37	1.37	1.20	0.64	1.20	2.40	1.20

that over the span of the data, the average number of incidents is 87% of the number seen in a typical January.

The number of incidents in each month is then multiplied by the appropriate weight factor. The resulting chart, with the calculations are given in Chart 19.7, and the interpretation is very similar to the raw hand tool incident frequency data, but the process changes are much clearer. The easiest change to identify is that commencing in month 33. From month 33 to the last month, there are six values less than any value in the previous 11 months and only three above the average, two of which are only just above.

The effect of inserting this process change can be seen in Chart 19.8. The first point to note is that the last month, although high, is not an out-of-control condition and should, pending further data, be treated as part of the current process. Secondly, the process change at month 11 is much clearer. Of the first 11 months, only two are above the average, the others being on or below. Inserting a process change at month 11 results in Chart 19.9, and it appears that we have three separate time periods, each of which is in a state of control.

In passing, we note that Chart 19.10, the cusum chart of the de-seasonalised data identifies very easily the two process changes.

The next step in the analysis is to attempt to identify the causes of these process changes and the reasons for seasonality. Once this investigation has been carried out, we can begin to analyse causes of hand tool incidents.

Relationships between variables

It is important to understand the relationships between variables where they exist. A very useful tool for doing so is the scatter diagram. Using our data we might hypothesise that the number of eye incidents is related to the number of wind blown incidents. Chart 19.11 demonstrates that there is a close relationship between the number of eye incidents and the number of wind blown incidents in the same month. There are 44 points plotted on the chart, but unfortunately because some points have the same values they overlie each other and so are not visible. It would be possible to carry out a statistical test to determine the degree of correlation, but that is beyond the scope of this book. The next step would be to review the incidents form to confirm whether most of the eye injuries were indeed due to wind blown foreign bodies in the eye. For example, if this was a workshop, analysis might show that these incidents occurred outside and that people were not wearing protective goggles. It would then be possible to estimate

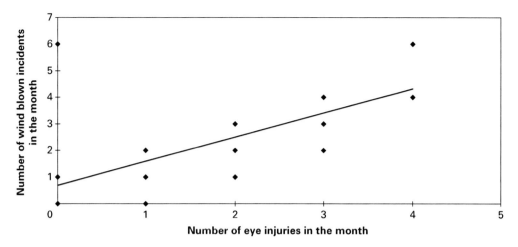

Chart 19.11 Scatter diagram: relationship between eye incidents and wind blown incidents

the reduction in both the number of eye incidents and the number of wind-blown incidents if goggles were worn outside. Analysis (especially of seasonality of wind-blown eye incidents) may show that they nearly all occur over the period of a few months during the dry season, and so the requirement to wear protective glasses could be confined to the dry season.

Splashes

During the study of the data, one observant person suggested that the number of incidents due to splashed chemical seemed to have reduced. The control chart was plotted, and a reduction was indeed obvious (the control chart is not reproduced here, but the change can be seen by reviewing the splash row in Chart 19.1). On investigation, it was found that a particularly corrosive chemical had been replaced by a newer safer alternative at the beginning of year 3. We noted that there was one incident due to splashes after this change that had been wrongly assigned to chemical splash. Note that this dramatic improvement is not readily discernable in the total incidents chart.

Comments

There is more work that can be done with this data which is rich in information. We have demonstrated the basic methods of analysis and shown that much can be learned by using a few simple charts.

The c or u chart?

Since the number of exposure hours is more or less the same from month to month, a c chart can be used to monitor the number of incidents. If the number of exposure hours per month varied by more than about 25% from the average, we would need to analyse the incident frequency (e.g. the number of incidents per million exposure hours) using the more complex u chart.

$\bar{X}$/R or $\bar{X}$/S chart?

An $\bar{X}$/range chart was drawn to investigate the existence of seasonality in the data. Equally, the range chart could have been replaced by a standard deviation (s) chart. The s charts are generally used for larger sample sizes, and R charts for smaller sample sizes.

The effect of process changes on the $\bar{X}$/Range chart

If there was a process change during the 4 years, it may show up as a seasonal effect in the $\bar{X}$/R chart. Considering, for example Chart 19.3, each plotted point is the average for the month taken over 4 years. If a sudden drop in process average were to occur in, for example, May of year 2 then the figures for the months May to December would be the average of one high value and two low values whilst the figures for January to April would be the of two high values and one low value.

The key is to be aware that this has happened, which is why we usually begin analyses with a chart that will identify whether a process change has occurred. Having identified the change, we would ideally exclude data from before the change before proceeding. However, this may lead to having too few data points for a seasonality analysis. If we believe that the process change will not affect seasonality, then the data before the change can be adjusted to take account of the change. In the case of specific incidents, if, for example, cut fingers had been reduced by 40% then the incident rates before the process change would be multiplied by $(1 - 0.40) = 0.60$.

Monitoring different processes separately

If it is found that there are two (or more) separate types of process, there are two possible ways of proceeding with monitoring:

- The processes can be monitored entirely separately, each with their own charts. In this case, it will probably be useful to check to see whether findings, conclusions and improvements in one process apply to the other. This approach would be appropriate if there were, for example, more outside activity in the summer and more inside activity in the winter, AND the types of injury were different in summer and winter.
- Data can be "adjusted" to remove the effect of one of the processes. In this case study, data can be "adjusted" to take into account the time of year. This approach would be appropriate if the same types of injury were occurring, but just more or less frequently depending on the month. This topic is discussed further in Part 4 where we deal with rare events.

What do we do about a seasonality factor if there is a process change?

In this case study the seasonality factor has been affected by process changes as shown in Chart 19.9. Whilst there are a variety of more or less complicated methods of dealing with this situation, one simple method is explained here, albeit with some added data complications, and the calculations are given in Table 19.2.

Process 1, January to November year 1:

- The number of incidents by month is given in the row "value".
- The overall average = 22 incidents/11 months = 2.0 incidents per month is given in the overall average row.
- The weight is then calculated as the number of incidents in the month divided by the overall average. (We could calculate the weight as the overall average divided by the number of incidents in the months, but this is not possible with this data as two months have zero incidents.)

Process 2, December year 1 to August year 3:

- Data was available for 21 months and so there is one value for some months and two values for the other months.
- The overall average is calculated as the average of the value average row = 2.75, so as to avoid the potential bias of including two values from months which have a higher (lower) incident rate than others.
- The weight is calculated as the number of incidents in the month divided by the overall average.

Weights for process 3 are calculated in the same way.

It is now possible to calculate an overall weight factor. We could simply average the three different weights, or as provided in the table, we could weight the weights according to the number of values used to calculate each one. For example, the weight for January is calculated as $(1.5 + 2 \times 0.91 + 1.60)/4 = 1.23$. The de-seasonalised incident frequency is calculated as the number of incidents divided by the weight factor.

Chart 19.12 shows the resulting control chart with the new weights applied. Fortunately, the identified process changes still appear intact. It is at comforting to see that the basic interpretation of the data holds whether we apply weights or not.

Is it worthwhile re-calculating the weights in this way? At the bottom of the table in Chart 19.12 we have added de-seasonalised incident rates from Chart 19.7 and below that the percentage difference between Charts 19.12 and 19.7 de-seasonalised incident rates. For September, the difference is 20%. For the other months, the differences are less than 10%. Despite these differences, the overall conclusions from both charts and from the raw data as charted in Chart 19.5 are similar. In conclusion, in this case study weighting adds a comfort feeling that we have done what we can to remove seasonality effects, however, doing so makes little difference to the interpretation.

We have identified and taken into account both seasonality and process changes. The resulting data could now be charted to show the changes to produce an updated version of Chart 19.9. Analysis of the chart would then be as normal, we would look for individual out-of-control points that may be due to special causes of variation and further process changes and other features.

Table 19.2 Weight factor calculations for hand tools

	Jan	Feb	Mar	Apr	May	Jun	Jul	Aug	Sep	Oct	Nov	Dec
Process 1												
Value (year 1)	3	3	4	2	1	1	0	3	4	1	0	
Overall average	2.0	2.0	2.0	2.0	2.0	2.0	2.0	2.0	2.0	2.0	2.0	
Weight	1.50	1.50	2.00	1.00	0.50	0.50	0.00	1.50	2.00	0.50	0.00	
Process 2												
Value (year 1)												3
Value 2 (year 2)	2	5	3	2	4	2	3	0	2	3	2	2
Value 3 (year 3)	3	4	6	2	1	2	3	5				
Value average	2.5	4.5	4.5	2.0	2.5	2.0	3.0	2.5	2.0	3.0	2.0	2.50
Overall average	2.75	2.75	2.75	2.75	2.75	2.75	2.75	2.75	2.75	2.75	2.75	2.75
Weight	0.91	1.64	1.64	0.73	0.91	0.73	1.09	0.91	0.73	1.09	0.73	0.91
Process 3												
Value 1 (year 3)			1	2	1	2	1		2	2	1	1
Value 1 (year 4)	3	4						0	7			
Value average	3.0	4.0	1.0	2.0	1.0	2.0	1.0	0.0	4.5	2.0	1.0	1.0
Overall average	1.88	1.88	1.88	1.88	1.88	1.88	1.88	1.88	1.88	1.88	1.88	1.88
Weight	1.60	2.13	0.53	1.07	0.53	1.07	0.53	0.00	2.40	1.07	0.53	0.53
Average weight	**1.23**	**1.73**	**1.45**	**0.88**	**0.71**	**0.76**	**0.68**	**0.83**	**1.88**	**0.89**	**0.42**	**0.78**

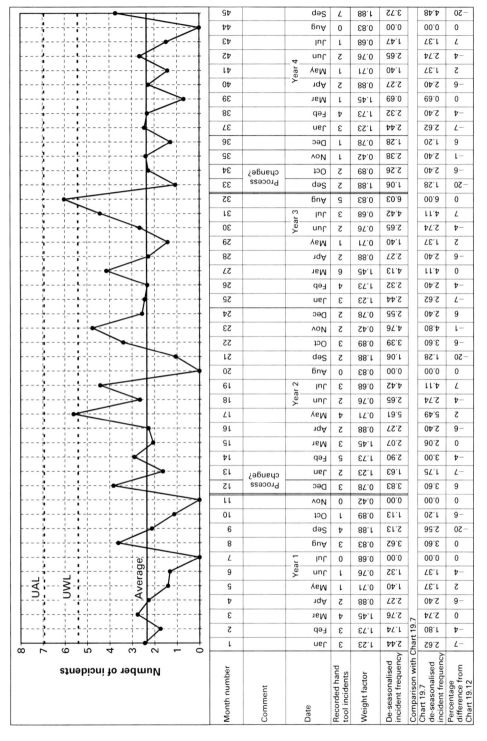

Chart 19.12 c chart: hand tool de-seasonalised incidents with revised weights; UAL: upper action limit; UWL: upper warning limit

Should the aim be to reduce the number of incidents?

In this case study we have focussed on reducing overall incident frequency. However, there are other metrics that we could use for selecting where to focus attention. Some organisations allocate a "risk" to each incident representing the either actual seriousness of the incident or potential effect of incurring the same incident again. For example, if the hand tool incidents only resulted in cuts or bruises, this may be considered insignificant compared to a fire which resulted in loss of life. To determine which incident type to focus on each incident is allocated a "risk" factor and the highest risk factors are singled out for attention first. It is possible to chart the risk factor rather than the number of incidents. One potential disadvantage of this approach is that where very serious incidents are few and far between, we do not have much information on which to develop and test theories as to causes, and so analysis is less likely to lead to effective solutions.

Total incident analysis may miss important changes in processes

When we analysed Chart 19.1 of all injurious incidents, we concluded that there was no process change. However, we have discovered two changes in the hand tool data, and one in the splash data and it is reasonable to assume that there are others within the data set as well. By analysing totals in this way we may miss changes that have occurred but are swamped by other data. The total incident chart is a useful indicator of the overall process and for predicting future performance. It should reflect changes in, for example, safety culture because this will affect all aspects of safety, but it will be slow to recognise changes that only affect one aspect of the total. For this reason, the overall charts are of limited use for improvement purposes, as we often need to stratify the data as demonstrated in this case study to improve aspects of the process.

This case study illustrates the need to take care when aggregating data, and the concepts apply to many types of data not just incidents. A very simple example will illustrate the issue. Suppose we receive between 80 and 120 complaints each month with an average of 100 complaints per month, split more or less evenly amongst 10 main causes. If there is an increase in one of the causes of complaint of 50%, that is, from 10 to 15, this would only increase the average total complaints by 5%, from 100 to 105, and it would take some months to recognise that something has happened.

Calculations

The c charts (Chart 19.1)

$$\bar{c} = \text{average} = \frac{\text{total number of incidents}}{\text{total number of months}} = \frac{384}{45} = 8.53.$$

$$s = \sqrt{\bar{c}} = \sqrt{8.53} = 2.92.$$

$$\text{UAL} = \bar{c} + 3s = 8.53 + 3 \times 2.92 = 17.3.$$
$$\text{LAL} = \bar{c} - 3s = 8.53 - 3 \times 2.92 < 0.$$
$$\text{UWL} = \bar{c} + 2s = 8.53 + 2 \times 2.92 = 14.4.$$
$$\text{LWL} = \bar{c} - 2s = 8.53 - 2 \times 2.92 = 2.7.$$

Cusum chart (Chart 19.2)

The target used is the average number of incidents per month = 8.53.

The difference between each month's incidents and the target is calculated as $c_i - 8.53$ for month i.

The cusum is the cumulative sum of the differences = $\Sigma(c_i - T)$.

$\bar{X}$/Range chart (Chart 19.3)

The table at the top of the chart gives the month, comments, average number of incidents in the month and the range.

In the table below the charts, the first four rows give the raw data of the number of incidents in each month. Note that there was no data for October to December for year 4. Below these rows are details of the calculations.

The first row gives the total number of incidents incurred in the month, and n is the number of years for which we have data ($n = 4$ in most cases, 3 for October to December).

$\bar{X}$ is the average number of incidents occurring in the month, and the range is the difference between the maximum and minimum number of incidents occurring the month.

$\bar{\bar{X}}$ is the overall average incident rate calculated by dividing the total number of incidents by 45, the number of months over which the incidents occurred = $384/45 = 8.53$.

$$\bar{R} \text{ is the weighted average range} = \frac{\sum n \times \text{Range}}{\sum n} = \frac{151}{45} = 3.36.$$

(the unweighted average range = $40/12 = 3.33$ is a good approximation.)

The control limits are calculated using the usual formula:

UAL_X is the UAL for the $\bar{X}$ chart = $\bar{X} + A_2\bar{R} = 8.53 + 0.729 \times 3.36 = 10.98$.
LAL_X is the LAL for the $\bar{X}$ chart = $\bar{X} - A_2\bar{R} = 8.53 - 0.729 \times 3.36 = 6.09$.
UAL_R is the UAL for the $\bar{R}$ chart = $D_4\bar{R} = 2.282 \times 3.36 = 7.66$.
LAL_R is the LAL for the $\bar{R}$ chart = 0 for $n = 3$ or 4.

Note that we have not taken into account that the number of values for October to December is only 3 when calculating the average range, but have used the appropriate values for A_2 and D_4. This is a reasonable simplification given the interpretation.

Summary

In this case study we have seen how to analyse a set of incident data where the incidents can be categorised in different ways. In this case the incidents were safety incidents, but a similar analysis method would apply to many other situations such as causes of breakdowns, failures/rejects, late deliveries/arrivals and missed deadlines.

We showed that it is important to categorise the data, for example, by cause of incident, and analyse each category separately because changes in the rates of occurrences due to individual categories may be swamped and missed when combining all categories into an overall incident rate.

We also discussed different methods of selecting which areas to target for improvement. We may, for example, wish to reduce the overall number of incidents, or the potential severity of incidents. One common approach is to attach a cost (or loss) to each incident and target those categories incurring the highest costs. We used Pareto charts to help identify key areas to target for improvement.

Using control charts we identified that the data not only had process changes but were also subject to seasonality and we gave one simple method of taking both process changes and seasonality into account during analysis.

20

Comparison of time-spent training across different facilities of an organisation

Charts used: X/MR, $\bar{X}$/range, $\bar{X}$/s and difference X/MR, histogram

Introduction

This study demonstrates how control charts can be used to compare the amount of training provided by different facilities in a large organisation. It also:

- Compares the differences between the range and standard deviation charts.
- Highlights the effect of changing the order of data on control limits.
- Demonstrates the uses of the difference chart.

Background

A large company has four subsidiaries (SS), we will call them Alpha, Beta, Gamma and Regions. Three of the SSs are themselves subdivided, for example, "Regions" is split into four further Business Units (BUs). Each BU has several facilities as shown in Figure 20.1. Since Alpha is not split into BUs it can be considered as both a BU and an SS.

Within the SSs Alpha, Beta and Gamma, the facilities are of a similar type, whereas the facilities within the Regions cover a diverse range of activities.

The organisation wants to review its training policies and practices and there is concern that the amount of training may vary between facilities, BUs and SSs. As one of the first fact-finding tasks, a study was carried out to determine if this was the case. The percentage of time-spent training at each facility in the previous calendar year was collected and charted.

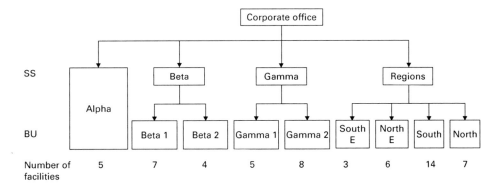

Figure 20.1 Corporate organisation chart

Analysis

X/moving range (Charts 20.1 and 20.2)

Chart 20.1 shows the un-interpreted data on an X/MR chart. The facilities have been numbered from 1 to 59 for quick reference, and the table also includes the BU, percentage of time-spent training and the moving range. The percentage of time-spent training was defined only for those people, staff or contract, for which the facility is responsible for providing training. For those people, the percentage of time-spent training is:

$$\% \text{ time} = \frac{\text{number of days time-spent training} \times 100}{\text{number of days worked by those eligible for training}}.$$

To calculate the number of days worked by those for whom the facility is responsible for training, it was suggested that facilities use the following formula:

$$\text{number of days} = \begin{bmatrix} \text{number of} \\ \text{staff on} \\ \text{site} \end{bmatrix} \times \begin{bmatrix} \text{average number} \\ \text{of days worked} \\ \text{per year by the} \\ \text{staff} \end{bmatrix} + \begin{bmatrix} \text{number of} \\ \text{contractors} \end{bmatrix} \times \begin{bmatrix} \text{average number} \\ \text{of days worked} \\ \text{per year by the} \\ \text{contractors} \end{bmatrix}$$

Though this figure will not be exact (e.g. because of varying staff and contractors numbers throughout the year) the result is deemed to be correct to within less than 1%, accurate enough for the analysis we wish to carry out.

The data is not plotted in any specific order, and we will not be analysing the data for trends. The data points have been joined with a line, broken between BUs, to make it easier to identify different BU.

It seems clear from the chart that some SSs have consistently low levels of training, for example, Alpha and Gamma 2 compared to other SSs such as Gamma 1. Some SSs such as Beta 2 and North East have large variation between their facilities.

Having determined that there is a difference between BUs, the next step is to investigate these differences further. To do so, we draw a control chart for each BU as shown in Chart 20.2. This chart suggests:

1. BUs Alpha, Gamma 2 and South have consistently low levels of training, with upper control limits less than 10.
2. SSs Beta 1 and North show out-of-control conditions. Beta 1 has two values above the upper action limit (UAL) in the X chart and one point above the UAL in the MR chart. North has one point above the UAL in both the X and the MR charts.
3. The remaining SSs, Beta 2, Gamma 1, South E, North East show little out-of-control evidence, thought South E especially has very few facilities reporting, only 3.

From the above analysis, we suspect that there are (significant) differences in the levels of training given between BUs, and in some cases perhaps also within BUs. As always, the step to take when identifying out-of-control values is to confirm the data are correct. Some facilities are reporting over 20% of time on training. Is this correct or, as is likely, have the figures been mis-reported, perhaps by a factor of 10? Do all facilities define

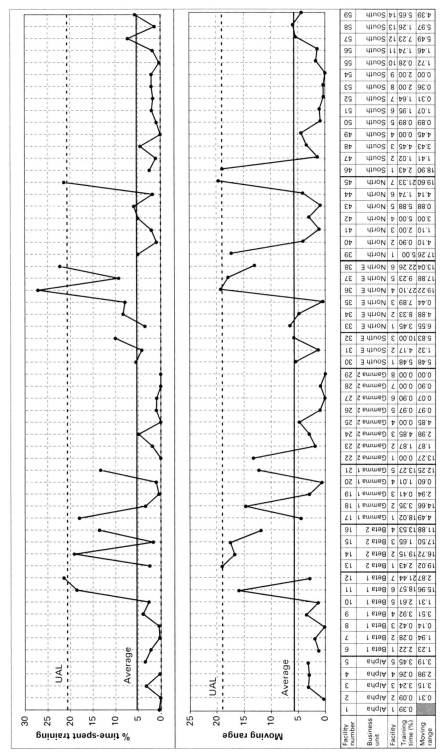

Chart 20.1 X/MR chart: percentage of time-spent training ordered randomly within BU; UAL: upper action limit

236

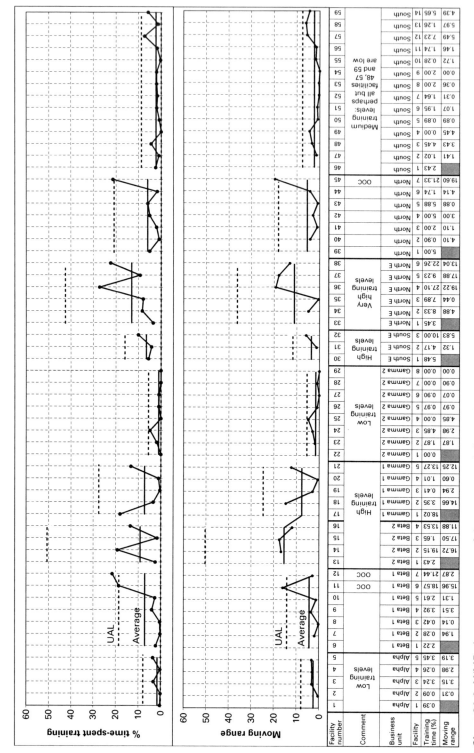

Chart 20.2 X/MR chart: percentage of time-spent training ordered randomly within BU with separate averages and limits for each BU; UAL: upper action limit; OOC: out of control

training in the same way (e.g. Do they include students on industry training schemes?). If the data are correct, we would want to understand why so much training is being provided. Similarly, some facilities are providing no training; is this correct? Again, we should investigate.

We re-chart the data excluding the out-of-control facilities in Chart 20.4 later.

Comparison of average training levels using average/range and average/standard deviation charts (Chart 20.3)

Continuing with the analysis, to compare average training rates between BUs, we can draw an average/range or an average/standard deviation chart. Both are given in Chart 20.3. We have included both in order to compare the results of the two charts. For this data we would normally use the s chart as the number of observations (facilities) varies.

The details of the calculations are given in the calculations section below. In both cases the individual facility training rates are provided in columns at the top of the chart. The calculations are provided in the next block and underneath are the charts. We continue assuming that the data are correct and without removing the apparent out-of-control facilities.

Comparing the average (top) charts of each pair, which have been drawn to the same scale, we see that the percentage of time-spent training and the average are the same in both cases, which we expect. The control limits are very similar, which we would hope to see, even though the calculations for them are different for the two charts. However, in most cases the control limits are slightly wider for the $\bar{X}$/s (standard deviation) chart than for the $\bar{X}$/R chart. This is comforting in that limits are similar whichever chart is used.

The interpretation for both average charts is that there is strong evidence that North E gives significantly more training than other BUs. However, if we look closer, we notice an interesting phenomenon: Alpha, Gamma 2 and South have remarkably similar training rates of between 1.1% and 2.3%. There is then a gap before Beta 1, Beta 2, Gamma 1, South E and North all have training rates between 6% and 9.2%, with North E alone at 13%. Perhaps this is chance, perhaps not, and it would be worthwhile checking to see if there is a connection from the training point of view between these BUs.

Turning to the R and s charts, again we see that the general pattern for R, s and the control limits is similar. It is not appropriate to plot the charts on the same scale, as though the range is related to the standard deviation (and indeed it is possible to estimate the standard deviation from the range), the relationship is not one-to-one. The differences to note are that the lower limit is just breached by South on the s chart, whereas it is only close to the limit on the R chart, similarly, Gamma 2 is nearer the lower action limit (LAL) on the s chart. However, other points are further away from the limits: for example, North E, South E. There is also a similar pattern in the variability charts as there is in the average charts in that Alpha, Gamma 2, and South, which had the lowest averages also have the low variability, but with the variability charts, South E is added to this group, whilst all the other BUs have much higher variability. With the exception of South on the s chart, there are no out-of-control conditions.

On these charts we have joined the points to aid interpretation, but as with the X/MR charts, it is not appropriate to look for runs or trends.

Average/range chart

	Alpha	Beta1	Beta2	Gamma 1	Gamma 2	South E	North E	North	South
Facility 1	0.39	2.22	2.43	18.02	0.00	5.48	3.45	5.00	2.43
Facility 2	0.09	0.28	19.15	3.35	1.87	4.17	8.33	0.90	1.02
Facility 3	3.24	0.42	1.65	0.41	4.85	10.00	7.89	2.00	4.45
Facility 4	0.26	3.92	13.53	1.01	0.00		27.10	5.00	0.00
Facility 5	3.45	2.61		13.27	0.97		9.23	5.88	0.89
Facility 6		18.57			0.90		22.26	1.74	1.95
Facility 7		21.44			0.00			21.33	1.64
Facility 8					0.00				2.00
Facility 9									2.00
Facility 10									0.28
Facility 11									1.74
Facility 12									7.23
Facility 13									1.26
Facility 14									5.65

Business unit	Alpha	Beta1	Beta2	Gamma 1	Gamma 2	South E	North E	North	South
n	5	7	4	5	8	3	6	7	14
Average	1.5	7.1	9.2	7.2	1.1	6.5	13.0	6.0	2.3
Range	3.36	21.16	17.50	17.61	4.85	5.83	23.66	20.43	7.23
Expected range if $n=7$	3.91	21.16	22.99	20.47	4.61	9.32	25.24	20.43	5.74
Overall average	5.3	5.3	5.3	5.3	5.3	5.3	5.3	5.3	5.3
Mean range	14.9	14.9	14.9	14.9	14.9	14.9	14.9	14.9	14.9
Mean range adjusted for n	13	15	11	13	16	9	14	15	19
UAL_X	13	11	14	13	11	15	12	11	10
LAL_X	–	–	–	–	–	–	–	–	–
UAL_R	27	29	26	27	29	24	28	29	31
LAL_R	0	1	0	0	2	0	0	1	5

Average/s chart

	Alpha	Beta 1	Beta 2	Gamma 1	Gamma 2	South E	North E	North	South
	0.39	2.22	2.43	18.02	0.00	5.48	3.45	5.00	2.43
	0.09	0.28	19.15	3.35	1.87	4.17	8.33	0.90	1.02
	3.24	0.42	1.65	0.41	4.85	10.00	7.89	2.00	4.45
	0.26	3.92	13.53	1.01	0.00		27.10	5.00	0.00
	3.45	2.61		13.27	0.97		9.23	5.88	0.89
		18.57			0.90		22.26	1.74	1.95
		21.44			0.00			21.33	1.64
					0.00				2.00
									2.00
									0.28
									1.74
									7.23
									1.26
									5.65

Business unit	Alpha	Beta 1	Beta 2	Gamma 1	Gamma 2	South E	North E	North	South
n	5	7	4	5	8	3	6	7	14
$n-1$	4	6	3	4	7	2	5	6	13
Average	1.5	7.1	9.2	7.2	1.1	6.5	13.0	6.0	2.3
Standard deviation, s	1.7	9.0	8.6	8.0	1.7	3.1	9.4	7.0	2.1
Variance	2.9	80.4	73.5	63.3	2.8	9.4	87.6	49.6	4.3
Overall average	5.3	5.3	5.3	5.3	5.3	5.3	5.3	5.3	5.3
Average, s	6.0	6.0	6.0	6.0	6.0	6.0	6.0	6.0	6.0
UAL_X	14	12	15	14	12	17	13	12	10
LAL_X	–	–	–	–	–	–	–	–	–
UAL_R	13	11	14	13	11	15	12	11	10
LAL_R	0	1	0	0	1	0	0	1	2

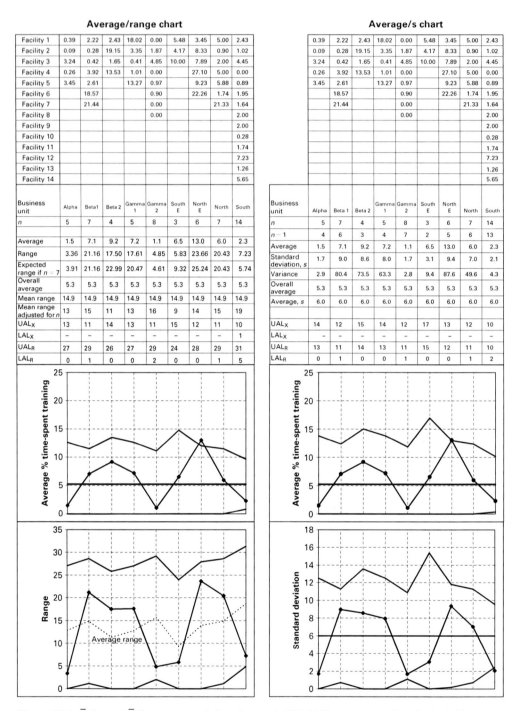

Chart 20.3 X̄/R and X̄/S charts: training for each BU. UAL: upper action limit; LAL: lower action limit

Excluding the out-of-control facilities from the analysis (Charts 20.4 and 20.5)

We now return to the data of Chart 20.2 and re-plot this with the out-of-control points for facilities Beta 1 (points 11 and 12) and North (point 45) removed from the calculations. The result is given in Chart 20.4. It is interesting to note that Beta 1 and North now join Alpha, Gamma 2, and South on having nearly identical training rates 1.1–3.4, with the other BU being unchanged. The two Beta 1 values removed were between 18.57% and 21.44%, which are very close to being 10 times the average Beta 1 rate of 1.9%. Similarly, the North value removed was 21.33, again quite close to being 10 times the average of 3.4 for the BU, and which is within the range of the other values (0.90–5.88%). This adds support to the theory that the rates reported might simply be too high by a factor of 10. This point has been laboured here to show the uses of control charts; of course, a phone call should be able to resolve the issue, but it is always useful to have an idea of where possible rogue data may have come from.

Redrawing the corresponding average/range and average/standard deviation charts having removed the three out-of-control points, Chart 20.5, accentuates that North E has significantly higher training rates than other facilities.

In conclusion, the control charts have:

- Identified three data values that need to be investigated.
- Concluded that North E is providing significantly more training than other BU.
- Suggested that there are two further levels of training:
  Low: which includes Alpha, Beta 1, Gamma 2, North and South;
  Medium: which includes Beta 2, Gamma 1 and South E.

Comments

Difference charts (Chart 20.6) (Page 242)

Another approach to analysing this data would be to consider the difference between the percentage of time-spent training for each facility and the average percentage of time spent for all facilities. The resulting X/MR chart is shown in Chart 20.6. The conclusions are, not surprisingly, the same as for Chart 20.2. The lines between BUs have been joined on Chart 20.6 to help compare the visual effect of including and removing them as in earlier X/MR chart.

Data order is important for X/MR charts (Chart 20.7) (Page 243)

It was purely fortuitous that the two high training percentages for Beta 1 occurred as the last two points in the Beta 1 data set. If *we change the order of the data, the conclusions may also change*. This is an unfortunate and potentially serious problem. Chart 20.7 is a re-print of Chart 20.2 but the high value of 21.44% has been swapped with an earlier value of 2.61%. The result is startling. BU Beta 1 no longer shows out-of-control conditions.

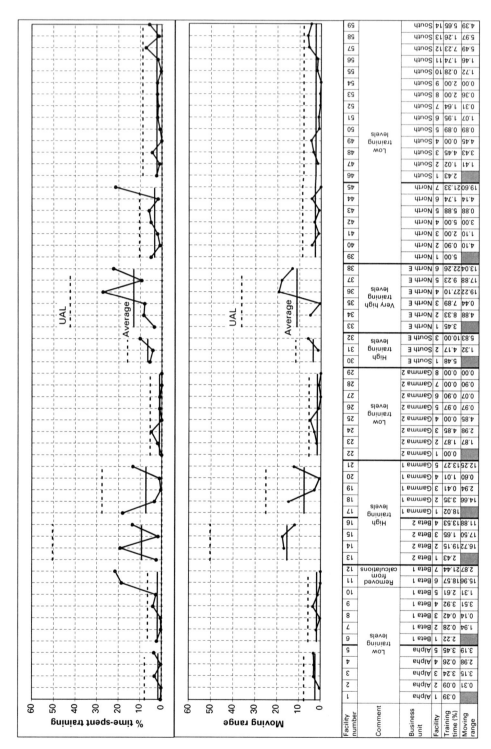

Chart 20.4 X/MR chart: percentage of time-spent training ordered by group, out-of-control facilities removed; UAL: upper action limit

Average/range chart

0.39	2.22	2.43	18.02	0.00	5.48	3.45	5.00	2.43
0.09	0.28	19.15	3.35	1.87	4.17	8.33	0.90	1.02
3.24	0.42	1.65	0.41	4.85	10.00	7.89	2.00	4.45
0.26	3.92	13.53	1.01	0.00		27.10	5.00	0.00
3.45	2.61		13.27	0.97		9.23	5.88	0.89
				0.90		22.26	1.74	1.95
				0.00				1.64
				0.00				2.00
								2.00
								0.28
								1.74
								7.23
								1.26
								5.65

Business unit	Alpha	Beta 1	Beta 2	Gamma 1	Gamma 2	South E	North E	North	South
n	5	5	4	5	8	3	6	6	14
Average	1.5	1.9	9.2	7.2	1.1	6.5	13.0	3.4	2.3
Range	3.36	3.64	17.50	17.61	4.85	5.83	23.66	4.98	7.23
Expected range if $n = 7$	3.91	4.23	22.99	20.47	4.61	9.32	25.24	5.32	5.74
Overall average	4.5	4.5	4.5	4.5	4.5	4.5	4.5	4.5	4.5
Range mean	9	9	9	9	9	9	9	9	9
Mean range adjusted for n	8	8	7	8	10	6	9	9	12
UAL_X	9	9	10	9	8	10	9	9	7
LAL_X	–	–	–	–	1	–	0	0	2
UAL_R	17	17	16	17	18	15	18	18	20
LAL_R	0	0	0	0	1	0	0	0	4

Average/s chart

0.39	2.22	2.43	18.02	0.00	5.48	3.45	5.00	2.43
0.09	0.28	19.15	3.35	1.87	4.17	8.33	0.90	1.02
3.24	0.42	1.65	0.41	4.85	10.00	7.89	2.00	4.45
0.26	3.92	13.53	1.01	0.00		27.10	5.00	0.00
3.45	2.61		13.27	0.97		9.23	5.88	0.89
				0.90		22.26	1.74	1.95
				0.00				1.64
				0.00				2.00
								2.00
								0.28
								1.74
								7.23
								1.26
								5.65

Business unit	Alpha	Beta 1	Beta 2	Gamma 1	Gamma 2	South E	North E	North	South
n	5	5	4	5	8	3	6	6	14
$n - 1$	4	4	3	4	7	2	5	5	13
Average	1.5	1.9	9.2	7.2	1.1	6.5	13.0	3.4	2.3
Standard deviation, s	1.7	1.5	8.6	8.0	1.7	3.1	9.4	2.1	2.1
Variance	3	2	74	63	3	9	88	4	4
Overall average	4.5	4.5	4.5	4.5	4.5	4.5	4.5	4.5	4.5
Average, s	4.7	4.7	4.7	4.7	4.7	4.7	4.7	4.7	4.7
UAL_X	11	11	12	11	10	14	11	11	8
LAL_X	–	–	–	–	–	–	–	–	1
UAL_R	10	10	11	10	9	12	9	9	8
LAL_R	0	0	0	0	1	0	0	0	2

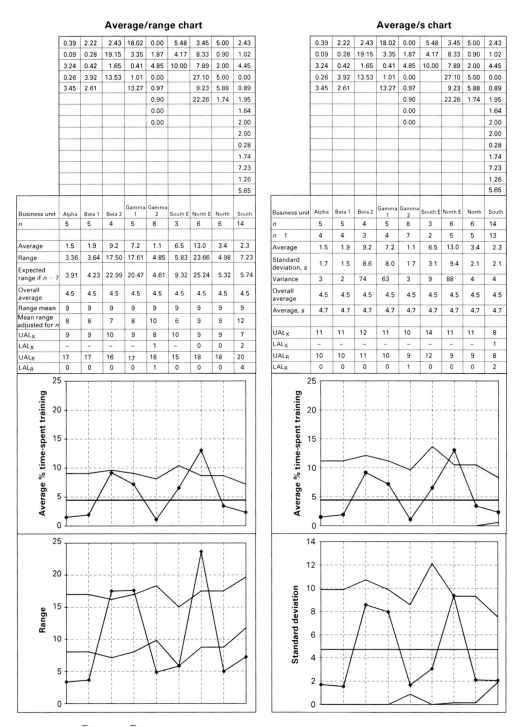

Chart 20.5 $\bar{X}$/R and $\bar{X}$/S chart: percentage of time spent training excluding out of control values. UAL: upper action limit; LAL: lower action limit

242

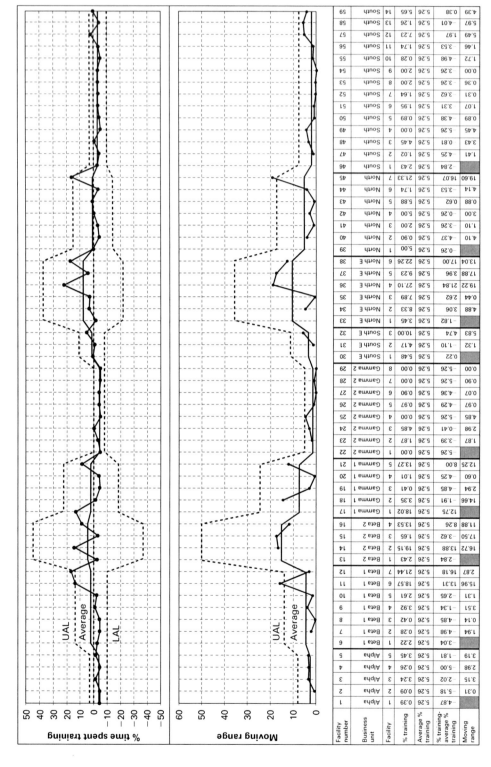

Facility number	Business unit	Facility	% training	Average % training	% training-average % training	Moving range
1	Alpha	1	0.39	5.26	-4.87	
2	Alpha	2	0.09	5.18	-5.18	0.31
3	Alpha	3	3.24	5.26	-2.02	3.15
4	Alpha	4	0.26	5.26	-5.00	2.98
5	Alpha	5	3.45	5.26	-1.81	3.19
6	Beta 1	1	2.22	5.26	-3.04	
7	Beta 1	2	0.28	5.26	-4.98	1.94
8	Beta 1	3	0.42	5.26	-4.85	0.14
9	Beta 1	4	3.92	5.26	-1.34	3.51
10	Beta 1	5	2.61	5.26	-2.65	1.31
11	Beta 1	6	18.57	5.26	13.31	15.96
12	Beta 1	7	21.44	5.26	16.18	2.87
13	Beta 2	1	2.43	5.26	-2.84	
14	Beta 2	2	19.15	5.26	13.88	16.72
15	Beta 2	3	1.65	5.26	-3.62	17.50
16	Beta 2	4	13.53	5.26	8.26	11.88
17	Gamma 1	1	18.02	5.26	12.75	
18	Gamma 1	2	3.35	5.26	-1.91	14.66
19	Gamma 1	3	0.41	5.26	-4.85	2.94
20	Gamma 1	4	1.01	5.26	-4.25	0.60
21	Gamma 1	5	13.27	5.26	8.00	12.25
22	Gamma 2	1	0.00	5.26	-5.26	
23	Gamma 2	2	1.87	5.26	-3.39	1.87
24	Gamma 2	3	4.85	5.26	-0.41	2.98
25	Gamma 2	4	0.00	5.26	-5.26	4.85
26	Gamma 2	5	0.97	5.26	-4.29	0.97
27	Gamma 2	6	0.90	5.26	-4.36	0.07
28	Gamma 2	7	0.00	5.26	-5.26	0.90
29	Gamma 2	8	0.00	5.26	-5.26	0.00
30	South E	1	5.48	5.26	0.22	
31	South E	2	4.17	5.26	-1.10	1.32
32	South E	3	10.00	5.26	4.74	5.83
33	North E	1	3.45	5.26	-1.82	
34	North E	2	8.33	5.26	3.06	4.88
35	North E	3	7.89	5.26	2.62	0.44
36	North E	4	27.10	5.26	21.84	19.22
37	North E	5	9.23	5.26	3.96	17.88
38	North E	6	22.26	5.26	17.00	13.04
39	North	1	5.00	5.26	-0.26	
40	North	2	0.90	5.26	-4.37	4.10
41	North	3	2.00	5.26	-3.26	1.10
42	North	4	5.00	5.26	-0.26	3.00
43	North	5	5.88	5.26	0.62	0.88
44	North	6	1.74	5.26	-3.53	4.14
45	North	7	21.33	5.26	16.07	19.60
46	South	1	2.43	5.26	-2.84	
47	South	2	1.02	5.26	-4.25	1.41
48	South	3	4.45	5.26	-0.81	3.43
49	South	4	0.00	5.26	-5.26	4.45
50	South	5	0.89	5.26	-4.38	0.69
51	South	6	1.95	5.26	-3.31	1.07
52	South	7	1.64	5.26	-3.62	0.31
53	South	8	2.00	5.26	-3.26	0.36
54	South	9	2.00	5.26	-3.26	0.00
55	South	10	0.28	5.26	-4.98	1.72
56	South	11	1.74	5.26	-3.53	1.46
57	South	12	7.23	5.26	1.97	5.49
58	South	13	1.26	5.26	-4.01	5.97
59	South	14	5.65	5.26	0.38	4.39

Chart 20.6 X/MR difference chart: percentage of time-spent training for each facility and the organisation average; UAL: upper action limit

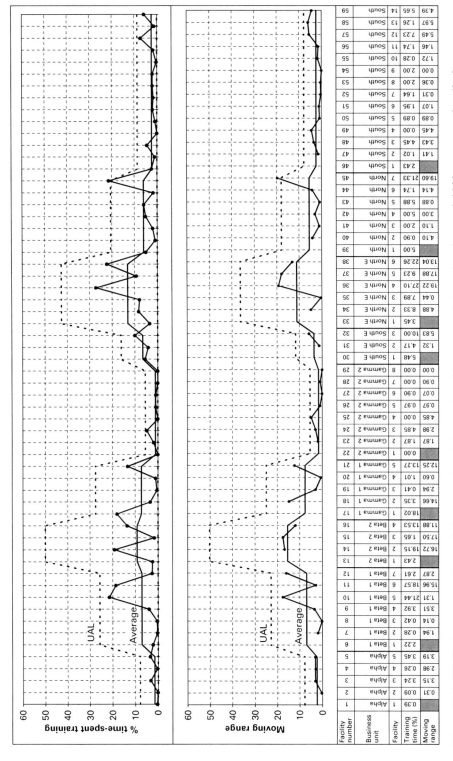

Chart 20.7 X/MR chart: percentage of time-spent training ordered by group, data order changed; UAL: upper action limit

The explanation is simple. The control limits is calculated as: $UAL = \bar{X} + 3s$ and the LAL and warning limits are calculated similarly. s is estimated as (average moving range)/1.128.

The moving range is simply the difference between successive training percentages. Each training percentage is used twice:

- once to calculate the moving range between it and the previous value,
- once to calculate the moving range between the next value and it.

The example below illustrates:

Data with high value at the end							
X	5	7	6	5	18	Sum of MR	Average MR	s
MR	–	2	1	1	13	17	17/4 = 4.25	4.25/1.128 = 3.8

Data with high value in the middle							
X	5	7	18	5	6	Sum of MR	Average MR	s
MR	–	2	11	13	1	27	27/4 = 6.75	6.75/1.128 = 6.0

The table demonstrates that the order of the data does influence the control limits. The more the data values there are, the less the influence this effect has. However, where there are few values, as in this case, the effect can be dramatic. It should be noted that:

- If the extreme values are separated, a high value of s will result, leading to the control limits being further away from the mean, and possible out-of-control conditions missed. If the resultant chart has points beyond the control limits, we can be certain that, we have identified that the data in the group do not come from the same underlying distribution.
- If the extreme values are grouped, a low value of s will be obtained, and the control limits will be nearer the mean. If the resultant chart has no points beyond the control limits, we can be certain that we have identified that the data in the group do come from the same underlying distribution.
- If, as in this case study, changing the order of the data changes the interpretation, then we need to find an alternative solution.

We present two possible solutions to this problem:

1. Perhaps the most pedantic solution is to calculate the average MR for all possible orderings of the data and use the average of these averages: a somewhat laborious task!
2. An alternative is to calculate the average moving range for a number of different random orderings and use the average of these as the basis for calculating the standard deviation.

We are not aware of any detailed work that has been done on this problem, but the most important point is to be aware of the fact that this can happen and be ready to look for potential problems.

Too much training?

In most uses of control charts it is possible to say that either high values are desirable (e.g. production rates, sales) or low values are desirable (e.g. rejects, accidents, lost sales, difference between plan and actual). However, in some cases, training being one, it is not appropriate to conclude that higher or lower values are better. It is not true to say that "more" training is good or "less" training is good. In situations like this, the concept of Taguchi's loss function provides a useful tool for thinking about these issues. For an explanation of Taguchi's loss function see, for example, Statistical Process Control by Oakland.

The optimum amount of training will partly depend on the attitude of the employer. Many employees strongly believe that a key responsibility of an employer is to encourage employees to develop themselves to their potential. Such employers will encourage and actively support training even when it is not directly related to the job. Other organisations view training as a necessary cost and will limit it to the absolute minimum. From the organisation's point of view, it is probably true that a lack of training leaves people unable to do their job in the most efficient and effective way. As the amount of (appropriate) training increases the payback to the organisation is at first high. This decreases as training increases until the cost of training equals the benefit to the organisation, after which the cost exceeds the benefit. Measuring the benefit of training is difficult as it includes not only the measurable effects (e.g. ability to do new tasks or ability to do them correctly and faster), but also the less quantifiable effects such as the attitude of the employee to the organisation. Training should be given in appropriate amounts, the perpetual trainee is not an employee but rather a sponsored student. The capacity of an organisation to train may also depend on the size. A small organisation with only a few employees may find it difficult to release staff for training for operational reasons.

Non-normality of data (Chart 20.8)

Chart 20.8 shows the histograms for all the data, as well as the South and Beta 1 BUs individually. The data are clearly non-normal, and, although there are very few values, the histograms also suggest that the distributions are not the same for each BU, as confirmed by the control chart analysis. Non-normality is common with this type of data and is discuss further in Part 4, but it is worth commenting here that the effect of non-normality is somewhat mitigated when using average charts. The distribution of means tends to normality as the number of observations included in the mean increases.

Other statistical analyses

A number of the analyses carried out in this and other case studies could also have used other statistical tools and tests. The purpose of this book is to explain how statistical process control (SPC) and control charts can be used to investigate a variety of different situations. Other tools would and should be used where appropriate and many will need the skills of a trained statistician.

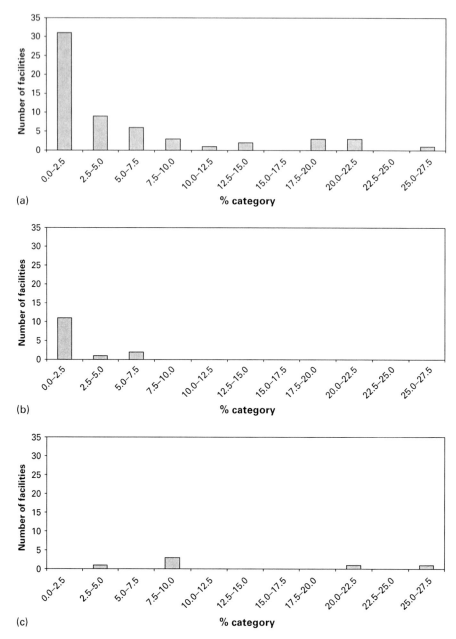

Chart 20.8 Histograms showing the non-normality of training percentage data: (a) all, (b) South and (c) North E

Calculations

$$\% \text{ time-spent training for a facility } = \frac{\text{number of days time-spent training} \times 100}{\substack{\text{number of days worked by all those} \\ \text{eligible for training}}}.$$

The average percentage of time-spent training for the whole organisation should be calculated as:

$$\% \text{ time } = \frac{\substack{\text{total number of days spent training} \\ \text{in the organisation} \times 100}}{\substack{\text{number of days worked by all those in the} \\ \text{organisation eligible for training}}}.$$

Unfortunately, the only data available were the individual facility percentages, and so the averages were calculated by averaging these percentages.

This would be acceptable if all facilities were of a similar size. As it happens, this is not the case. The result is that small facilities will weight the results more than their size warrants, and large facilities will weight the results less than their size warrants.

As an example, consider three facilities, A, B and C with the data as given in the table below:

Facility	Number of hours training	Number of hours worked (thousands)	Time-spent training (%)
A	10	100	10
B	10	200	5
C	10	400	2.5

Calculating the overall average by using the percentages gives $= (10 + 5 + 2.5)/3 = 5.8\%$.

The true average is $(10 + 10 + 10)/(100 + 200 + 400) = 0.043 = 4.3\%$.

X/MR Chart (Chart 20.2), Beta 2

The average percentage of time-spent training

$$\bar{x} = \frac{2.43 + 19.15 + 1.65 + 13.53}{4} = 9.19$$

The average moving range,

$$\overline{MR} = \frac{16.72 + 17.5 + 11.88}{3} = 15.37.$$

$$\text{UAL}_X = \bar{x} + \frac{3\overline{MR}}{1.128} = 50.1.$$

$$\text{UAL}_{MR} = 3.27\overline{MR} = 50.3.$$

Calculations for the other BU use the same formula.

$\bar{X}$/Range (Chart 20.3)

The methodology is explained in more detail earlier in the case study. The calculations below relate to the Alpha BU and the constants A_2 and D_4 can be found in the Appendix.

The overall average, $\bar{\bar{x}}$ = average of all facility training percentages = 310/59 = 5.3%.

The average number of facilities, $\bar{n}$ = 59 facilities/9 BUs = 6.55, rounded to 7.

To calculate the average range we need to estimate the ranges for each group to do this we calculate:

$$\text{expected range } (n = 5) := \frac{\text{range} \times d_{2,7}}{d_{2,5}} = \frac{3.36 \times 2.704}{2.326} = 3.91.$$

The mean range is calculated as the average of the expected ranges = 14.9.

The mean range now needs to be adjusted back to the sample size, n. To do this we calculate:

$$\text{mean range adjusted if } (n = 5) := \frac{\text{mean range} \times d_{2,5}}{d_{2,7}} = \frac{14.9 \times 2.326}{2.704} = 12.8$$

$\text{UAL}_X = \bar{\bar{x}} + A_2 \times$ mean range adjusted for n = 5.3 + 0.729 $\times$ 12.8 = 14.6.

$\text{LAL}_X = \bar{\bar{x}} - A_2 \times$ mean range adjusted for n = 5.3 $-$ 0.729 $\times$ 12.8 < 0.

For the range chart:

$\text{UAL}_R = D_4 \times$ mean range adjusted for n = 2.11 $\times$ 12.8 = 27.0.

As D_3 is zero, the LAL_R = 0.

Note that the average range varies depending on the value of n.

Standard deviation (s) (Chart 20.3)

The calculations for the average percentage of training are the same as for the Average chart given above and gives the average as 5.3%. All the calculations below relate to the South E BU. The values for the constants can be found in the appendix:

n is the number of observations (=3).

$n - 1$ is used to calculate the standard deviation.

x_i are the individual percentage (=5.48, 4.17, 10.00).

$\bar{x}$ is the average (=6.55).

s is the standard deviation is calculated using the usual formula for each facility.

$$\text{variance} = s^2 = \frac{\sum (x_i - \bar{x})^2}{(n - 1)}$$

$$= \frac{(5.48 - 6.55)^2 + (4.17 - 6.55)^2 + (10.00 - 6.55)^2}{(3 - 1)} = 9.36.$$

And $s = \sqrt{9.36} = 3.1$

The average standard deviation is calculated as:

$$\overline{s} = \sqrt{\frac{(n_1 - 1)s_1^2 + (n_2 - 1)s_2^2 + \cdots + (n_9 - 1)s_9^2}{n_1 + n_2 + n_3 + \cdots + n_9 - 9}}$$

$$= \sqrt{\frac{4 \times 2.9 + 6 \times 80.4 + \cdots + 13 \times 4.3}{50}} = 6.0.$$

$$\text{UAL}_X = \overline{\overline{x}} + A_3\,\overline{s} = 5.3 + 1.954 \times 6.0 = 17.$$
$$\text{LAL}_X = \overline{\overline{x}} - A_3\,\overline{s} = 5.3 - 1.954 \times 6.0 < 0.$$

And for the s chart:

$$\text{UAL}_R = B_4\,\overline{s} = 2.57 \times 6.0 = 15.$$
$$\text{LAL}_R = B_3\,\overline{s} = 0 \times 6.0 = 0.$$

Summary

In this case study we used control charts to compare the amount of training provided by different facilities in a large corporation. Using control chart we were able to identify three suspect data items, and conclude that there were differences in the training provided between BUs.

We demonstrated that when data have no pre-determined sequencing, that the order in which the data are entered into a control chart may affect the conclusions derived from the chart.

A difference chart was used as an alternative to compare the percentage of training provided by each facility and the average for the organisation.

We also explained the use of the range and standard deviation charts. We concluded that when applied to the data in this case study there were minor differences in the results but the overall conclusions were the similar.

PART 4

Implementing and Using SPC

In Parts 2 and 3 of this book we aimed to persuade the reader that statistical process control (SPC) can bring many benefits to organisations. We purposefully kept the theory to a minimum providing just enough to enable an appreciation of the benefits of SPC without requiring the reader to first become familiar with and understand the statistical theory and formulae behind control charting.

In Part 4 we provide further information on SPC theory.

We begin, in Chapter 21, by providing a more detailed explanation as to how to interpret a control chart. We explain the importance of the standard deviation and how it is used in interpreting process performance. We discuss in detail how to discover whether a process is in a state of control, and explain the reasons behind the interpretation guidelines.

The next three chapters, 22–24, are linked. The first explains how to select the appropriate control chart in different situations. The second explains in detail the application, formulae and how to develop each chart. The third chapter introduces cumulative sum charts. This type of chart can be very complex and the aim is to cover their main uses and theory. In addition to the basic cusum chart, we discuss the more complex weighted cusum chart.

Chapter 25 discusses a number of issues that arise from time to time when charting. We discuss:

- how to determine the average number of observations required to identify a process change after it has occurred,
- how to check for normality of data, and what to do if the data are non-normally distributed,
- the problem of autocorrelation, how to identify it and what to do about it,
- how to analyse rare event data,
- how to analyse data that can be divided into groups.

In Chapter 26, we explain several of the common tools used along with control charts for data analysis and process improvement, in particular we discuss how to interpret the histogram.

Chapter 27 suggests how to set up and implement an effective performance measurement and monitoring system, while the following chapter provides some ideas on metrics that can be monitored.

21 Understanding and interpreting a control chart

Introduction

In Part 1 of this book we outlined some basic theory of control charting. In this chapter we provide more information to help deepen the understanding of some of the concepts and theory.

We explain the standard deviation and why it is key to control charting. We also explain some of the statistics behind the chart interpretation rules. Whilst the theory is appealing to those who like to see some sort of "proof" that the ideas behind control charts are valid, it is important to remember that these techniques are successful not because of the theory, but simply because experience has shown that they work. In statistics (and many if not all other sciences), the theory aims to explain what we experience. It is not perfect, but it is a useful approximation to the truth and helps us to make appropriate process decisions consistently and effectively.

The normal distribution and the standard deviation

As explained in Chapter 1, if we gather data and plot a histogram, it will have a shape similar to that in Figure 21.1. It is characterised by having one peak, and is symmetrical about the average. Data like this frequently follow a very well understood statistical distribution called the normal distribution.

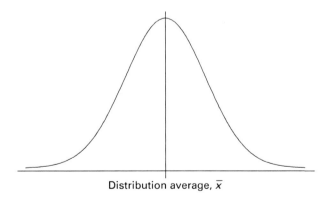

Distribution average, $\bar{x}$

Figure 21.1 The normal distribution

There are two ways in which a normal distribution can change: location and spread (Figure 21.2). The basic shape of a normal distribution cannot change: that is, if the shape changes it is no longer a normal distribution. The location is determined by the

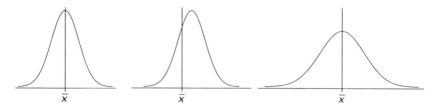

Figure 21.2 A normal distribution can only change location ... or spread

distribution average, denoted $\bar{x}$, and the spread, or variability, by the standard deviation, denoted s. The formulae for both are given in Chapter 1, and we explain in a little more detail where the standard deviation comes from.

Suppose we have taken measurements of a characteristic, for example, goals scored, and have obtained the five values, 1, 2, 3, 4, 5 (denoted x_1, x_2, x_3, x_4 and x_5 in Table 21.1). The mean of these five values is obtained by adding them to give 15 and dividing by the number of values, 5, to give an average of $\bar{x} = 3$. We do not want the variability to be affected by changes in the mean. For example, if the values had been 51, 52, 53, 54, 55 we would want the measure of variability to be the same as for the values 1, 2, 3, 4, 5. To achieve this, we subtract the mean from each value, given in the column $(x - \bar{x})$. This shows how much each value varies from the mean. Unfortunately, if we add the values of $(x - \bar{x})$, we get 0, because the negative values cancel out the positive. (This is inevitable since we subtracted the mean from each value.) However, what is of interest is the size of the variation, not whether the variation is positive or negative. One way of overcoming this problem is to square the differences, to obtain $(x - \bar{x})^2$. We can then calculate the average of these squared values, giving 2 in our example. However, since we squared the values, we now need to take the square root, which gives $\sqrt{2}$. This, in essence, is how the standard deviation is calculated.

To write this as a formula, let us consider each calculation that we did:

- we subtracted the mean from each value, $(x - \bar{x})$,
- squared them, $(x - \bar{x})^2$,
- added them up, $\sum (x - \bar{x})^2$,
- divided by the number of values, denoted n, and took the square root:
  $$\sqrt{\sum (x - \bar{x})^2 / n}.$$

This, with one small alteration is the formula for the standard deviation. The powerhouse of statistics tells us to use $(n - 1)$ rather than n as a divisor. The explanation is beyond the scope of this book, but is explained in many basic statistics book. The resulting formula is:

$$s = \sqrt{\frac{\sum (x - \bar{x})^2}{(n - 1)}}.$$

This is the formula used widely in statistics, and is very useful in helping to understand the concept of the standard deviation. However, from the point of view of statistical process control (SPC), it assumes the data come from a single normal distribution and is greatly inflated if this is not the case. The group of charts called attributes charts

Table 21.1 Calculating the standard deviation

Observation	x	$(x - \bar{x})$	$(x - \bar{x})^2$
x_1	1	-2	4
x_2	2	-1	1
x_3	3	0	0
x_4	4	1	1
x_5	5	2	4
Sum = 15		Sum = 0	Sum = 10
$\bar{x}$ = 15/5 = 3			Mean = 10/5 = 2

frequently follow well understood but non-normal distributions and have their own formulae for calculating the standard deviation. To protect against non-normality and extreme values in the X chart, we calculate the standard deviation using the differences between successive values (i.e. moving ranges). The formulae are explained in Chapter 22.

The importance of the standard deviation

For data that do follow the normal distribution we know some very useful facts. If we calculate the standard deviation of the data, s, then (Figure 21.3):

- about two-thirds (68%) of the data values will lie within 1 standard deviation of the mean;
- about 95% of the data values will lie within 2 standard deviations of the mean;
- about 99.7% of the data values will lie within 3 standard deviations of the mean.

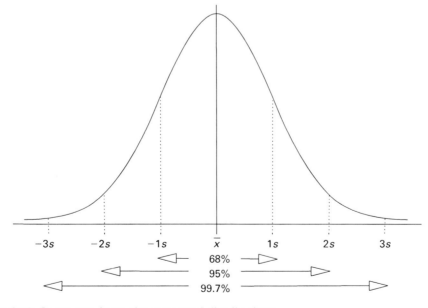

Figure 21.3 Some use facts about normal distributions

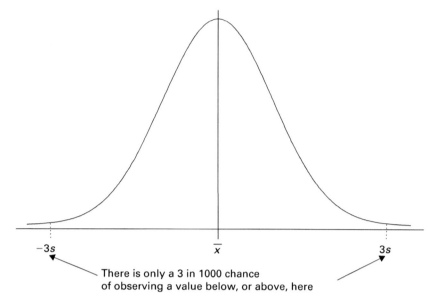

There is only a 3 in 1000 chance
of observing a value below, or above, here

Figure 21.4 Probability of observing a point outside the control limits

We can infer from this that the probability of obtaining a value more than 3 standard deviations away from the mean is only 3 in 1000 (Figure 21.4). Therefore, if we do obtain a value outside these limits we conclude that it has probably not come from the same distribution as that from which the mean and standard deviation were calculated. To translate this to monitoring a process, if a process is in a state of control and we obtain a value outside the control limits, we conclude that the process has probably changed: that is, we have an out-of-control condition.

Why do we use 3 standard deviations to determine the control limits and not 2 or 4 or something else? Experience has shown that using the 3 standard deviation rule is a good compromise between reacting when we should not react (e.g. because on average, 3 in 1000 observations will be outside the control limits by chance alone) and not reacting when we should.

Definition of a process in a state of control

We are now in a position to define what we mean by a process in a status of statistical control. If the data from a process are:

- randomly distributed, that is there are no obvious (i.e. predictable) patterns;
- distributed randomly around the mean with approximately half the data lying above the mean and half below;
- distributed with less points the further away from the mean we move;
- within certain limits (i.e. within 3 standard deviations from the mean).

Then the process is said to be in a state of statistical control.

Conversely a process that does not follow these rules is said to be not in a state of statistical control, or "out of control".

The aim of control charts is to analyse data to determine whether a process is or is not in a state of statistical control.

Interpreting a control chart

With control charts we are looking for data values which do not come from the same distribution as the rest of the process data.

In Part 1 we explained the structure of the control chart as consisting of a run chart with warning and control limits. We also gave the key indicators of an out-of-control process. We repeat those here with further comments:

- One point outside the control limits.

  The probability of obtaining a value outside the control limits by chance is only 3 in 1000 (0.3%).
- Eight or more consecutive values on the same side of the mean (Figure 21.5). Some people recommend 7 or even (rarely) 6.

  For an in control process the probability of recording a value on the same side of the mean as the previous value is 50% or $\frac{1}{2}$. The probability of the next value being on the same side is again $\frac{1}{2}$. Therefore the probability of both values being on the same side as the first value (i.e. three consecutive values on the same side) is $\frac{1}{2} \times \frac{1}{2} = \frac{1}{4}$. Using the same argument, the probability of eight consecutive points being on the same side of the mean is $\frac{1}{2} \times \frac{1}{2} \times \frac{1}{2} \times \frac{1}{2} \times \frac{1}{2} \times \frac{1}{2} \times \frac{1}{2} = 1/128$, or about 0.78%.
- A run of seven alternating high/low values (Figure 21.6).

  The reasoning behind this rule is similar to the argument above: the probability of an observation being lower than the previous observation is (very approximately) $\frac{1}{2}$, and the following one being higher is again $\frac{1}{2}$ and so on.
- Two consecutive observations between the warning and control limits (Figure 21.7).
- Some people like to use warning limits. Where these are used they can be a guide to determine whether too many points are far away from the average. Only about 5 in

Probability of any one of these points being on same side of the average as A is 1/2
The probability that all 7 are on the same side as A is 1/128

Figure 21.5 The probability of a run of observations being on the same side of the average; UAL: upper action limit; LAL: lower action limit

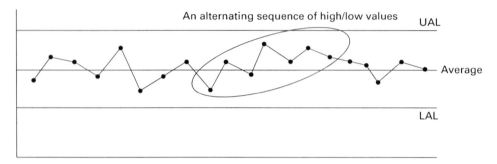

Figure 21.6 The probability of alternating observations; UAL: upper action limit; LAL: lower action limit

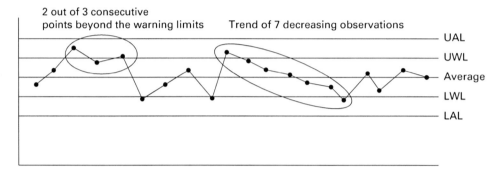

Figure 21.7 Other indicators of special causes of variation; UAL: upper action limit; LAL: lower action limit; UWL: upper warning limit; LWL: lower warning limit

every 100 observations should lie beyond the warning limits, so if two are observed consecutively, it is taken as a signal of an aberration in the process. Many analysts consider that two out of any three consecutive points beyond the warning limits indicates a change in the process.
- A trend consisting of seven consecutive increasing (or decreasing) observations (Figure 21.7).
- Any other non-random pattern.

When getting to grips with analysing control charts it is important to realise:

- That the guidelines given above are precisely that: guidelines. They are indicators that there is a special cause of variation, and for this reason I hesitate to call them rules as many texts do.
- The problem with applying rules strictly is illustrated in Figure 21.8a.
    - Is there a signal if we have, for example, only one observation beyond the warning limit, but several very close to it?
    - Is there a signal if rather than seven consecutive decreasing observations, some are at the same value?
    - Figure 21.8b does not break the "rules" but is hardly a random set of data!

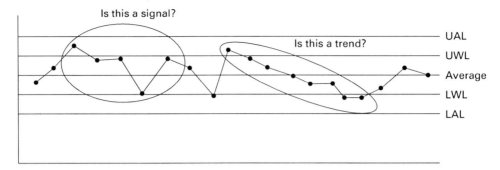

Figure 21.8(a) Rule: Are these indicators of the presence of special causes?; UAL: upper action limit; LAL: lower action limit; UWL: upper warning limit; LWL: lower warning limit

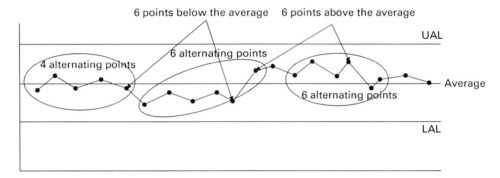

Figure 21.8(b) Random set of data: Are these indicators of the presence of special causes?; UAL: upper action limit; LAL: lower action limit

- It is recommended that these guidelines be more rigorously applied when first using control charts. After some time you will begin to sense something is not quite right – and the guideline becomes more of a confirmation tool than a rule to be rigidly followed.
- Sometimes we will see a false positive signal (the chart tells us that there is a special cause of variation present when there is not) or there will be a special cause present and we will miss it. In organisations today we generally over-react: we believe there is a special cause present when there is not. The action manager want to take action, wants to find out what went wrong, why and how to fix it. Unfortunately, the response is often a knee jerk reaction, an investigation, or an opportunity to blame someone. This is tampering. The control chart aims to minimise the error of inappropriately reacting to data. Far less often we fail to react when we should. When this happens it is usually because we are not measuring the right thing, we are using the wrong tool and it is not providing the correct information, or we just cannot read the information being presented to us. What the control chart enables us to do is to focus our energies where they will have most positive effect: investigating ONLY those data values that warrant investigation, and improving stable process to achieve higher levels of performance.

- With experience, the interpretation guidelines can be modified to take into account the risk associated of reacting to a signal when there is none, and not reacting to a signal when there is. For example, when monitoring mortality rates after a procedural change in a hospital, one might begin an investigation after only four or five points above the average rather than waiting for the usual seven, because the "cost" of missing a signal is so great compared to the cost of investigating.

Possible causes of control chart signals

The out-of-control conditions, or signals, in a control chart not only warn us that something unusual is going on in the process, they can also give clues as to what is going on, and so help us identify problems. Each organisation will build up its own specific list with experience, but a few general pointers are given in Table 21.2.

A note on the British and American control chart limits

This book glosses over the differences between the British and American conventions for drawing and interpreting control charts as in practice the differences have little effect. For those who may be interested the differences are outlined below.

The British system is to set:

- Action limits at ±3.09 standard deviations from the mean.
- Warning limits at ±1.96 standard deviations from the mean.

These limits correspond to 2 in 1000 probability of an observation being outside the action limits by chance, and a 2 in 40 probability of an observation being outside the warning limits by chance IF the underlying distribution is normal.

The American system is to set:

- Control limits at ±3 standard deviations from the mean which corresponds to a 3 in 1000 probability of an observation being outside the limits by chance.

In practice, distributions may well not be exactly normal and processes will seldom be strictly in control. We must not lose sight of the fact that control charts are successful not because they follow some theoretical model, but because they work in practice. The theory is only a way of modelling life and explaining mathematically what we experience.

In practical terms, it make little difference if we set our action/control limits at 3 or 3.09 standard deviations away from the mean, and the warning limits at 1.96 or 2 standard deviations away from the mean.

Summary

- Data from stable processes frequently follow a normal distribution.
- A normal distribution is characterised by having one peak, and being symmetrical about it's average (Figure 21.1).
- Normal distributions can vary in only two ways (Figure 21.2):
  - location (measured by the average);
  - variability (measured by the standard deviation).

Table 21.2 Possible causes of control chart signals

Signal	Possible causes
Single point outside the control limits followed by data showing no abnormalities	Measurement or recording error Temporary event such as stand-in personnel/sickness
Sudden jump in values	Sudden change in process (e.g. damaged equipment, new staff, manager, procedure, supplier, target, bonus payments)
Gradual drifting of data	Wear and tear on machinery or measuring equipment
Cycling high/low	The output of more than one process is being monitored examples include: two shifts, two machines, two suppliers, two people. Over control (tampering): after a high value the process is being adjusted and then providing a low value, etc.
Repeating pattern	Same as cycling but with more than two processes involved, (e.g. several machines, shifts)
Too many points near specification or target limits, but few if any over the limit	Measurements being "adjusted" so that they fall within the specification or target
Substantially more than two-thirds of the data lie within 1 standard deviation either side of the mean	Measurements from different processes with different average are being grouped together and averaged thereby removing the more extreme values. Data have been "manipulated" (i.e. more extreme values are not being recorded or recorded incorrectly)
Substantially less than two-thirds of the data lie within 1 standard deviation either side of the mean	Over control (tampering): after a high value the process is being adjusted and then providing a low value, etc. Measurements are being taken from different subgroups with very different averages. For example, different shifts, groups of people or suppliers.
In daily, weekly or monthly data, recurring patterns	Often difficult to spot, and usually worth checking anyway: – Time of day trend (values change in a systematic way throughout the day) – Day trend (e.g. Mondays always high, Fridays low) – End of week/month values high/low in order to meet targets
Single point outside the control limit (charts monitoring variability)	An increase in variation could indicate: – a change in process average – for an $\bar{X}$ or S chart, depending on the make up of the sample, it could indicate that one stream or source has changed

- For a normal distribution we know that (Figure 21.3):
  - about two-thirds (68%) of the data values will lie within 1 standard deviation of the mean;
  - about 95% of the data values will lie within 2 standard deviations of the mean;
  - about 99.7% of the data values will lie within 3 standard deviations of the mean.
- Processes that are in a state of control have data that are;
  - randomly distributed, that is there are no obvious (i.e. predictable) patterns;
  - distributed randomly around the mean with approximately half the data lying above the mean and half below;
  - distributed with less points the further away from the mean we move;
  - within certain limits (i.e. within 3 standard deviations from the mean).
- Processes whose data do not follow this pattern are said to be not in a state of control (i.e. they are changing).
- There are guidelines for analysing data that are plotted on a control chart to help determine whether a process is or is not in a state of control.

22 Selecting the appropriate control chart

Introduction

As can been seen in the case studies in this book there are a variety of different types of control charts. Selecting the appropriate chart for any particular set of data can be difficult for those new to SPC. Sometimes, the choice of chart makes little difference to the conclusions drawn, whilst in other situations it does. In this chapter we give some guidance as to how to choose the appropriate chart to analyse a set of data, and some of the relative advantages of different types of chart. The decision chart, Figure 22.1, will be used as a guide map. Further information is given in the next chapter on how to draw the charts, here the emphasis is on how to select which chart to draw. The case study in Chapters 14, 16 and the Rods Experiment in Part 5 of the book in particular use a variety of charts to analyse the same basic set of data and help to contrast the application of different charts.

If in doubt as to which chart is appropriate, it is possible to draw and compare all those that you think may be appropriate. Frequently the conclusions will be the same. If the conclusions are different, there is an opportunity to investigate and learn why they are different. Remember, the purpose of charting is the insight the chart brings, and not the chart itself.

Note:

- The $\bar{X}$ (pronounced "x bar") chart is also known as the "averages" or "means" chart.
- The R chart is also known as the range chart.
- The s chart is also known as the standard deviation chart.
- The X chart is also known as the individuals or I chart.
- The MR chart is also known as the moving range chart.
- The MR chart is usually plotted with the X chart, and the two together are often referred to as if they are one chart: the X/MR chart.

Usually when we refer to the R chart in this book, it should be understood that the s chart is an alternative, and usually preferable when samples sizes are large or when calculations are automated.

Referring to Figure 22.1, the first decision is to determine what type of data we are dealing with.

Variables or attributes data?

For charting purposes, data can be split into two types: variables and attributes. Variables data usually result from measurements; you can choose the precision (i.e. number of decimal places recorded) and for this reason they are sometimes called continuous data. Typical examples of variables data include length, weight, time, volume, porosity, chemical concentration and cost. Attributes data (also known as counts data) are so-called

264

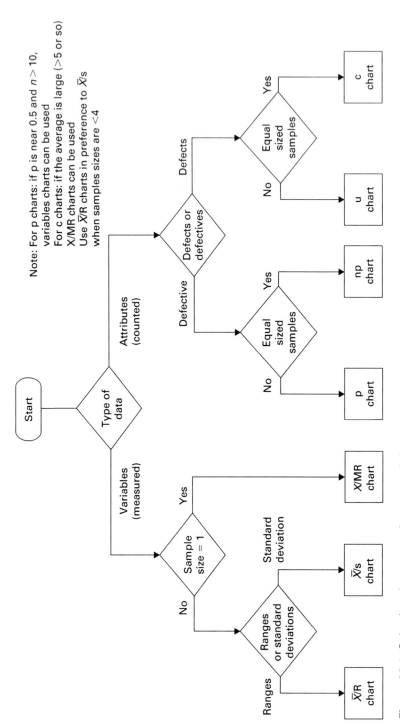

Note: For p charts: if p is near 0.5 and $n > 10$, variables charts can be used
For c charts: if the average is large (>5 or so) X/MR charts can be used
Use $\bar{X}$/R charts in preference to $\bar{X}$/s when samples sizes are <4

Figure 22.1 Selecting the appropriate control chart

because they are based on whether an item has an attribute or not. The data are always counted and hence are whole numbers; for this reason they are sometimes called discrete data. Typical examples of attributes data include numbers of complaints, accidents, fires, rejects, orders, errors, flaws (e.g. in a square meter of carpet).

The distinction is important because attributes data do not usually follow a normal distribution (we discuss which distributions they follow below). Attributes data are usually charted using c, u, np or p charts (see Figure 22.1). The most common charts for charting variables data are X, moving range, $\bar{X}$, range or s charts as appropriate.

There are other types of data such as categories (e.g. examination grades). This type of data is turned into counts or variables data for analysis. For example, in opinion surveys we may record the percentage of people that prefer item A from several alternatives, or the percentage of people that ticked (checked) a certain answer to a multiple choice question. We could use control charts to monitor, for example:

- The number or proportion or percentage of people preferring item A in successive surveys.
- The number or proportion or percentage of people in different locations preferring items A to B.
- The number or proportion or percentage of people from different age groups preferring items A to B.

Similar analyses could be carried out for the multiple choice questions.

In practice, however, the differences can become blurred as, for example, when the counts values are high, we can often treat counts data as if it were variables data, as discussed below.

We may monitor some situations by variables or counts or both. Consider a pharmaceutical process in which painkillers are produced. We could chart the weight of an active ingredient in each tablet (variables data) or the number of tablets with too little/much active ingredient (counts data).

There are advantages and disadvantages to both approaches (see Figure 22.2):

- If we chart the weight of active ingredient we will be able to deduce if there is a trend, cycling or other non-random variation occurring much more readily than if we

Variables	Attributes
Usually the result of a measurement	Are counts data
Contains more information	Contains less information
Is more difficult/expensive to collect/analyse	Is less difficult/expensive to collect/analyse
	Under certain conditions, we can use variables charts to analyse attributes data
	Process may not be in control, but may not be picked up by attributes charts
Examples: distances, weights, costs, percentages, chemical concentrations	Examples: number of failures, rejects, accidents, fires, spills

Figure 22.2 Summary comparing variables and attributes data

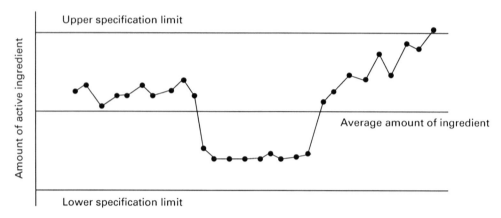

Chart 22.1 Run chart of the amount of active ingredient of a substance from samples taken over a period of time

only record whether there is too little active ingredient. For example, Chart 22.1 is a plot of the amount of active ingredient of a substance from samples taken over a period of time. If the sample value lies with the specification limits, the sample passes the test. The first and only failure is the last value which is above the upper specification limits. However, we can see from the chart that the process is not in a state of control and could have predicted that the process would produce a result above the upper specification limit before the last point. However, if we had only recorded whether the sample was within the specification limits there would have been no knowledge that the process was not in a state of control and no warning that it was about to produce a value outside the specification limit.

● As variables data give us more information than attributes data, in some sense we can say we need fewer measurements to get a similar "amount" of information, and so can often draw conclusions earlier than if we record attributes data.

● Variables data are usually more difficult/expensive to collect and analyse. To record whether a patient survives a year or more after being discharged from hospital is much easier than measuring how long they survived for.

● The decision to count an item as a "pass" or a "fail" can be a personal judgement. Similarly deciding whether an event has occurred may sometimes be left up to the person collecting the data, and each person may have his own interpretation, as illustrated in the case study in Chapter 6. The marking of examination questions is a typical example and examination boards go to great lengths to try to ensure consistency of marking. Whenever a result is left to judgement of an individual (or even a group) there is room for inconsistency. The advantage of a measurement is that it is usually more consistent.

Defects or defectives data?

Referring to the decision chart in Figure 22.1, if we are using attributes data then we next need to know whether we have defects or defectives. The differences between these two types of data are summarised in Figure 22.3.

Defectives	Defects
Whole item classified as having a particular attribute	Number of incidents counted
Examples	Examples
The number of forms completed correctly	Number of errors on a form
Number of surgical operations that went smoothly	Number of problems during a surgical operation
Pass/fail an inspection	Number of flaws in inspected items
3% of all orders are not completed according to contract	There were four non-compliances with the contract
	Often thought of as "rates"
Can have a "negative" event, for example: a patient can get better or not get better, an insurance claim can be paid or not paid, a defendant can be found guilty or not guilty	A "negative" event has no meaning, for example: We know how many car accidents there were, but we do not know how many car accidents did not happen We know how many insurance claims there were but we do not know how many insurance claims there were not
Has an upper limit of 100% "failures"	No upper limit
	Provides more information than defectives
Analysed with a p or np chart	Analysed with a c or u chart

Figure 22.3 Defectives vs. defects

The terminology is unfortunate, and probably comes from the development of these charts in the manufacturing industry. With defectives data we determine whether an item has a particular attribute; that is, if it "passes" or "fails" a test. It may "pass" or "fail" a test for one reason or one of several. For example, a person could pass as being "ill" simply because they have a cold, or because they have a range of conditions (bone fractures hypothermia and brain damage).

Another guide for distinguishing between defects and defectives is that defects can be thought of as a "rate", such as incident or failure rates.

Defective data is binary: it can only have one of two values. An item can pass or fail an inspection, a form can have errors or not have errors, a person can be in paid employment or not in paid employment. In a sense, we have a "positive" or "negative" outcome. However, with defects data a "negative" event has no meaning. We can count the numbers of errors on a form, but we cannot count the numbers of errors that were not on the form; we can count the number of non-compliances during an audit, but we cannot count the number of compliances.

We can contrast defects and defectives by considering completed forms. If the form is incorrectly completed it is defective; however, it may have several defects (e.g. there may be two occurrences of missing information, and one of incorrect information giving three defects in total).

Where there is a choice between collecting defectives or defects data, there are advantages and disadvantages to each:

- Identifying whether a form is defective or not is quick and easy. As soon as we identify the first defect the form is labelled as defective. If we monitor the number of defects we have to keep "inspecting" and keep a count of and log the number of defects.
- There is more information in defects data. For example, when collecting the number of defects, we could log the number of problems with each question/part of the form, this would give us useful information when re-designing and improving the form.

In some situations, there may not be a choice of which type of information to collect. The number of accidents, fire alarms and criminal acts can only be considered as defects data.

The distinction between defects and defectives is important because defective data follow a statistical distribution called a binomial distribution and defects data follow a different distribution called a Poisson distribution, and the formulae for calculating the standard deviation are different.

Equal or variable size samples?

The next level in Figures 22.1 asks whether we have equal sample sizes, and the answer determines the chart that we should use.

In all attributes data, it is important to consider the "opportunity" for an event to be recorded. For example, we may be recording the number of accidents per month in the work place (defects). However, if the number of hours worked varies from month to month, there is less opportunity for an accident to occur when the number of hours worked is reduced. Similarly, the opportunity for rejects decreases if the number of items inspected falls. To account for this we need to "normalise" the data by reporting for example, the number of accidents per thousand hours worked and the number of rejects per 50 items inspected.

Choosing between the p and np chart

Figure 22.4 shows which types of chart to use for different types of attributes data. For defectives charts, if the number of items scrutinised remains the same, then an np chart is used, otherwise the p chart is used. Note that if the number of items remains the same, the p and np chart will give exactly the same conclusions. For defects charts, the c chart is used where the opportunity for an event remains the same and the u chart otherwise.

An example should help to clarify the issue. Consider the spinning of a coin. If we spin a coin 10 times and record the number of heads and keep repeating this experiment we could plot the number of heads in each 10 spins as an np chart. A typical set of results is given below:

Experiment	1	2	3	4	5
Number of spins	10	10	10	10	10
Number of heads	4	5	5	4	6
Proportion of heads	0.4	0.5	0.5	0.4	0.6

Sample	Constant	Varying
Defectives	**np chart** Examples Number of rejected invoices per 30 inspected Number of patients suffering relapses per 50 treated	**p chart** Examples Proportion of rejected invoices per week Number of patients suffering relapses per month
Defects	**c chart** Number of errors in 30 invoices Number of accidents per million hours exposure (hours per month can vary) Number of flaws in a square metre of carpet	**u chart** Number of errors per invoice per week Number of accidents per month (same hours worked each month) Number of flaws per meter of carpet inspected (area inspected changes from carpet to carpet)

Figure 22.4 Charts for defectives and defects

Alternatively, we could allow the number of spins to vary, as in the results below:

Experiment	1	2	3	4	5
Number of spins	10	15	8	16	10
Number of heads	4	5	5	4	6
Proportion of heads	0.4	0.33	0.63	0.25	0.6

Clearly charting the number of heads would be misleading. We need to chart the proportion of heads and to do this we use a p chart.

Suppose the number of spins is normally 20, but occasionally we lose a result so that we have only recorded 19 results, or sometimes we record 21 results by mistake. Is it still necessary to use the more complicated p chart rather than the np chart? What if the number of spins is always 20 ± 1?

Two (slightly different) guidelines have been developed to help decide when we can use the np rather than the more complicated p chart:

1. If the number of items inspected does not vary by more than 25% from the average (i.e. 75 to 125 in the above example) then we can use the np chart.
2. If the maximum number of items inspected is less than 1.5 times the minimum number inspected we can use the np chart.

What we need to consider is the cost in terms of effort to use the p chart weighed against the saving of using the simpler np chart.

As in all cases of doubt, we suggest trying both to determine which should be used on a long-term basis.

Choosing between the c and u chart

As Figure 22.4 shows, there is a similar decision process regarding the c and the u chart, as for the p and np charts. If we are counting the number of accidents per month in an organisation where the workforce is stable, we could compare the numbers directly. However, if the workforce is varying, perhaps due to expansion, we need to take account of the size of the workforce. This is usually done by calculating the number of accidents per, for example, million hours worked. Where the area of opportunity, as it is often called, does not vary we can use a c chart to compare numbers of the event that we are monitoring, and if the area of opportunity varies, we use the u chart.

There is also the same issue regarding the point at which we should use a u chart rather than a c chart, and the guidelines are very similar to those for choosing between the p and np chart, that is:

1. If the area of opportunity (number of hours worked in the example above) does not vary by more than 25% from the average, then we can use the c chart.
2. If the maximum area of opportunity is less than 1.5 times the minimum area of opportunity, we can use the c chart.

The c and the np charts are simplified variants of the u and p charts where the background to the measuring regime remains constant. Therefore, it is always possible to use the u and p charts instead of the c and np charts. The advantages of the c and np charts are that the calculations are simpler and they also do not have variable limits, which add to the complexity when explaining the use of charts to others.

Choosing between X and $\bar{X}$ charts

Referring to the left-hand side of Figure 22.1, the X chart is used when we collect samples one at a time and there is no logical grouping of data. The $\bar{X}$ chart is used when the data can be logically grouped. The choice will depend on what precisely what we are monitoring and often the choice will be clear. Some examples should help to clarify the situation.

- On a production line we may collect five samples every hour. These could grouped to form one sample and we monitor the means ($\bar{X}$).
- An application illustrated in some of the case studies is comparing the number of accidents occurring in different months of the year. If we have 4 years of data, for each of the 12 months, we can compare the average number of accidents each month using an $\bar{X}$ chart.
- Monthly hours of downtime for a piece of equipment, sales and production rates are data that would be monitored using an X chart.

The Rods experiment in Part 5 of this book further explains the relationship between the X and $\bar{X}$ charts.

There are some important points to bear in mind when worth making (Figure 22.5).

- The X chart is much more sensitive than the $\bar{X}$ chart to individual measurement abnormalities and non-normality. This is not surprising since in the $\bar{X}$ chart, for every point plotted we are averaging several pieces of data. It is a statistical fact that when the data are non-normal, then the distribution of means will be closer to a normal

X chart	$\bar{X}$ chart
Charts individual values	Charts means
Less robust to non-normally distributed data	More robust to non-normally distributed data
More sensitive to individual measurement aberrations	Less sensitive to individual measurement aberrations
Standard deviation = s	Standard deviation = $s/\sqrt{n}$
Less sensitive to shifts in process average	More sensitive to shifts in process average

Figure 22.5 Comparison of X and $\bar{X}$ charts

distribution than the individual values. This is proved in the central limit theorem, which we do not discuss in here, but can be found in basic statistics texts.

- If the standard deviation for a set of data, $x_1, x_2, x_3, \ldots, x_n$ is s, then the standard deviation for the mean, $\bar{x}$, is $s/\sqrt{n}$. Therefore, whilst the control limits on an X chart are set at the mean $\pm 3s$, the control limits for the $\bar{X}$ chart will be $\bar{x} \pm 3s/\sqrt{n}$, that is they are closer to the mean. The chart is, therefore, much more sensitive to shifts in process average (see the Section 1 on Average Run Length for more information).
- The application of the chart needs to be considered. If, for example, we are interested more in whether the average is changing we would use the $\bar{X}$ chart, if the individual values are important, the X chart should be used.
- With batch processes, if the batch is believed to be homogenous (perhaps demonstrated by previous experiments) then the X chart is likely to be appropriate. This is particularly important if the cost or difficulty of measurement is high.
- One way of viewing the $\bar{X}$ chart is that it answers the question "Is the variability within a sample small compared to the variability between samples?".

Monitoring the mean and variability

There are two key aspects in a data set that we are interested in monitoring: the mean (or other measure of location) and the variability. For example, if we are monitoring the percentage of active ingredient in a product, we want to know whether the mean level of ingredient is increasing, decreasing, showing some other non-random fluctuation or staying the same. We also need to know if the variability is changing. The average level of ingredient may be consistent at, say, 30%, but the variability may increase from $\pm 0.2\%$ to $\pm 2\%$. To monitor both the mean and variability we usually need two charts because the one can change independently of the other.

When using the X chart to monitor the mean, the MR (moving range) chart is used to monitor the variability, and when the $\bar{X}$ chart is used to monitor the mean, the variability is monitored by either the range or s chart.

However, for attributes charts (c, u, np, p) we only need one chart as it responds to changes both in mean and variability simultaneously. For example, for the c chart, $s = \sqrt{\text{average}}$.

Use of X/MR charts in place of c, u, np and p charts

The variables charts, particularly the X/MR charts, are often used in place of the attributes charts (c, u, np, p). Unfortunately, the published guidelines as to when the X/MR can or should be used vary. Some books suggest that if the average of the data being plotted is greater than 5 then an X/MR chart can be used. Others suggest that the average only has to be greater than 1 and still others (Wheeler) suggest that we should always use X/MR charts unless we know that the data follow the specific statistical distribution that the attributes charts assume. The argument for this last guideline is that the X/MR chart is more robust than the attributes charts to deviations from the assumed distributions. Conversely, Carey has found that in his experience using the X/MR chart resulted in missing special causes that were found on the appropriate attribute chart.

The difference between the c, u, np, p charts and the X/MR chart is that the limits for the attributes charts are based on the theoretical distributions associated with counts data (for the c and u chart this is the Poisson distribution; and for the np and p charts, the binomial distribution). If the data follow these distributions, then the appropriate attribute chart with its limits should be used; however, it is argued by Wheeler that in this case the X/MR chart will mimic these limits. If the data do not follow the theoretical distribution, then the attribute chart limits will be inaccurate. However, because the X/MR limits are empirical and do not rely on these distributions being followed, the X/MR chart will be the better chart to use.

The advantage of using the attributes charts is that only one chart is needed, whereas with the X/MR charts two are needed.

If in doubt, both the attribute and the X/MR charts should be drawn. If there is no discrepancy then we need to give the matter no further thought, and if there is a discrepancy we need to consider carefully why it is there and what it is telling us.

Median and mid-range charts

A median chart is a simple alternative to the $\bar{X}$ chart for monitoring averages. It is particularly useful because once the initial chart and limits have been developed, no further calculations are required, unlike the $\bar{X}$ chart where the mean must be calculated every time samples are taken. In the median chart it is the median, the middle value in an ordered set of data, that is monitored. The main disadvantage is that the median does not take account of the variability of the data. In the example below, the two medians are the same though the mean is different:

Sample	Observed values	Mean	Median
1	12, 14, 16, 18, 20	16.0	16.0
2	14, 15, 16, 20, 22	17.4	16.0

It is also possible, as in sample 1, that the median and mean are the same.

This relatively little-used chart is useful either where calculating means is difficult, or where the data are very unlikely to be skewed (in which case the median and mean are likely to be similar) or if we wish to treat extreme values with suspicion, for example, because the measurement method is liable to error.

One advantage of the median chart is that every point is plotted and so we get a visual impression of the variability of the data.

The median chart is used with odd numbers of samples (e.g. 3, 5, 7) so that the median can be easily identified and highlighted on the chart. If there are an even number of values in the sample then there is no middle value and the value plotted is the average of the two middle values. For example, if the observed values are 10, 13, 15 and 22 then $(13 + 15)/2 = 14$ is plotted.

Another little used alternative to the $\bar{X}$ chart is the mid-range chart. The mid-range is calculated as the average of the highest and lowest values. In sample 2 above, the mid-range is $(14 + 22)/2 = 18$. This chart is likely to be used if we are interested in extreme values.

Median moving range charts

A little used alternative to the moving range chart is the median moving range chart. If the process is liable to extreme values that we are suspicious of, perhaps because the measurement method is liable to error, this will result in inflated moving range values. Inflated moving range values will lead to wide control limits and the risk of missing signals.

A potential solution in this scenario is the median moving range chart. The only difference between this and the moving range chart is that the control limits are calculated from the median of the moving ranges rather than the average of the moving ranges, and so reduces the effect of inflated moving range values.

Difference charts

Difference charts are very useful when we are frequently adjusting the process average. An obvious example comes from short run production runs where we may gather just a few data values before resetting the line to produce a product with different average. The X/MR control chart will only tell us what we already know – that the process average has changed. In these situations we monitor the difference of the measured item from the target value and plot them in time order on an X/MR chart, that is:

$$\text{plotted value} = \text{measured value} - \text{target value.}$$

Difference charts for means can also be used in place of the usual $\bar{X}$ charts in the same way that they are used in place of X/MR charts, that is:

$$\text{plotted mean} = \text{measured mean} - \text{target.}$$

Having calculated the differences, the calculations and the interpretations are the same as for the non-differenced data.

Z charts

Difference charts assume that the variability remains the same when the process average is adjusted. This is likely to be true in some situations, for example, producing different length spindles on a machine are likely to be within a certain tolerance whatever the spindle length. However, in other situations (such as for project delivery vs. actual date) it is likely that variability does change, and in these situations the Z chart can be used. The transformation in this case is:

$$\text{transformed value} = \frac{\text{value} - \text{target}}{s}.$$

This effectively transforms the data to be a normal distribution with a standard deviation of 1. We usually set the control limits at ±3 standard deviations from the mean, and because the standard deviation is 1, the limits are always ±3 units from the mean.

The $\bar{Z}$ charts can also be used in place of the usual $\bar{X}$ charts in the same way that Z charts can be used in place of X/MR charts.

The process for deciding whether to use a difference chart or a Z chart is straightforward: if analysis of the variability (usually by analysing the MR, range or s charts) suggests that the variability does not change, then we can use a difference chart, otherwise we need to use a Z chart.

See the rods experiment for an example.

R or s charts?

When monitoring averages the $\bar{X}$ chart is used and the variability is monitored with R (range) charts. R charts are used because the calculations are easy. In theory the standard deviation, s, chart is a better indicator of process variability, but the calculations are more difficult. Where the sample size is small (some texts recommend nine values) it makes little practical difference which is used, but above nine the s chart is always recommended. If calculations are automated, then the s chart can always be used in place of the R chart. It is worth noting that the s chart is less sensitive than the R chart at identifying special causes of variation when only one of the observations in the sample is the cause of the out-of-control condition.

When the sample size, n, varies the s chart has the great advantage that the calculations are much easier than with a range chart.

Cumulative sum (cusum) charts

Cusum charts can be used for the same data as any of the individuals charts (X, c, u, np, p). They are usually used in addition to other charts and are particularly powerful at identifying changes in mean. Though they are extremely useful, they are more difficult to interpret.

Selecting the most powerful chart

There are situations where we are able to select what data is collected and how it is reported. For example, when monitoring the active ingredient in medication, we could at each sampling:

- take several samples, measure the amount of active ingredient in each sample and chart the results on an $\bar{X}$/R chart;
- take one sample, measure the amount of active ingredient and chart the results on an X/MR chart;
- take several samples, measure the amount of active ingredient in each sample. We could then chart the number of samples outside the specification limits as an np chart if the sample size remains the same, or a p chart if the sample size varies.

The question then arises as to what would be the best-monitoring scheme and chart to use.

Generally we would like to use the chart that will identify an out-of-control conditions soonest after they have occurred. Based on the discussions above, our preferred charts would be:

1. The $\bar{X}$/s would be out first choice, slightly better than $\bar{X}$/R (depending on sample size).
2. The X/MR or similar charts such as the median/MR chart.
3. c or u chart.
4. The np and p are the slowest at identifying process changes.

Summary

In this chapter we have discussed how to select the appropriate chart, with alternative, depending on the type of data that we are analysing.

The variables charts, $\bar{X}$/R, $\bar{X}$/s are the most power at detecting changes in process average and where we have the option of grouping data, we would normally choose one of them. Where we have variables data but it is not grouped, the X/MR or one of its substitutes (e.g. median/R chart) would be our choice.

The attributes charts are used for monitoring data that is counted. If the counts are binomial (e.g. the result can be termed either a "pass" or "fail", "true" or "false") then we use the np if the sample size remains constant, or the p chart if it varies.

If the attributes are counting the number of occurrences, for example the number of errors on a form, or the number of flaws in a square meter of metal, number of accidents, then we use a c chart if the area of opportunity for an event occurring remains the same, or the u chart if it does not.

In addition to these charts, there are a number of lesser-used charts, such as the Z, difference, median which have specific applications.

Finally, the cumulative (cusum) chart is an extremely powerful chart for identifying changes in process average. It is usually used with the X, c, u, np or p charts. The disadvantage of the cusum chart is that it is more difficult both to draw and interpret.

23 Procedures and formula for drawing control charts

Introduction

In Chapter 22 we discussed how to select the appropriate chart for a given set of data. In this chapter we discuss the applications of each of the charts and how to draw them. Whilst there is some overlap between the content in this and the previous chapter, it has the benefit that it is not necessary to keep referring back and forth from one to the other.

In this chapter we give an easy reference step-by-step guide for drawing charts. Additional examples of calculations for most charts are illustrated in the case studies.

There are two terms commonly applied to outer limits of a control chart: action limits and control limits. The subtle difference between them is explained in Chapter 21, and we use them interchangeable throughout this book. Both terms are unfortunate. Some people object to the word "control" because we are not actually controlling these limits, it is the observed values that determine them. The word "action" is objected to because it implies we should take action. Similarly, some people object to the term warning limits, the inner set of limits which are often used on control charts. However, no other terminology is in common use, and inappropriate as they may be we are stuck with them.

The following charts are described in some detail in this chapter:

- $\bar{X}/R$ (range)
- s (standard deviation)
- Median/range
- Difference
- Z
- X/MR (individuals/moving range)
- Moving mean/moving range
- p
- np
- c
- u
- Multivariate charts.

The cumulative sum (cusum) chart is somewhat more complex and an introduction to it is given in the next chapter.

Frequency of measurements

Often the measurement frequency will be determined by some event, such as a daily, weekly or monthly report. When we have the choice of frequency of measurement, there are several issues to bear in mind:

- The aim of process measurement is usually to detect signals in the data that indicate changes in the process over time or single extreme values. Measurement should be taken frequently enough to catch process changes or extreme values that may be occurring in the process. However, if they are taken too frequently they may be auto-correlated (i.e. not independent of each other). This is particularly likely with manufacturing data. There are tests for auto-correlation, which are outside the scope of this book, but a simple scatter diagram such as the one used in the analysis of hospitality suites in Chapter 16 data is a good starting point. Auto-correlation is discussed in a little more detail in Chapter 25.
- The effort and cost of collecting and analysing the data needs to be balanced against the impact of missing a signal.
- To set up a control chart, most texts recommend a minimum of 20–30 data points. By that time the estimate of the mean and standard deviation, and hence control limits, should be quite accurate. This does NOT mean that no useful information can be gleaned from less data, and in some of the case studies in this book useful conclusions were drawn with far fewer, but 20–30 data points is a reasonable aim.
- A useful approach to determine the frequency of data collection is to begin by collection data frequently and reduce the frequency if the process appears relatively stable.

Setting up charts

Before we can set up a control chart we need to have already collected some data. As stated previously, a minimum of 20 values is desirable, but with as few as six data values we can get a rough estimate of the average and control limits. This allows us to start recording data on a chart with a reasonable scale. The more data we start with, however, the less likely it is that we will have to re-draw the chart because the scale is inappropriate. If the charts are being developed using a computer package, scaling and re-drawing the chart will probably not be an issue as the package will probably select and adjust scales as appropriate.

It is recommended that additional information be annotated on the chart to help trace back any queries or to help with future analysis. Information such as time, person recording the data, equipment used for measurements, any unusual events, calculations, etc., could be appropriate.

PART I VARIABLES CHARTS

The $\bar{X}$/R charts

Application

The $\bar{X}$/R charts are used when data can be logically grouped, such as taking a group of four samples from a production line every hour. Grouping should always be in such a way that the variability of data within the group is thought to be less than the variability between groups. For example, the variation between these four samples taken every hour, is likely to be less than the variation from hour to hour.

Other typical applications include:

- Several samples are taken and analysed in a local area, and the results grouped and then compared with results from a different area.
  The area could be:
  - one sheet of glass (or metal or plastic, etc.), different areas would be different sheets of glass (or metal or plastic, etc.).
  - a farmers field or a county and different areas would be different fields or counties.
- Comparing performance of different "analysts" where we take n samples of each analyst's results (the analysts could be chemical, medical, etc.).
- Comparing locations with respect to the carrying out particular tasks. For example, sampling five surgical operations from each of 10 hospitals. The groups are the five operations and we are comparing hospitals to ascertain if there is a difference between hospitals with respect to the time taken to complete the operation.

The $\bar{X}$ chart monitors the mean of these groups and the R chart monitors the variability.

Gathering data

We will have chosen $\bar{X}$/R charts because our data are naturally grouped. However, we may be able to choose the sample size, n and the frequency of measurement. The following should be borne in mind:

- The tacet assumption behind the $\bar{X}$/R chart is that the variation within each sample is small compared with the variation between samples, and to an extent this is what the control chart is testing: is the variability between samples large compared to the variability within samples?
- In general, the larger the sample size, n, the better we will be able to discern differences between groups. Unfortunately, the larger n the more expensive data collection is likely to be. It is difficult to give guidelines that will be appropriate to all situations, but a useful rule of thumb is to start with $n = 4$ or 5. As the sample size increases, the control limits for the chart will narrow, resulting in the chart being more sensitive to process changes.
- Sample sizes should remain the same if at all possible. However, after the control chart is established it may be appropriate to change sample sizes, for example, if the

process is in a state of control or if the variability within samples is very small we may reduce the sample size.

- The process itself may determine the sample size. For example, if there are four processes running in parallel (e.g. four people, groups and machines) then sample sizes of 4 (or a multiplier of 4) are probably appropriate. However, even here care needs to be taken as we need to be aware that there are now three sources of variation:
  - within the people/groups/machines taken at a point in time;
  - between people/groups/machines taken at a point in time;
  - variation over time.

Whilst it is often clear that the $\bar{X}$/R is the appropriate chart to use in a particular situation, the selection of sample sizes and sampling frequency can be difficult. It requires careful thought and perhaps the help of a statistician.

Chart structure (Chart 23.1)

The $\bar{X}$/R chart consists of two parts, the $\bar{X}$ chart and the R chart, and the $\bar{X}$ chart is conventionally drawn above the R chart. As with all charts it is usual to include a table giving the raw data and some other information. This table can be above or below the charts, and in this case consists of the shift number, which identifies when the data was taken, space for comments, shift identifier, day number and the five observations. Below this are the two calculated rows, the mean and range.

It is not usual to show the remaining calculations but are included here for reference and include:

- n, the number of observations per shift;
- $\bar{X}$, the average of the samples (this is also included in the table that accompanies the chart);
- $\bar{\bar{X}}$, the average of all $5 \times 13 = 65$ observations;
- upper action limit (UAL) and lower action limit (LAL) for the $\bar{X}$ chart;
- $\bar{R}$, the mean range;
- UAL and LAL for the range chart.

Setting up and interpreting the chart (Chart 23.1)

The data are from a manufacturing company. There were concerns about the amount of rejected product and an investigation was begun. As a first step it was decided to analyse the data from one production run which lasted $k = 13$ shifts. During every shift five samples were taken, analysed and the results recorded. This chart is one of a series that was used to investigate concerns about the quantity of waste being produced. The investigation, using only the tools explained in this book identified that a key source of variation in the results was due to how different analysts carrying out the analyses.

The steps in setting up the chart are as follows:

1. Complete the "observations" part of the table.

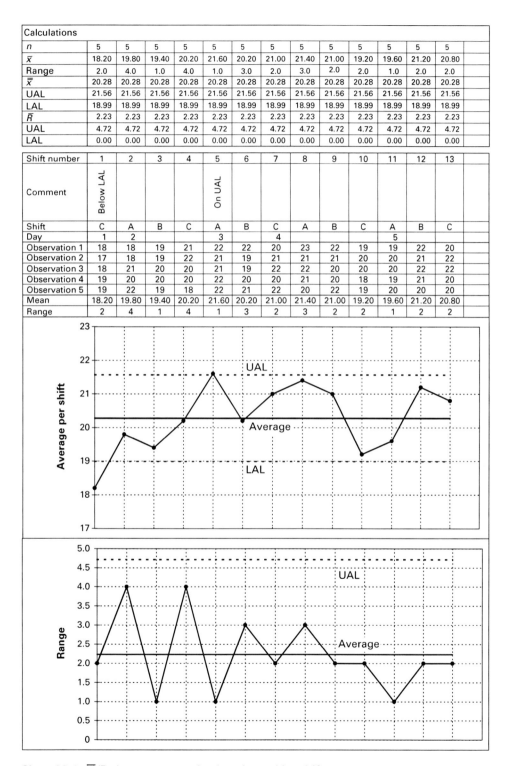

Calculations													
n	5	5	5	5	5	5	5	5	5	5	5	5	5
$\bar{x}$	18.20	19.80	19.40	20.20	21.60	20.20	21.00	21.40	21.00	19.20	19.60	21.20	20.80
Range	2.0	4.0	1.0	4.0	1.0	3.0	2.0	3.0	2.0	2.0	1.0	2.0	2.0
$\bar{\bar{x}}$	20.28	20.28	20.28	20.28	20.28	20.28	20.28	20.28	20.28	20.28	20.28	20.28	20.28
UAL	21.56	21.56	21.56	21.56	21.56	21.56	21.56	21.56	21.56	21.56	21.56	21.56	21.56
LAL	18.99	18.99	18.99	18.99	18.99	18.99	18.99	18.99	18.99	18.99	18.99	18.99	18.99
$\bar{R}$	2.23	2.23	2.23	2.23	2.23	2.23	2.23	2.23	2.23	2.23	2.23	2.23	2.23
UAL	4.72	4.72	4.72	4.72	4.72	4.72	4.72	4.72	4.72	4.72	4.72	4.72	4.72
LAL	0.00	0.00	0.00	0.00	0.00	0.00	0.00	0.00	0.00	0.00	0.00	0.00	0.00

Shift number	1	2	3	4	5	6	7	8	9	10	11	12	13
Comment	Below LAL				On UAL								
Shift	C	A	B	C	A	B	C	A	B	C	A	B	C
Day	1	2			3		4				5		
Observation 1	18	18	19	21	22	22	20	23	22	19	19	22	20
Observation 2	17	18	19	22	21	19	21	21	21	20	20	21	22
Observation 3	18	21	20	20	21	19	22	22	20	20	20	22	22
Observation 4	19	20	20	20	22	20	20	21	20	18	19	21	20
Observation 5	19	22	19	18	22	21	22	20	22	19	20	20	20
Mean	18.20	19.80	19.40	20.20	21.60	20.20	21.00	21.40	21.00	19.20	19.60	21.20	20.80
Range	2	4	1	4	1	3	2	3	2	2	1	2	2

Chart 23.1 $\bar{X}$/R chart: measured value charted by shift

2. Calculate the mean

$$\bar{x} = \frac{x_1 + x_2 + x_3 + \cdots + x_n}{n}$$

for each shift.

In this case $n = 5$, and the mean for the first set of observations:

$$\bar{x}_1 = \frac{18 + 17 + 18 + 19 + 19}{5} = 18.2.$$

The other means, $\bar{x}_2$ to $\bar{x}_{13}$ are entered below the observations.

3. Calculate the range for each shift = maximum − minimum.

For shift 1 this is

$$r_1 = 19 - 17 = 2.$$

The ranges are entered below the means.

4. Calculate the mean range

$$\bar{R} = \frac{r_1 + r_2 + r_3 + \cdots + r_k}{k}.$$

For this data

$$\bar{R} = \frac{2 + 4 + 1 + \cdots + 1 + 2 + 2}{13} = \frac{29}{13} = 2.23.$$

5. Calculate the mean for ALL x values

$$\bar{\bar{x}} = \frac{\text{sum of all observations}}{\text{number of observations}}.$$

For this data

$$\bar{\bar{x}} = \frac{(18 + 17 + 18 + 19 + 19) + \cdots(\cdots 22 + 20 + 20)}{65} = 20.08$$

6. Calculate the control limits for the $\bar{X}$ chart

$$\text{Upper control limit (UCL)}_x = \bar{\bar{x}} + A_2\bar{R}$$
$$\text{Lower control limit (LCL)}_x = \bar{\bar{x}} - A_2\bar{R}.$$

A_2 are constants depending on the sample size, n, given in the tables in the appendix. For this data:

$$\text{UCL}_x = 20.28 + 0.58 \times 2.23 = 21.56$$
$$\text{LCL}_x = 20.28 - 0.58 \times 2.23 = 18.99.$$

If required add the warning limits:

$$\text{Upper warning limit (UWL)}_x = \bar{\bar{x}} + (2/3) A_2 \bar{R}$$
$$\text{Lower warning limit (LWL)}_x = \bar{\bar{x}} - (2/3) A_2 \bar{R}.$$

7. Calculate the control limits for the $\bar{R}$ chart

$$\text{UCL}_R = D_4 \bar{R}$$
$$\text{LCL}_R = D_3 \bar{R}.$$

D_3 and D_4 are constants depending on the sample size, n, given in the tables in the appendix. For this data:

$$\text{UCL}_R = 2.11 \times 2.23 = 4.71$$
$$\text{LCL}_R = 0 \text{ since } D_3 = 0 \text{ for values of } n < 7.$$

If using the British system, the action limits are:

$$\text{UAL}_R = D_5\,\bar{R}$$
$$\text{LAL}_R = D_6\,\bar{R}.$$

And the warning limits are:

$$\text{UWL}_R = D_7\,\bar{R}$$
$$\text{LWL}_R = D_8\,\bar{R}.$$

Values of the constants A_2, D_3, D_4, D_5, D_6, D_7 and D_8 are given in the appendix.
8. Determine the scale of the charts.
   To ensure that the control limits fit on the scale of the chart, determine the maximum of the UAL_x and the highest observed mean, and choose a suitable maximum value above these.
     With this data $\text{UAL}_x = 21.56$ and the maximum mean is 21.60 and we have chosen 22 as the maximum for scaling to give a little space at the top of the chart.
     Similarly, select the lower value for scaling by determining the minimum of the UAL_x and the lowest mean and choose a suitable value below these.
     With this data $\text{LAL}_x = 18.99$ and the maximum mean is 18.20 and we have chosen 15 as the minimum for scaling to give a little space at the bottom of the chart. Repeat for the range chart.
9. Plot the data and add lines for $\bar{\bar{x}}$ and the limits on the $\bar{X}$ chart, and add lines for $\bar{R}$ and the control limits on the R chart.
10. Interpret the R chart. Since interpretation of the $\bar{X}$ chart relies on the constant variability, the R chart is analysed first. The usual guidelines apply; that is, a point outside the control limits, trends, runs and any predictable pattern.
     However, note that as the group size, n, decreases the likelihood of runs below the average increases slightly, so some analysts suggest that for $n < 6$, a run of eight points below the average is required to signal a decrease in process variability.
     Also beware that if the data are not in any logical order (e.g. if we are comparing analysts) the guidelines regarding runs and cycling, etc. become meaningless and are not used.
     In Chart 23.1 there appears to be regular cycling for the first nine values. This is unlikely to be by chance, and should be investigated.
11. Interpret the $\bar{X}$ chart. Once the R chart has been analysed, any out-of-control conditions have been analysed and the data excluded from analyses and the charts redrawn (if appropriate) analyse the $\bar{X}$ chart. The usual interpretation rules apply.
     In Chart 23.1, we have determined that the range chart is not in a state of statistical control, so interpreting the $\bar{X}$ chart is somewhat academic. However, we see that it is also not in a state of control. The first value is below the LAL, the fifth is on or slightly above the UAL. In addition, there seems to be a shortage of points near the average!

Comments

Change of sample size

It is a statistical fact, and intuitive, that as the number of observations in a sample increases, the range will increase. Consider recording dice throws. The maximum range is $(6 - 1) = 5$. If the dice is only thrown twice the chance of obtaining a range of 5 is small (it is actually equal to the probability of throwing a 1 followed by a 6 or a 6 followed by a $1 = (1/6 \times 1/6) + (1/6 \times 1/6) = 1/18$. However, if we were to throw the dice 20 times we would be very surprised if we did not throw at least one 1 and one 6.

Since the control limits are based on the ranges, and the ranges are dependent on the sample size, if the sample size changes we need to recalculate the control limits.

This situation would occur, for example, if a process was in a state of control and it was decided to reduce the number of observations per sample. The standard deviation for the new data can still be based on the historic data using the formula:

$$s' = \bar{R}/d_2$$

where d_2 is read off the tables in the appendix for the previous value of n. $\bar{R}$ is then recalculated using

$$\bar{R} = s'd_2$$

where d_2 is read off the tables in the appendix for the new value of n.

For example, if in the current example we reduced the sample size from 5 to 3.

$$s' = 2.23/2.33 = 0.96.$$

And the new $\bar{R}$ is calculated as:

$$\bar{R} = 0.96 \times 1.69 = 1.62.$$

The limits are then recalculated using the usual formula, and the line representing $\bar{R}$ reduced to 1.62 from where the new measurement system begins.

Varying sample size (Chart 23.2)

In Chart 23.1, the sample size, n is the same for each sample. In reality, the number of samples varied widely from one sampling to the next. The complete data and chart are given in Chart 23.2.

The organisation of the chart is the same as for Chart 23.1, except that the individual observations are placed in a separate table at the top of the chart for easy reference. A simple way of charting data using the procedure above is:

1. Calculate the average number of observations, $\bar{n} = 10.9$. This is close to 11, and 11 will be used in calculation.
2. For each of the $k = 13$ samples, estimate what R would have been if 11 measurements had been taken.

The new range based on $\bar{n}$ = the old range,

$$R \times \frac{d_{22}}{d_{21}} = 3.0 \times \frac{3.17}{2.97} = 3.2$$

	1	2	3	4	5	6	7	8	9	10	11	12	13	
Observation 1	18	18	19	21	22	22	20	23	22	19	19	22	20	
Observation 2	17	18	19	22	21	19	21	21	21	20	20	21	22	
Observation 3	18	21	20	20	21	19	22	22	20	20	20	22	22	
Observation 4	19	20	20	20	22	20	20	21	20	18	19	21	20	
Observation 5	19	22	19	18	22	21	22	20	22	19	20	20	20	
Observation 6	18	22	19	18	24	20	21	19	22	20	20	20		
Observation 7	19	22	20	18	24	20	24	21	20	21	20	22		
Observation 8	18	21	20	19	18	22	22	21	21	18	21	24		
Observation 9	20	20	21	20	20	22	22	22	21	21	18	21		
Observation 10		21	20	19	19	22	21	21	20	18	20	20		
Observation 11			21	21	20	23	22	20		20	19	22		
Observation 12			19	19	22		22	21		19		19		
Observation 13										19				
Observation 14										21				
n	9	10	12	12	12	11	12	12	10	14	11	12	5	
$\bar{n}$	10.9	10.9	10.9	10.9	10.9	10.9	10.9	10.9	10.9	10.9	10.9	10.9	10.9	
$\bar{x}$	18.44	20.50	19.75	19.58	21.25	20.91	21.58	21.00	20.90	19.50	19.64	21.17	20.80	
Range	3.0	4.0	2.0	4.0	6.0	4.0	4.0	4.0	2.0	3.0	3.0	5.0	2.0	
Range based on $\bar{n}$	3.2	4.1	1.9	3.9	5.8	4.0	3.9	3.9	2.1	2.8	3.0	4.9	2.7	
$\bar{\bar{x}}$	20.39	20.39	20.39	20.39	20.39	20.39	20.39	20.39	20.39	20.39	20.39	20.39	20.39	
UAL	21.51	21.45	21.36	21.36	21.36	21.40	21.36	21.36	21.45	21.29	21.40	21.36	21.89	
LAL	19.26	19.32	19.42	19.42	19.42	19.37	19.42	19.42	19.32	19.49	19.37	19.42	18.88	
$\bar{R}$	3.56	3.56	3.56	3.56	3.56	3.56	3.56	3.56	3.56	3.56	3.56	3.56	3.56	
$\bar{R}$ revised	3.33	3.45	3.65	3.65	3.65	3.56	3.65	3.65	3.65	3.45	3.82	3.65	3.65	
UAL	6.05	6.13	6.27	6.27	6.27	6.21	6.27	6.27	6.13	6.39	6.21	6.27	5.51	
LAL	0.61	0.77	1.03	1.03	1.03	0.91	1.03	1.03	0.77	1.25	0.91	1.03	0.00	

Number	1	2	3	4	5	6	7	8	9	10	11	12	13
Comment	Below LAL						Above UAL			Below LAL			
Shift	C	A	B	C	A	B	C	A	B	C	A	B	C
Day	1	2		3			4			5			
Mean	18.44	20.50	19.75	19.58	21.25	20.91	21.58	21.00	20.90	19.50	19.64	21.17	20.80
Range	3	4	2	4	6	4	4	4	2	3	3	5	2

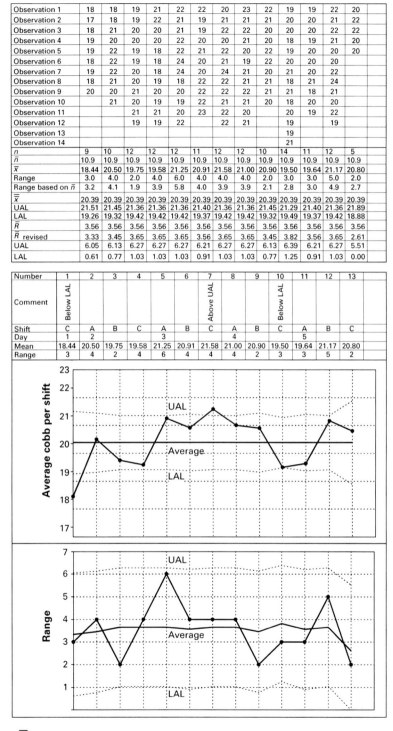

Chart 23.2 $\bar{X}$/R chart: measured value charted by shift – all data

for the first sample, where d_{21} is the value of the constant d_2 for $n = 9$ (i.e. the actual number of observations in the first sample) and d_{22} is the value of the constant d_2 for $n = 11$ (i.e. the average number of observations). Note that where $n < \bar{n}$ the range based on $\bar{n}$ will increase, where $n = \bar{n}$ it will be the same and otherwise it will decrease.

3. Calculate $\bar{R}$ based on the revised R values. This is the usual calculation for means and is 3.56 for this data.
4. Calculate a revised $\bar{R}$ for each sample to take account of the varying sample sizes.

$$\bar{R} \text{ revised} = \bar{R} \times \frac{d_{21}}{d_{22}} = 3.56 \times \frac{2.97}{3.17} = 3.33$$

for the first sample, where d_{21} and d_{22} have the same values as in step 2 above.

Charting and interpretation now continues as normal.

In this case $\bar{n} = 10.9$ which is close to 11 and this was the value used to look up d_2 in the tables. When n and $\bar{n}$ are large, rounding makes little difference to the results. However, as n or $\bar{n}$ falls below about 7 we may decide to interpolate between the two nearest values. This could be done either by interpolating a (non-linear) line to the data, or, more simply by calculating a ratio. For example, if

d_2 for $n = 3 = 1.693$
d_2 for $n = 4 = 2.059$

so a value of d_2 for $n = 3.3$ could be estimated as $1.693 + 0.3 \times (2.059 - 1.693) = 1.800$.

Such changes may often have minimal effect on the interpretation of the chart, and, particularly if sample sizes do not vary much, it may be reasonable to assume all sample sizes are the same, at least for an initial interpretation.

The s (standard deviation) chart

Application

The s (standard deviation) chart can be used in place of the R chart when sample sizes rise above about 9 or if calculations are automated or particularly if the sample size varies from group to group. The R chart is usually used for small sample sizes or when calculations are manual.

The process is the same for the s chart as for the R chart, but some of the formulae are different.

In place of calculating R for each group of observations, we calculate and chart the standard deviation (s), with the formula:

$$s = \sqrt{\frac{\sum (x_i - \bar{x})^2}{n - 1}}$$

where x_i are the individual observations, $\bar{x}$ is the average for the group and n is the sample size.

The control limits for the $\bar{X}$ chart are:

$$UCL_X = \bar{\bar{x}} + A_3 \bar{s}$$
$$LCL_X = \bar{\bar{x}} - A_3 \bar{s}.$$

And for the s chart:

$$UCL_s = B_4 \bar{s}$$
$$LCL_s = B_3 \bar{s}.$$

If using the British system, the action limits are:

$$UAL_s = \bar{\bar{x}} + B_5 \bar{x}$$
$$LAL_s = \bar{\bar{x}} - B_6 \bar{s}.$$

And the warning limits are:

$$UWL_s = \bar{\bar{x}} + B_7 \bar{R}$$
$$LWL_s = \bar{\bar{x}} - B_8 \bar{R}.$$

Values of the constants A_3, B_3, B_4, B_5, B_6, B_7 and B_8 are given in the appendix, where $\bar{s}$ is the average of the individual sample standard deviations. The average value of $\bar{s}$ may be estimated by the usual formula:

$$\bar{s} = \sqrt{\frac{(n_1 - 1)S_1^2 + (n_2 - 1)S_2^2 + \cdots + (n_k - 1)S_k^2}{(n_1 + n_2 + \cdots + n_k - k)}}.$$

There are examples of the s chart in the case studies, and none are reproduced here as they are similar to the R chart and it is likely that only automated s charts would be used.

The median/R chart (Chart 23.3)

Application

The median chart is a simple to use alternative to the $\bar{X}$ chart. It would usually be used by those not wanting to carry out the calculations required for the $\bar{X}$ chart, or for the reasons discussed in Chapter 22. It also has the advantage that it can be used without the R chart (though the R chart is included here) as each value in each sample is plotted. However, it is not as efficient as the $\bar{X}$/R or $\bar{X}$/s charts at identifying out-of-control conditions. It is normally only used when group sizes are less than about 10.

The charting procedure is similar to that for the $\bar{X}$/R charts with the following exceptions (Chart 23.3):

- Each individual point is plotted (in Chart 23.3 they are shown as dots).
- The median, $\tilde{X}$ is plotted and highlighted in some way (in Chart 23.3 it is plotted as a square). Where there are an odd number of values, the median is the middle value; where there is an even number of values, the median is calculated as the average of the middle two values.
- The average median, $\bar{\tilde{X}}$ is calculated and plotted as the average on the median chart.
- The limits for the median chart are:

$$UCL_X = \bar{\tilde{X}} + A_4 \bar{R}$$
$$LCL_X = \bar{\tilde{X}} - A_4 \bar{R}.$$

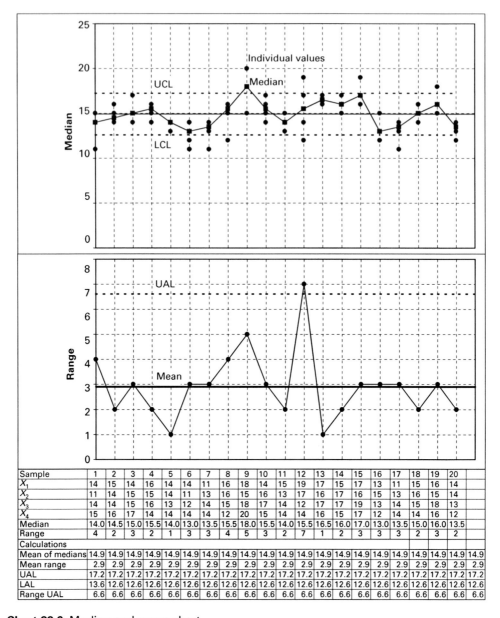

Chart 23.3 Median and range chart

The warning limits can be set at:

$$\text{UWL}_x = \overline{\overline{X}} + (2/3)A_4\,\overline{R}$$
$$\text{LWL}_x = \overline{\overline{X}} - (2/3)A_4\,\overline{R}$$

and for the range chart:

$$\text{UAL}_R = D_4\,\overline{R}$$
$$\text{LAL}_R = D_3\,\overline{R}$$

where A_4, D_3 and D_4 are constants given in Appendix A.

Difference charts

In some situations we know that the mean is changing. A typical example comes from manufacturing where machines are setup for short runs and only a few measurements are taken before re-setting for a different product. The ultimate changing process could be considered to be project work where every project is different. For example, installing office telephone systems or treating individual patients with different needs. In these cases it is possible to monitor the difference between the observed outcome and the planned outcome by using the formula:

$$\text{difference} = \text{observed valued} - \text{planned value}$$

or

$$\text{difference} = \frac{\text{observed} - \text{planned}}{\text{planned}},$$

which could be multiplied by 100 to convert to a percentage.

The X/MR chart of the differences is plotted and interpreted in the normal manner. However, in this case there are likely to be some negative figures and negative control limits should be drawn.

Difference charts are applicable to both $\bar{X}$/R (or s charts) and X/MR charts.

The assumption behind difference charts is that the variability does not change with the planned value. In many cases this will be a reasonable assumption. However, if it is not, as perhaps would be the case for project work where larger projects may have larger variances than shorter ones, we need a Z chart. The relationship between variability and planned value can be investigated using a scatter diagram.

An example of the difference chart is given in the Rods Experiment in Part 5.

Further information and examples can be found in Wheeler's excellent book Short Run SPC.

Z charts

Z charts are used in place of difference charts where the variability is not constant. They plot values of Z where

$$Z = \frac{\text{observed value} - \text{planned value}}{\text{standard deviation}}.$$

That is, it converts the raw data into a normal distribution with a standard deviation of 1.

The standard deviation is easy to calculate when applied to $\bar{X}$/R charts as the standard deviation can be calculated for each group. When applied to X/MR data the standard deviation is calculated in the normal way as $\bar{R}/1.128$, as described in the next section.

Further information and examples can be found in Wheeler's excellent book Short Run SPC.

X/MR charts

Application

"X" charts, also known as individuals charts, are used for monitoring averages of data from processes when sampling groups is not appropriate. Under many circumstances they can also be used for monitoring counts data. The MR chart monitors the variability. The X and MR charts are usually used together and for this reason are treated as if they were one chart. In most situations it is useful to be aware of the distribution of the data: firstly, to be aware of any deviations from normality and secondly, as it may give hints to causes of variation in the data, as seen for example in the first case study in Chapter 7. Some charts are designed to include a histogram for analysing the distribution, as shown in Chart 23.4. The histogram is shown above the calculations and is drawn with the bars horizontal so that it is best viewed by turning the chart through 90 degrees.

The X/MR chart is:

- less sensitive at identifying out-of-control conditions than the $\bar{X}$/R charts,
- far less robust to non-normality of data than $\bar{X}$/R charts.

However, it:

- has a wider application than $\bar{X}$/R charts,
- is very simple to set up and use.

Typical applications include:

- Any measured value where measures are taken one at a time.
- Equipment downtime/non-productive time/availability.
- Time to complete tasks (e.g. process invoices, register patients, complete a design, produce a report/plan, respond to queries/orders).
- Delays in time and or/related costs due to, for example, materials shortages, late delivery, breakdowns, illness.
- Quotients.
- Sales, production rates.
- Physical characteristics of products from a production line (where sampling is not appropriate).
- Examination marks.

Chart structure (Chart 23.4)

X/MR charts are usually structured with the X chart above the MR chart and the data and information tables underneath. In this case, as is common with many charts, there is space on the chart for calculations. Some charting packages also have a facility for drawing a histogram alongside the chart.

291

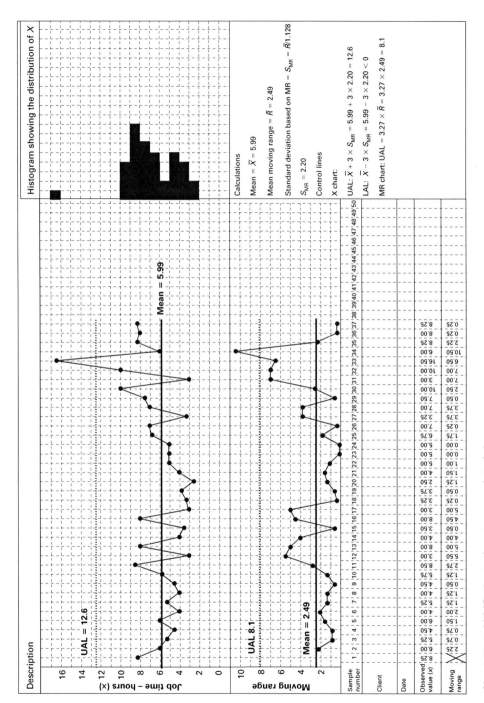

Chart 23.4 X/MR chart: time to complete jobs

Setting up and interpreting the chart

The data in Chart 23.4 are the times required to complete packages of similar jobs for clients. The company was concerned that the profitability of projects seemed to vary greatly and as part of the analysis the time to complete the tasks was plotted.

The steps for drawing an X/MR chart are:

1. Enter the individual data values, x_i on the chart.
2. Calculate the moving ranges.

The moving range corresponding to the ith observation is $x_i - x_{i-1}$.

Note that there is no moving range for the first observation.

The sign of the moving range should be ignored. For example, if the moving range is -2.8, then record 2.8. This is because we are only interested in the size of the difference.

3. Calculate the mean of all the individual data values,

$$\bar{x} = \sum_{i=1}^{n} \frac{x_i}{n} = \frac{8.25 + 6.00 + 5.25 + \cdots + 8.25}{37} = 5.99.$$

4. Calculate the mean of the moving ranges $\bar{R}$.

For this data the average $= \dfrac{2.25 + 0.75 + \cdots + 0.25}{36} = 2.49.$

5. Calculate the standard deviation based on the mean moving range,

$$S_{MR} = \frac{\bar{R}}{1.128} = \frac{2.49}{1.128} = 2.2$$

(1.128 is Hartley's constant to convert the mean moving range to an estimate of the standard deviation).

6. Calculate the control limits for the X chart as:

$$UAL_x = \bar{x} + 3 \times S_{MR} = 5.99 + 3 \times 2.20 = 12.6$$
$$LAL_x = \bar{x} - 3 \times S_{MR} = 5.99 - 3 \times 2.20 < 0.$$

Note that sometimes the lower limit will be less than zero, in these cases the limit is ignored (except in the difference chart where both values and limits can be less than zero as described above).

For the moving range chart there is only an upper limit which is calculated as:

$$UAL_{MR} = 3.27\bar{R} = 3.27 \times 2.49 = 8.1.$$

There is no lower limit for moving range charts.

7. Determine the scale of the charts, taking into account the action limits and the maximum and minimum values. Scaling limits for the moving range are chosen in the same way.
8. Plot the x values and moving ranges, and draw lines representing averages and control limits.
9. If required draw a histogram to show the distribution of x.
10. Interpret the chart. Interpret the moving range chart first to ensure that the variability is in control. However, the moving ranges are non-normally distributed and

auto-correlated (because successive moving ranges have a value in common). For these reasons the guidelines for interpreting runs are different. Some sources suggest that 14 points below the mean are required before a process change is indicated. Next interpret the X chart and histogram using the usual guidelines. However, because of the relative insensitivity of the X chart, it is useful to include the warning limits on the chart and some texts recommend drawing additional horizontal lines at ±1 standard deviations from the mean, and if any four of five consecutive points fall outside the 1s limits, a special cause of variation is suspected.

Chart 23.4 is clearly not in a state of control:

- One outlier in both the MR and X charts at observation number 33.
- Three very high moving ranges at observations 30–32.
- A disproportionally high number of moving ranges above the average moving range from observations 11–17.
- Eight consecutive values below the average on the X chart beginning with observation 17 (the reader might like to recalculate the average excluding observation 33 to determine whether these eight points are still below the average).
- The histogram has two peaks and an outlier, and is definitely not normally distributed.

The fact that there are two peaks in the histograms suggests that there may be two different processes. One theory is that the "similar jobs" are not actually similar. We could review the jobs to identify whether there are any obvious links between the jobs and the timings.

Comments

The usual method of calculating the standard deviation is not used here because it is susceptible to outliers and non-normality. Instead the standard deviation is derived from the moving range.

If the moving range has some very high values, these will inflate the estimate of the standard deviation and hence the control limits will be too wide. A possible solution is to use a median moving range chart. To decide whether there is a problem, calculate the standard deviation using both methods and compare the results.

Moving mean/moving range charts

There are procedures for charting moving means (averages) and ranges. The data (x values) are collected and entered onto an X/MR chart in the normal way and one additional row is added to calculate the moving mean. The span, n, of the moving mean is determined and the means of n consecutive values calculated as shown in the table below using as an example span of 4, and the range over the same span is entered in the moving range row.

Observation	3	5	4	4	6	5	3	etc.
Moving mean				4	4.75	4.75	4.5	
Moving range				2	2	2	3	

For $n = 4$, the first $(n - 1) = 3$ moving means and ranges are blank. The first moving mean is the average of 3, 5, 4, 4 = 5, and the first moving range is the maximum minus the minimum of these observations, that is, $5 - 3 = 2$.

The argument for using such charts is that by charting the average of several consecutive values we will "calm" the chart, thus we will not react to single individual results and this should help identify trends. However, selecting the span and interpreting such charts is not easy; and the risks have already been demonstrated. If you must use these charts, it is strongly suggested that a competent statistician be involved. However, my own experience suggests that cusum charts are a better alternative.

PART 2 ATTRIBUTES CHARTS

Chart structure

All attributes charts have the same basic structure. Only one chart is required to monitor both mean and the variability. Conventionally the data are included in a table below the chart, and often there is space on the chart for calculations.

Since attribute data are integers, interpretation around the limits is a little less clear cut. For example, if the UAL is 6.01 and an observation of 6 is received this will cause more concern than if the UAL were 6.99.

p charts

Application

p charts are used to analyse proportion of items or events that fall into a specific category. Typical applications include:

- Proportion of rejects/passes/failures after an inspection (e.g. of products and forms).
- Proportion of late deliveries/arrivals/payments.
- Proportion of jobs that are vacant.
- Proportion of rejects due to wrong item delivered, damaged item, no longer required, etc.
- Proportion of patients suffering infection.

Gathering data

For convenience, and in keeping with tradition, we use the following terminology: data are collected in samples of n items; each item is inspected and those that fall into the category of interest are called rejects. In addition to the usual requirements of data collection, there are several recommendations regarding the p (and np) charts:

- Sample sizes should be large, >50, to be able to detect even moderate shifts in mean.
- The average proportion of rejects, $n\bar{p}$ should be >5.

- Whilst variation from sample to sample is allowed, the smaller the sample size the slower the chart will be to identify out-of-control conditions. For this reason, whenever possible sample sizes should vary as little as possible.
- The frequency with which samples are inspected should make sense from the process point of view, for example, proportion of patients suffering infection after surgery per week or per month.
- The "reject" rate should not vary during the sampling. For example, if rejects are being collected shift by shift, than the reject rate should not vary during the shift.

Setting up and interpreting the chart (Chart 23.5)

The data are rejects of supplied pipes on site and were collected as part of a study to investigate causes of rejects. As inspections were also carried out when the pipes were in the foundry it was possible to trace each item back and confirm that the reject was due to damage during transportation and/or storage. The proportion of rejects was reported on a monthly basis:

1. At the end of each month, the number of items inspected and the number of rejects are recorded. The data are shown in Chart 23.5. There is little choice over the number inspected as every pipe is inspected but the number required each month varies widely from around 200 to nearly 2000.
2. The proportion of rejects is calculated and entered on the chart. For the pipe rejects, the percentage is recorded. Failure rates are rather low at an average of under 1%.
3. Calculate the mean failure rate, $\bar{p}$:

$$\bar{p} = \frac{\text{total number of rejects}}{\text{total number inspected}} = \frac{34}{2240} = 0.0071.$$

4. Calculate the average sample size, $\bar{n}$:

$$\bar{n} = \frac{\text{total number inspected}}{\text{total number of samples}} = \frac{11,620}{14 \text{ months}} = 830 \text{ per month.}$$

5. Calculate the standard deviation (s) for each sample:

$$s = \sqrt{\frac{\bar{p}(1 - \bar{p})}{n}}.$$

For sample no. 9 which has the smallest sample size, $n = 195$, and hence will have the largest value of s:

$$s = \sqrt{\frac{0.0071(1 - 0.0071)}{195}} = 0.0060.$$

Note that the standard deviation depends on n. The smaller n is, the larger is the value of s.

6. Calculate the control limits:

$$\text{UAL} = \bar{p} + 3s$$
$$\text{LAL} = \bar{p} - 3s.$$

Description	In the 6 months prior to the chart there were
Rejects of inspection on site.	34 rejects from 2240 inspected Failure rate = 34/2240 = 0.0152 = 1.52% From June year 1 on there were 82 rejects from 11,620 inspected Reject rate = 82/11,620 = 0.0071 (i.e. 0.71%) Since there is a significant improvement in reject rate from 1.52% to 0.71%, the earlier data has been ignored

Calculations

		UAL	LAL
$\bar{p}$ = average rejection rate = .0071	$622 < n < 1037$	0.0158	
$\bar{n}$ = average sample size = 11,620/14 months = 830	$375 < n < 623$	0.0183	
$s = \sqrt{\bar{p}(1-\bar{p})\,\bar{n}} = \sqrt{0.0071(1-0.0071)/830} = 0.0029$	$n = 319$	0.0211	
only if $622 < n < 1037$	$n = 195$	0.0250	
Otherwise use specific values of n	$1036 < n < 1725$	0.0138	0.0003
Limits are $\bar{p} \pm 3s$	$n = 1785$	0.0130	0.0011

$$\text{UAL} = 0.0071 + 3 \times 0.0029 = 0.0158$$
$$\text{LAL} = 0.0071 - 3 \times 0.0029 < 0$$

Comment			Year 1						Year 2											
Month	J	J	A	S	O	N	D	J	F	M	A	M	J	J						
Number rejects (x)	7	13	1	8	6	5	2	6	4	3	13	6	7	1						
Number inspected (n)	551	1785	443	1121	680	674	675	849	195	319	896	777	1459	1196						
Reject rate (%) $100*(x/n) = p$	1.27	0.73	0.23	0.71	0.88	0.74	0.30	0.71	2.05	0.94	1.45	0.77	0.48	0.08						
Sample number	1	2	3	4	5	6	7	8	9	10	11	12	13	14	15	16	17	18	19	20

Chart 23.5 p chart: rejects on site

In Chart 23.5 taking the sample with $n = 195$ and $s = 0.0060$ the limits are:

$$\text{UAL} = 0.0071 + 3 \times 0.0060 = 0.025$$
$$\text{LAL} = 0.0071 - 3 \times 0.006 < 0.$$

That is, it is possible that there will be no rejects with this sample size.

Since the chart is plotting reject rates as a percentage, these limits need to be multiplied by 100. Warning limits can be added at $\bar{p} \pm 2s$.

7. Choose the scale of the chart, taking into account the highest and lowest values and the control limits.
8. Plot the data and the control limits.
9. Interpret the chart. All the usual guidelines apply. Chart 23.5 is in a state of control. However it has some interesting features which illustrate the importance of taking the sample size into account when analyzing this type of data:
   - Sample 9 is very high and if we had not calculated the UAL separately for this point we would have concluded that it was abnormally so. However, with only 195 pipes inspected, we would need a reject rate of 2.5% which equates to five rejects before the point would exceed the UAL.
   - Samples 2 and 11 have the most number of rejects, 13. However while the reject rate is average for sample 2, for sample 11 is very close to the UAL. Also note that month 1 has seven rejects, which is just over half the number of rejects of month 2, but the reject rate for month 2 is lower than for month 1. This amplifies the importance of monitoring reject rates rather than number of rejects.

Also note that although there are only 14 samples, the chart is already providing useful information.

Comments

If the calculations are being done by hand there are a number of short cuts:

1. The control limits change with every value of n because s is a function of n. However, the rate of change of the limits is relatively slow so a common guideline for reducing calculations is to calculate the control limits for $\bar{n}$ and use these limits for values of n within 25% of $\bar{n}$, that is, for $0.75\,\bar{n} < n < 1.25\,\bar{n}$ the same limits can be used.

   In the worked example, $\bar{n} = 830$ therefore for
   $0.75 \times 830 < n < 1.25 \times 830$,   that is, $622 < n < 1037$,
   we can use the limits as calculated for $n = 830$. This gives an UAL of 0.0158 and a LAL of 0. If any points are very near the limits, the exact limit based on n can be calculated.

   Similarly it would be possible to calculate limits for other ranges of n to further reduce time spent on calculations as shown on the chart.

Apart from the time saving, another advantage of calculating limits for bands of n is that the person completing the chart can be told to select the limit corresponding to a value of n and does not have to carry out the calculations themselves.

2. Calculate the value of n for which the LAL is zero, that is,

$$s = \bar{p} - 3\sqrt{\frac{\bar{p}(1 - \bar{p})}{n}} = 0 \quad \text{that is,} \quad \bar{p} = 3\sqrt{\frac{\bar{p}(1 - \bar{p})}{n}}.$$

Squaring both sides gives:

$$\bar{p}^2 = \frac{9\bar{p}(1 - \bar{p})}{n}.$$

Multiplying both sides by n and dividing by $\bar{p}^2$ gives:

$$n = \frac{9(1 - \bar{p})}{\bar{p}}.$$

For any sample size larger than $(9(1 - \bar{p})/\bar{p})$ the LAL = 0 and does not need to be calculated. In addition, for any sample where the reject rate, p, is less than the average, $\bar{p}$, the upper limit does not need to be calculated (as the upper limit cannot be less than the average).

For Chart 23.5 this gives:

$$n = \frac{9(1 - 0.0071)}{0.0071} = 1259.$$

Therefore, for any occasion where $n < 1259$ the LAL will be zero and do not need to be calculated. On Chart 23.5 this applies to samples 2 and 13.

If the failure rate for any month $< \bar{p}$ (i.e. 0.71%) then we do not need to calculate the upper limit either, and similarly when $p > \bar{p}$, it is not necessary to calculate the lower limit.

In summary, in order to minimise the number of manual calculations:

1. Calculate the limits for the values of n within 25% of $\bar{n}$.
2. For each point near the limit recalculate the limit for that point based on the appropriate value of n.
3. For values of n outside the $0.75 < \bar{n} < 1.25$ rule, determine which if any limits need to be calculated and add them to the chart.

The calculation section on Chart 23.5 gives limits for bands of n between 357–623 and 1036–1725.

np charts (Chart 23.6)

Application

np charts are used in exactly the same applications as the p chart. The differences are:

- The charted figure is the number of rejects, not the proportion.
- The sample size should remain constant from sample to sample. (In reality some variation is allowed, some texts recommend that all sample sizes should always be within 25% of the average sample size.)

As a result, the calculations for the np chart are much simpler and for this reason it is preferred where sample sizes are constant.

Setting up and interpreting the chart (Chart 23.6)

The setting up and interpretation of the np chart mirrors the p chart, with the obvious simplification that no allowance needs to be made for different sample sizes. As an example, Chart 23.6 is the np chart of the number of invoices paid late from samples of 50 taken each week.

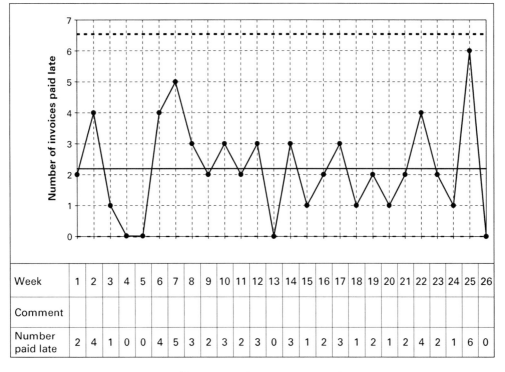

Week	1	2	3	4	5	6	7	8	9	10	11	12	13	14	15	16	17	18	19	20	21	22	23	24	25	26
Comment																										
Number paid late	2	4	1	0	0	4	5	3	2	3	2	3	0	3	1	2	3	1	2	1	2	4	2	1	6	0

Chart 23.6 np chart: number of invoices paid late, sample size 50

The calculations for the np chart are different to the p chart and are as follows:

$$n\bar{p} = \frac{\text{total number rejects}}{\text{number of samples}}.$$

$$n\bar{p} = \frac{57 \text{ invoices paid late}}{26 \text{ weeks}} = 2.2 \text{ invoices paid late per 50 inspected.}$$

The standard deviation is calculated as:

$$s = \sqrt{n\bar{p}\left(1 - \frac{n\bar{p}}{n}\right)} = \sqrt{2.2\left(1 - \frac{2.2}{50}\right)} = 1.45 \text{ for Chart 23.6.}$$

The control limits are calculated in the usual way as

$$\text{UAL} = n\bar{p} + 3s = 2.2 + 3 \times 1.45 = 6.6$$
$$\text{LAL} = n\bar{p} - 3s = 2.2 - 3 \times 1.45 < 0.$$

Warning limits can be added at $n\bar{p} \pm 2s$.

Whilst Chart 23.6 shows no points above the UAL or runs above/below the average, there is a lot of cycling between weeks 8 and 16.

c charts

Application

The c charts are used to monitor the number of occurrences of an event. It requires that the opportunity for events remains the same from observation to observation. Its use is best illustrated by examples of typical applications which include:

- Number of accidents per week/month where the number of people exposed to accidents remains the same.
- Number of flaws in an inspected item (e.g. carpet and bubbles in a pane of glass) where the amount inspected remains the same.
- Number of defects in an inspection lot. (Note that the np and p charts simply determine if an item passes or fails, but do not take into account that there could be several defects on each item. If we count the number of defects we use a c chart in place of an np chart, and an u chart in place of a p chart.)
- Number of failures/breakdowns/alarms.
- Number of non-conformances during an audit.

As c and u charts collect more information than a p or np chart, where there is an option they are preferred.

The c chart assumes that:

- Events occur independently of each other.
- The probability of an event occurring is proportional to the area of opportunity. (For example, when counting the number of errors in a report, if we double the number of pages we check, we expect to double the number of errors found.)

Setting up and interpreting the chart (Chart 23.7)

The procedure for setting up and interpreting c charts is similar to the p chart. Chart 23.7 shows the number of a specific type of loss incident, c, recorded on a monthly basis over $n = 35$ months. The formulae for the c chart are very simple:

$$\text{The average } \bar{c} = \frac{c_1 + c_2 + \cdots + c_n}{n} = \frac{60 \text{ incidents}}{35 \text{ months}} = 1.71 \text{ incidents per month}.$$

$$\text{The standard deviation, } s = \sqrt{c} = \sqrt{1.71} = 1.31$$

The control limits are:

$$\text{UAL} = \bar{c} + 3s = 1.71 + 3 \times 1.31 = 5.64$$
$$\text{LAL} = \bar{c} - 3s = 1.71 - 3 \times 1.31 < 0.$$

Warning limits can be added at $\bar{c} \pm 2s$.

The usual interpretation guidelines apply. Chart 23.7 has 2 months out of control: 11 and 16. Apart from that there are no other obvious out-of-control signals. The next step would be to investigate the causes of these high months, and recalculate the mean and limits excluding these months. The reader might like to do this as an exercise.

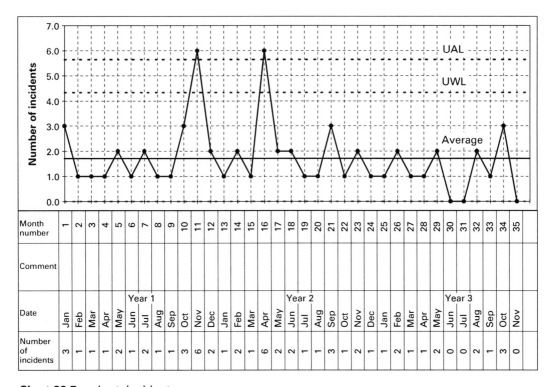

Chart 23.7 c chart: incidents

u charts

Application

The u chart is used in the same situations as the c chart when the opportunity for events varies from observation to observation. Examples include:

- Number of incidents per million exposure hours worked per week/month, where the exposure hours vary.
- Number of flaws per unit inspected, when the number of units inspected varies
- Number of defects per lot where the lot size varies.

Setting up and interpreting the chart (Chart 23.8)

The procedure for setting up and interpreting u charts is similar to the p chart. In the analysis for Chart 23.7 it was assumed that the exposure hours during which a loss could occur were the same for each month. This was not, in fact, the case as the number of hours varied greatly, which changes the chart markedly. We have re-plotted the data as a u chart taking into account the exposure hours.

In Chart 23.8, the number of incidents per month, c, and the number of exposure hours, n, in millions are both shown. The incidents per million man hours, u, is calculated as c/n and entered on the chart.

The formula for calculating the average incident rate, $\bar{u}$ for k observations, is:

$$\bar{u} = \frac{c_1 + c_2 + c_3 + \cdots + c_k}{n_1 + n_2 + n_3 + \cdots + n_k} = \frac{60 \text{ incidents}}{13.92 \text{ million hours}} = 4.31 \text{ incidents per million hours.}$$

The standard deviation is given by:

$$s = \sqrt{\frac{\bar{u}}{n}} = \sqrt{\frac{4.31}{0.23}} = 4.33, \text{ for the first observation.}$$

The control limits are calculated in the usual way as:

$$\text{UAL} = \bar{u} + 3s$$
$$\text{LAL} = \bar{u} - 3s.$$

For the first month the limits are:

$$\text{UAL} = 4.31 + 3 \times 4.33 = 17.3$$
$$\text{LAL} = 4.31 - 3 \times 4.33 < 0.$$

Warning limits can be added at $\bar{u} \pm 2s$.

The interpretation of the chart identifies points of interest which are not obvious in Chart 23.7. Months 11 and 16 are still above the UAL, but in addition, from around month 22, the incident frequency appears to fall. Nine of the following ten points are below the average which includes the first 2 months with zero incidents in the whole dataset. It is also interesting to note that in month 7, $u = 20$, and is within the control limits even

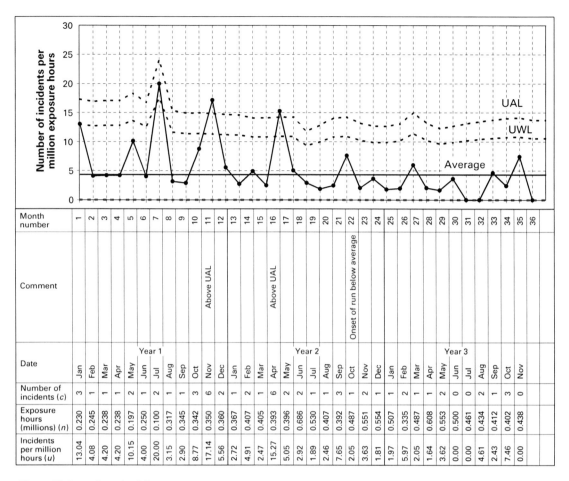

Chart 23.8 u chart incidents

Month number	1	2	3	4	5	6	7	8	9	10	11	12	13	14	15	16	17	18	19	20	21	22	23	24	25	26	27	28	29	30	31	32	33	34	35	36
Comment											Above UAL					Above UAL						Onset of run below average														
Date (Year 1 / Year 2 / Year 3)	Jan	Feb	Mar	Apr	May	Jun	Jul	Aug	Sep	Oct	Nov	Dec	Jan	Feb	Mar	Apr	May	Jun	Jul	Aug	Sep	Oct	Nov	Dec	Jan	Feb	Mar	Apr	May	Jun	Jul	Aug	Sep	Oct	Nov	
Number of incidents (c)	3	1	1	1	2	1	2	1	1	3	6	2	1	2	1	6	2	2	1	1	3	1	2	1	1	2	1	1	2	0	0	2	1	3	0	
Exposure hours (millions) (n)	0.230	0.245	0.238	0.238	0.197	0.250	0.100	0.317	0.345	0.342	0.350	0.360	0.367	0.407	0.405	0.393	0.396	0.686	0.530	0.407	0.392	0.487	0.551	0.554	0.507	0.335	0.487	0.608	0.553	0.500	0.461	0.434	0.412	0.402	0.438	
Incidents per million hours (u)	13.04	4.08	4.20	4.20	10.15	4.00	20.00	3.15	2.90	8.77	17.14	5.56	2.72	4.91	2.47	15.27	5.05	2.92	1.89	2.46	7.65	2.05	3.63	1.81	1.97	5.97	2.05	1.64	3.62	0.00	0.00	4.61	2.43	7.46	0.00	

though the plotted value is higher than months 11 and 16 that are both above the UAL. The reason that month 7 should not be considered abnormally high is that the exposure hours were relatively low.

Comments

The same comments apply regarding the ways to minimise the calculations as for the p chart. For example, the average number of exposure hours per month, $\bar{n}$ is calculated as:

$$\bar{n} = \frac{n_1 + n_2 + n_3 + \cdots n_k}{k} = \frac{13.92 \text{ million hours}}{35 \text{ months}} = 0.40 \text{ million hours per month}$$

And so for every month where the exposure hours lie between $0.75\bar{n} < n < 1.25\bar{n}$ we can use the same limits as for $\bar{n}$.

That is, for

$$0.75 \times 0.40 < n < 1.25 \times 0.40$$
$$0.30 < n < 0.50,$$

the limit of:

$$\mathrm{UAL} = \bar{u} + 3\sqrt{\frac{\bar{u}}{n}} = 4.31 + 3\sqrt{\frac{4.31}{0.40}} = 14.15$$

can be used and LAL < 0.

As with the p chart further simplifications can be made. For example, having identified the value of n for which the LAL $= 0$ (i.e. $\bar{u} = 3\sqrt{\bar{u}/n}$, giving $n = 9\bar{u}$) then for observations where n is less than this value (2.09 in this case) we do not need to calculate the LAL.

Multivariate charts

The multivariate chart is an extension of the attributes charts. In addition to identifying how many counts have occurred the multivariate chart shows the reason for the count occurring. Chart 23.9 gives an example of a multivariate np chart. Tenders are analysed in batches of 100 and the main reason for the rejection identified as one of price, delivery, specification, etc. The total number rejected is also recorded.

Development, analysis and interpretation of multivariate charts is exactly the same as for the type of chart with which it is being used.

Chapter 19 gives an example of a multivariate chart for accidents where the causes of accidents are categorised in two ways: by the immediate cause of accident (e.g. falling object) and the part of the body injured.

Multivariate charts are very useful for recording additional information that will probably be used in process improvement and/or investigations.

Description

Tenders that are rejected are analysed and the main cause for rejection identified and charted. Tenders are analysed in batches of 100.

x = number of defectives per sample
n = sample size = 100
$\bar{p}$ = average proportion defectives = total number of defectives/total number items inspected

$$= \frac{\Sigma x}{\Sigma n} = 109/(20 \times 100) = 0.0545$$

$$s = \sqrt{n\bar{p}(1-\bar{p})} = \sqrt{100 \times 0.0545(1-0.0545)} = 2.27$$

Limits:

$$\text{UAL} = n\bar{p} + 3 \times s = 100 \times 0.0545 + 3 \times 2.27 = 12.26$$

$$\text{LAL} = n\bar{p} - 3 \times s < 0$$

Comment	1	2	3	4	5	6	7	8	9	10	11	12	13	14	15	16	17	18	19	20	21	22	23	24	25	26	27
Price		1	3		2		1	1		1		3	1		2		2		2	1							
Delivery		1	1	2	3		1	3	5	1			2	3			1	1	3	1							
Specification	1	1	2		1	1	2	1		1	1	2	1		1		1		1	3							
Track record		1	2	1	3	1	2	1		1	1		2		2		1										
Too remote			1		1	2		1		1				1		2	1	1	1	2							
Other			2	2		1		1							1	1				1							
Total number of rejects (x)	1	4	11	5	10	5	6	8	5	5	2	5	6	4	6	3	6	2	7	8							
Date																											
Batch number	1	2	3	4	5	6	7	8	9	10	11	12	13	14	15	16	17	18	19	20	21	22	23	24	25	26	27

Chart (y-axis: Number of rejects (x), scale 0–12):

UAL = 12.26

Mean = 5.45

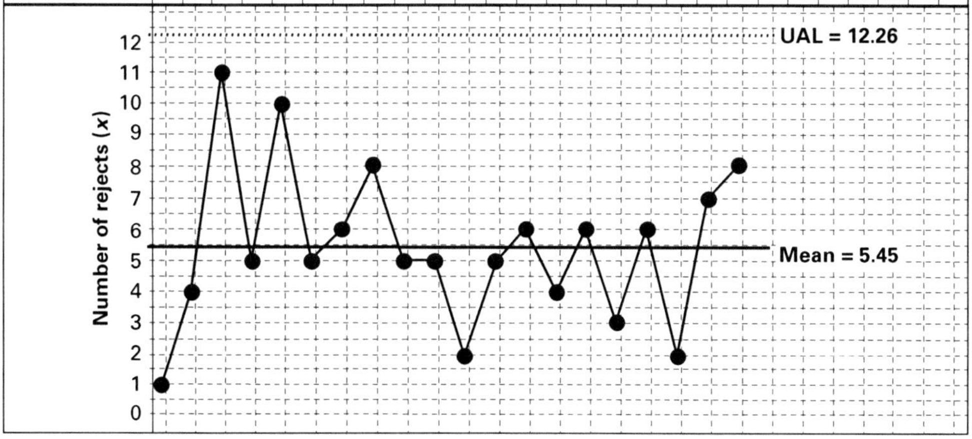

Chart 23.9 Multivariate np chart: rejected tenders

24 An introduction to cusum (cumulative sum) charts

Introduction

A cusum chart is a plot of the cumulative differences between successive values and a target value. The key features of a cusum chart are that they:

- are extremely good at identifying changes in process mean;
- can be applied to both variables and attributes charts including, for example, ranges and standard deviations.

The disadvantages are that they:

- are not very powerful at identifying other out-of-control signals;
- are more difficult to set up and use than other types of control chart.

Unlike other charts they:

- make use of all historic data: that is, each value on the cusum is a function of all previous data points;
- are interpreted by analysing the slope of the chart.

In this book we provide an introduction to cusum charts with the aim of demonstrating their power. Many people may wish to use them to identify changes in process average, and use other charts or statistical techniques for further investigation.

Basic cusum charts

A simple golfing example (Table 24.1, Charts 24.1–24.4)

A simple example will clarify how the chart is used. On an 18-hole golf course each hole has a "par" score depending on the difficulty of successfully getting the ball from the tee into the hole. Table 24.1 gives the par and score for each of the 18 holes and the score achieved by a better golfer than myself, along with the (score – par) and cusum.

Chart 24.1 is a run chart of the number of strokes taken for each hole. Not surprisingly, this does not tell us much. The process seems to be in control, and uneventful except perhaps for the score at hole 12, which has the highest value, 6. We could refine the chart by recognising that the target (par) for each hole is different, and chart both the par and the actual score as in Chart 24.2. Unfortunately, this chart still does not make it obvious as to how well we are doing. We could use a different chart, as in Chart 24.3, to plot (score – par). This also does not give a quick indication as to what is happening.

Table 24.1 Golfing scores

Hole	Par	Score (number of strokes)	Score – par	Cusum = Σ(score – par)
1	4	4	0	0
2	3	4	1	1
3	4	3	−1	0
4	4	3	−1	−1
5	4	4	0	−1
6	3	3	0	−1
7	4	3	−1	−2
8	3	4	1	−1
9	5	4	−1	−2
10	4	4	0	−2
11	4	3	−1	−3
12	5	6	1	−2
13	3	3	0	−2
14	4	3	−1	−3
15	4	4	0	−3
16	3	3	−1	−4
17	4	5	1	−3
18	5	4	−1	−4

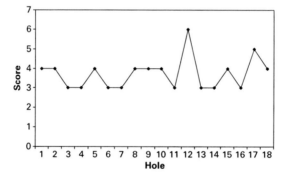

Chart 24.1 Run chart of golf scores

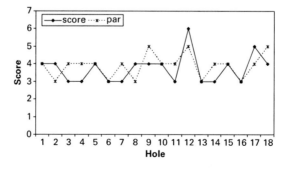

Chart 24.2 Run chart of par and actual golf scores

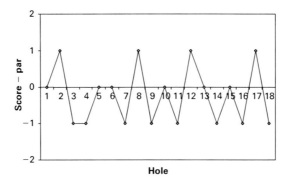

Chart 24.3 Run chart of differences (score − par)

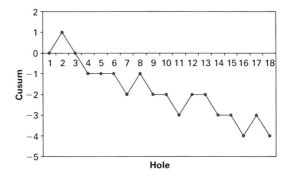

Chart 24.4 Cusum chart of (score − par)

Generally we are interested in how well we are performing. Using the (par) as the target, we know that sometimes we are on par, sometimes above par and sometimes below par, but is there a trend? Are we usually above or below par? Did our performance change part way round the course? One way of answering this question is to compare the number of scores above, below and on par. There are 8 holes below par, 6 on par and 4 above par. Unfortunately, whilst this suggests that we may generally be below par, it does not take into account how many strokes away from par we are, nor do we have a method of determining if the process changed and if so when. To answer this question we can look at the cumulative differences between par and our score. The calculations are given in Table 24.1 and the resulting chart is Chart 24.4. The chart trends steeply down and this tells us that our average score is below par, and because there are no changes in trend, we conclude that our performance has not changed, apart from random variation (for details on how to interpret a cusum chart see below).

Setting up a cusum chart (Chart 24.5)

The next example is taken from an organisation that was concerned about the amount of downtime on a critical piece of equipment. The metric of interest is the number of

Chart 24.5 Cusum chart: downtime per week

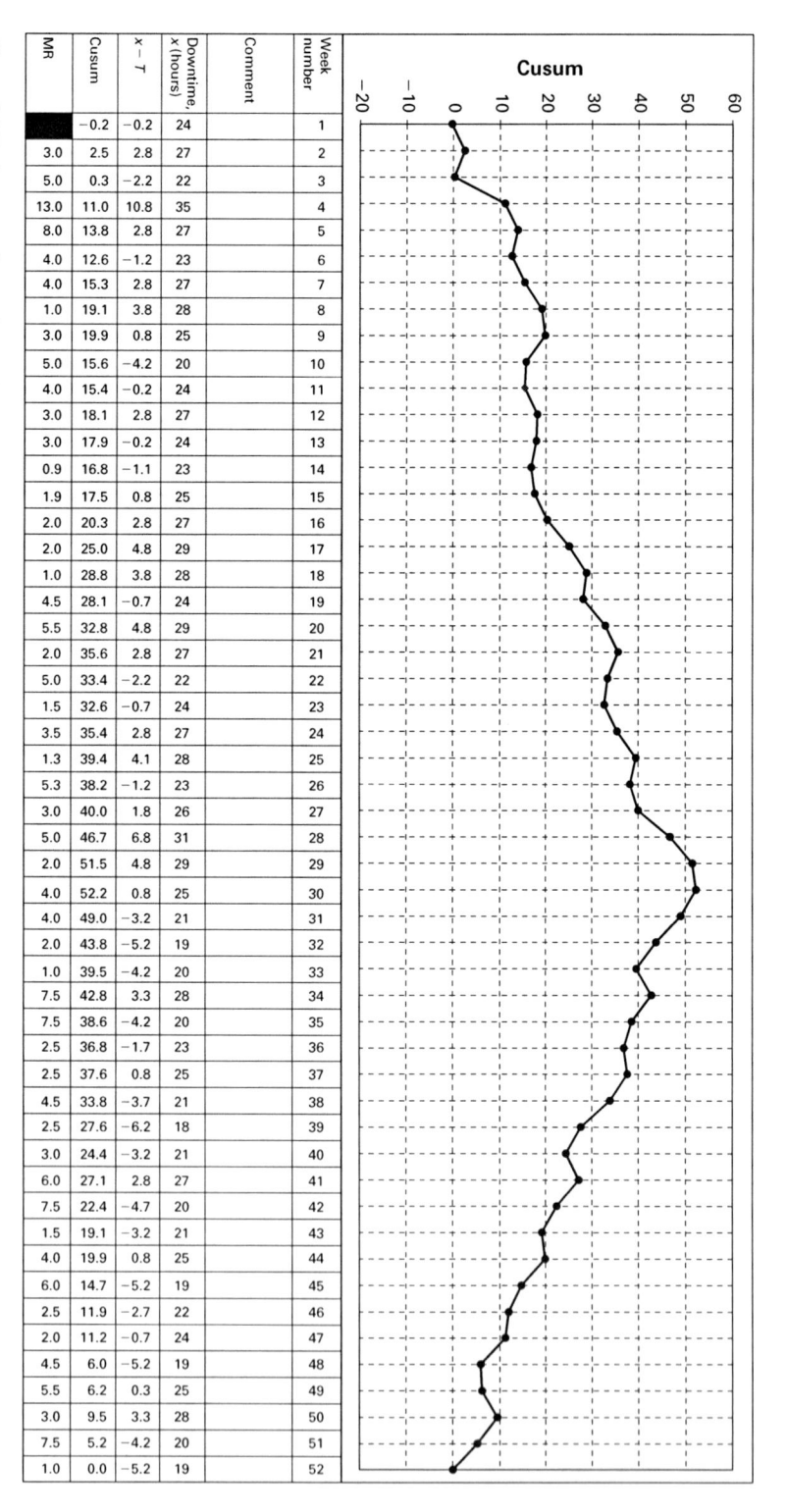

MR	Cusum	x − T	Downtime, x (hours)	Comment	Week number
	−0.2	−0.2	24		1
3.0	2.5	2.8	27		2
5.0	0.3	−2.2	22		3
13.0	11.0	10.8	35		4
8.0	13.8	2.8	27		5
4.0	12.6	−1.2	23		6
4.0	15.3	2.8	27		7
1.0	19.1	3.8	28		8
3.0	19.9	0.8	25		9
5.0	15.6	−4.2	20		10
4.0	15.4	−0.2	24		11
3.0	18.1	2.8	27		12
3.0	17.9	−0.2	24		13
0.9	16.8	−1.1	23		14
1.9	17.5	0.8	25		15
2.0	20.3	2.8	27		16
2.0	25.0	4.8	29		17
1.0	28.8	3.8	28		18
4.5	28.1	−0.7	24		19
5.5	32.8	4.8	29		20
2.0	35.6	2.8	27		21
5.0	33.4	−2.2	22		22
1.5	32.6	−0.7	24		23
3.5	35.4	2.8	27		24
1.3	39.4	4.1	28		25
5.3	38.2	−1.2	23		26
3.0	40.0	1.8	26		27
5.0	46.7	6.8	31		28
2.0	51.5	4.8	29		29
4.0	52.2	0.8	25		30
4.0	49.0	−3.2	21		31
2.0	43.8	−5.2	19		32
1.0	39.5	−4.2	20		33
7.5	42.8	3.3	28		34
7.5	38.6	−4.2	20		35
2.5	36.8	−1.7	23		36
2.5	37.6	0.8	25		37
4.5	33.8	−3.7	21		38
2.5	27.6	−6.2	18		39
3.0	24.4	−3.2	21		40
6.0	27.1	2.8	27		41
7.5	22.4	−4.7	20		42
1.5	19.1	−3.2	21		43
4.0	19.9	0.8	25		44
6.0	14.7	−5.2	19		45
2.5	11.9	−2.7	22		46
2.0	11.2	−0.7	24		47
4.5	6.0	−5.2	19		48
5.5	6.2	0.3	25		49
3.0	9.5	3.3	28		50
7.5	5.2	−4.2	20		51
1.0	0.0	−5.2	19		52

hours downtime per week. Below the cusum chart are the data that would normally be included with the chart. The steps for setting up the chart are:

1. Enter the downtime onto the chart (Chart 24.5).
2. Calculate the average.
   The average hours downtime per week, $\bar{x}$.
   $\bar{x} = 1260$ hours downtime/52 weeks $= 24.2$ hours per week.
3. Select the target (T).
   Any value can be selected as the target. However, when selecting the target there are two considerations:
   (a) It is easier to interpret the chart if on average the slope is nearly horizontal (to achieve this select the target equal to the average, 24.2 in this case).
   (b) If appropriate, select the target to be an appropriate value near the average, for example, in the golf example, selecting a target at par was appropriate.
   In the downtime example, we select the average.
4. Calculate the difference between the recorded values and the target, $x - T$, and enter them on the chart.
5. Calculate the cusum values, $\Sigma(x - T)$, for each week and enter them on the chart. The first cusum $= -0.2$, the second $= -0.2 + 2.8 = 2.5$ and the third $= 2.5 - 2.2 = 0.3$, etc.
6. Find the maximum and minimum cusum as this will determine the limits of cusum scale. For the downtime data the values are -0.2 and $+52.5$.
7. Determine scaling.
   Unless a suitable scaling convention is adopted the cusum chart may be difficult to interpret. At one extreme the slopes may be very flat and at the other extreme trivial changes may look dramatic as shown in Chart 24.6. The convention that has been widely adopted is that one observation along the horizontal axis should cover approximately the same distance as 2 standard deviations, $2s$, on the vertical axis.

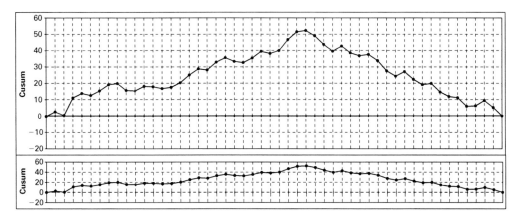

Chart 24.6 The effect of changing the scale on a chart

s is determined in the usual way as $S_{MR} = \bar{R}/1.128$, where $\bar{R}$ is the average moving range. The moving ranges, ignoring the sign, are calculated and entered on the chart. For the downtime data, the sum of the 51 moving ranges is 194.4 giving:

$$\bar{R} = \frac{194.4}{51} = 3.81$$

$$s = \frac{3.81}{1.128} = 3.38$$

$$2s = 6.76.$$

We round this down to a more convenient scaling of 5, and so the distance of 1 week on horizontal axis will be the same distance as 5 on the vertical axis.
8. Plot the cusum.
9. Interpret the chart.

Interpreting the cusum chart (Charts 24.7 and 24.8)

To interpret the cusum we are interested in slope and changes in the slope. The difficulty comes not in deciding what the slope is, but rather whether a change in slope is significant.
The rules as shown in Chart 24.7 are that if the cusum:

- Is horizontal, the process average equals the target.
- Slopes downwards (negative slope), the process average is less than the target.
- Slopes upwards (positive slope), the process average is greater than the target.
- The steeper the slope, the greater the difference between the target and the process average.
- Changes slope, the process average has changed (Chart 24.8).
- Is a curve, the process average is continually changing.

Other interpretation clues are:

- The process change occurs at the point where the slope changes.
- A jump in the chart signals a single very high/low observation. However, the cusum chart is not very effective at identifying single abnormally high or low values.

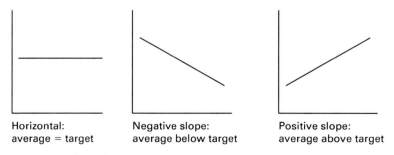

| Horizontal: | Negative slope: | Positive slope: |
| average = target | average below target | average above target |

Chart 24.7 Interpretation of a cusum chart

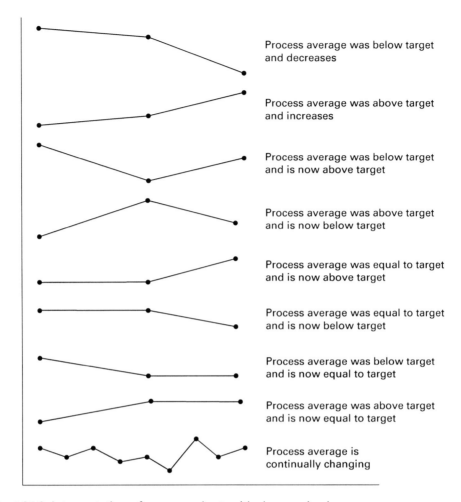

Process average was below target and decreases

Process average was above target and increases

Process average was below target and is now above target

Process average was above target and is now below target

Process average was equal to target and is now above target

Process average was equal to target and is now below target

Process average was below target and is now equal to target

Process average was above target and is now equal to target

Process average is continually changing

Chart 24.8 Interpretation of a cusum chart: with changes in slope

Interpreting the cusum chart is far more difficult than for the other control charts. There are two basic methods of determining whether a change in slope is random variation or not. The first is to construct a mask out of paper or card and place the mask on the chart, the other is to construct decision lines on the chart. Decision lines and masks are constructed to identify changes of a specific size of shift. We explain the construction of both a mask and decision lines for general-purpose analysis, that is, for identifying shifts of 1s.

Constructing and using a mask (Figures 24.1 and 24.2)

To make a mask drawn on transparent film (or cut out of paper) Figure 24.1:

1. Calculate the quantity 5s (i.e. 5 × standard deviation).
2. Measure the distance (e.g. in mm) on the vertical axis of the cusum chart that represents 5s.

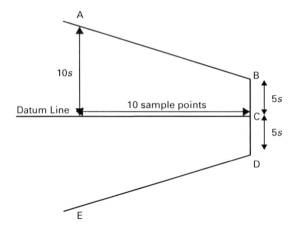

Figure 24.1 Cusum mask

3. Draw a horizontal line on the transparent film (datum line).
4. From the right-hand end of this line draw a vertical line up and down for a distance equal to $5s$ (CB and CD).
5. Measure the distance (e.g. in mm) on the horizontal axis of the cusum chart that represents 10 observations.
6. Measure a distance equivalent to 10 observations to the left of C along the datum line.
7. Mark off two points at $10s$ vertically above and below, A and E.
8. Join the points AB and DE.

If you are constructing a paper mask, cut out the paper around ABDE.

To use the mask, place it on the chart (Figure 24.2) so that DB runs parallel to the vertical axis with the limbs pointing to the left, as constructed. Slide the mask horizontally over the chart placing C of the mask on each point plotted. If at any point the mask cuts the cusum, then the average has shifted from the target by the equivalent of $1s$ or more.

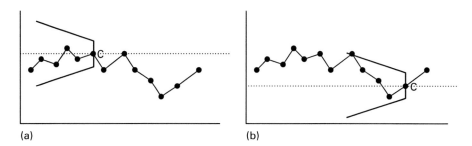

(a)   (b)

Figure 24.2 Cusum (a) does not cross the mask implying the process average has not changed and (b) crosses the mask implying the process average has changed by more than $1s$

Note that the coordinates of the plotted points are not of themselves of much interest. This is because each point represents the cumulative difference between the actual data value and a target since the beginning of the chart.

However, it is easy to calculate the process average for a span of data between any two points i and j, $\bar{x}_{i+1,j}$ The formula for doing so is:

$$\bar{x}_{i+1,j} = \frac{C_j - C_i}{j - i} + T$$

where C_j is the cusum for week j and T is the target value.

As an example, we see from Chart 24.5 that there appears to be a process change around week 29 and we can use the above formula to calculate the average incident rates before and after this point.

For the data up to week 29, $i = 0$ so $i + 1 = 1$ and $j = 29$ giving:

$$\bar{x}_{1,29} = \frac{51.5 - 0}{29 - 0} + 24.2 = 26.0$$

note that $C_0 = 0$.

To check that this is correct, the total downtime for the first 29 month = 754.4. Dividing by 29 weeks gives $754.4/29 = 26.0$ incidents per week.

Similarly, for the data from week 30, $i + 1 = 30$ and $j = 52$ and

$$\bar{x}_{30,52} = \frac{0 - 51.5}{52 - 29} + 24.2 = 22.0.$$

Comments

- The turning point was taken to be week 29. However, the cusum for week 30 is higher than that for week 29; should week 30 be selected as the turning point? The key to interpreting a cusum chart is the slope, and since the slope of the data from point zero is greatest at week 29, that week is taken as the turning point. However, the purpose of the chart is to gain insight. It is quite likely that whatever caused the process change did not occur exactly at the end of ANY particular week, and even if it did, changes are seldom abrupt. The message from the chart is that there is strong evidence that a change in process average occurred at around week 29, and the most likely week for the change is week 29.
- The construction of masks is tedious and time consuming; however, it is recommended that a few be constructed to develop understanding and gain familiarity with the method.
- There are different types of masks used for different situations. For more information see the British Standard BS5703.

Constructing and using decision lines (Chart 24.9)

A common method of highlighting changes on the cusum chart is to construct decision lines. The steps are:

1. Identify the suspected change in process average.
   In Chart 24.9, this is week 29.
2. Identify the beginning of the earlier process.
   In Chart 24.9, there seems to be no earlier process change, so the first section of the chart for comparison is taken to be the first week.

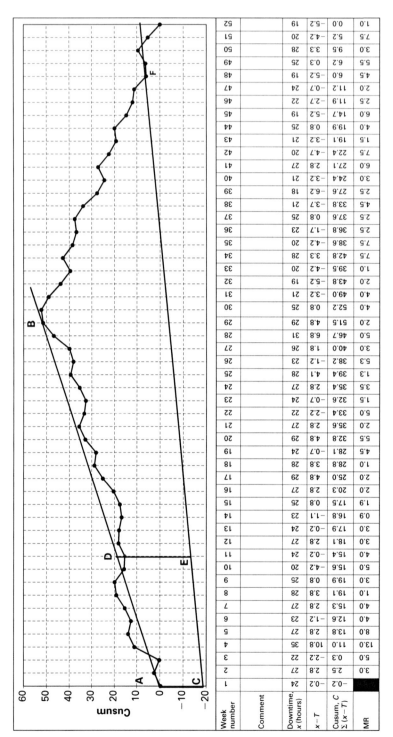

Week number	Comment	Downtime, x (hours)	x − T	Cusum, C Σ(x−T)	MR
1		24	−0.2	−0.2	
2		27	2.8	2.5	3.0
3		22	−2.2	0.3	5.0
4		35	10.8	11.0	13.0
5		27	2.8	13.8	8.0
6		23	−1.2	12.6	4.0
7		27	2.8	15.3	4.0
8		28	3.8	19.1	1.0
9		25	0.8	19.9	3.0
10		20	−4.2	15.6	5.0
11		24	−0.2	15.4	4.0
12		27	2.8	18.1	3.0
13		24	−0.2	17.9	3.0
14		23	−1.1	16.8	0.9
15		25	0.8	17.5	1.9
16		27	2.8	20.3	2.0
17		29	4.8	25.0	2.0
18		28	3.8	28.8	1.0
19		24	−0.7	28.1	4.5
20		29	4.8	32.8	5.5
21		27	2.8	35.6	2.0
22		22	−2.2	33.4	5.0
23		24	−0.7	32.6	1.5
24		27	2.8	35.4	3.5
25		28	4.1	39.4	1.3
26		23	−1.2	38.2	5.3
27		26	1.8	40.0	3.0
28		31	6.8	46.7	5.0
29		29	4.8	51.5	2.0
30		25	0.8	52.2	4.0
31		21	−3.2	49.0	4.0
32		19	−5.2	43.8	2.0
33		20	−4.2	39.5	1.0
34		28	3.3	42.8	7.5
35		20	−4.2	38.6	7.5
36		23	−1.7	36.8	2.5
37		25	0.8	37.6	2.5
38		21	−3.7	33.8	4.5
39		18	−6.2	27.6	2.5
40		21	−3.2	24.4	3.0
41		27	2.8	27.1	6.0
42		20	−4.7	22.4	7.5
43		21	−3.2	19.1	1.5
44		25	0.8	19.9	4.0
45		19	−5.2	14.7	6.0
46		22	−2.7	11.9	2.5
47		24	−0.7	11.2	2.0
48		19	−5.2	6.0	4.5
49		25	0.3	6.2	5.5
50		28	3.3	9.5	3.0
51		20	−4.2	5.2	7.5
52		19	−5.2	0.0	1.0

Chart 24.9 Cusum chart: downtime per week, using decision lines

3. From the point identified in step 2, draw a straight line through the suspected process change point (week 29 in Chart 24.9, line AB).
4. From the point identified in step 2, draw a vertical line above (or below depending on whether we are testing to see if the process has increased or decreased) a distance of $5s$.
   In Chart 24.9, $5s = 5 \times 3.38 = 16.9$ and the first cusum value is -0.2, so the line, AC extends down to $-0.2 - 16.9 = -17.1$.
5. From the point identified in step 2, count 10 observations along the chart and draw a vertical line above (or below) a length of $10s$ from the line AB: $10s = 10 \times 3.38 = 33.8$. The line is labelled DE.
6. Draw a line from C through E and continue it. If it cuts the chart, then we conclude that the process has changed. In this case the line cuts the cusum at F.

Comments

- In Chart 24.9, week 4 is an out-of-control point, as would be seen if an X/MR chart were been drawn. There is no simple way of seeing this on the cusum chart. For this reason, the cusum chart is usually used in conjunction with other charts. (It would, of course, be possible to calculate the upper action limit (UAL) using the formula for the X/MR chart and check by eye that no individual observations are above this value.)
- Constructing decision lines can be quite quick with a little practice, but is still laborious.
- A key use of the cusum chart is to identify where a process might have a change in average, and then to return to the other control charts to see if this is indeed mirrored there.

Weighted cusum charts

When to use weighted cusum charts

Weighted cusums are an extension of cusums used wherever the opportunity for the metric being monitored to vary is different for each observation. Thus, the weighted cusum is used in the same situations as p and u charts as well as for many variables situations. For example, the downtime chart recorded hours of downtime per week and so the opportunity for downtime remained constant, however, if the number of items of equipment being monitored varied from week to week, perhaps because we are monitoring hired items and the number of items varies with workload, then a weighted cusum chart would be needed.

The basic concepts behind the weighted cusum chart are the same as for the cusum chart, but the formulae and charting are more difficult.

As an example of the weighted cusum we return to the loss incident data used to illustrate the c and u charts in the previous chapter. Since the number of exposure hours varies each month we should use a weighted cusum chart rather than a cusum chart.

Chart format

As usual the data and calculations are provided below the table. In a weighted cusum chart we plot the cumulative cusum against the cumulative hours (in the case of

Chart 24.10). For example, after 10 months we have monitored a total of 2.50 million hours, and the cumulative sum of incidents is +5.22. As will be immediately obvious, the horizontal distance between the points varies. This is because each month has a varying number of exposure hours.

Setting up and interpreting a weighted cusum chart

1. Enter the number of incidents, x, and the number of exposure hours, w, onto the chart.
2. Determine the target value. In this case the average number of incidents per million exposure hours was used, and was calculated as:

$$T = \frac{\text{total number of incidents}}{\text{total number of exposure hours}} = \frac{60 \text{ incidents}}{13.924 \text{ million hours}}$$
$$= 4.31 \text{ incidents per million hours.}$$

3. For each month calculate the expected number of incidents wT and enter the data on the chart. For month $i = 1$, $w = 0.23$ hours so $wT = 0.23 \times 4.31 = 0.991$ incidents. This tells us that at a rate of 4.31 incidents per million man hours, we expect 0.991 incidents this month which had 0.23 exposure hours.
4. For each month calculate the difference between the actual and the expected incidents, $x - wT$.
   For month 1, $x = 3$ incidents and $wT = 0.991$ giving a value of $3 - 0.991 = 2.009$. Enter the data into the appropriate row below the chart.
5. For each month calculate the cumulative number of hours, Σw and enter the data into the appropriate row below the chart.
6. For each month calculate the weighted cusum; that is, the cumulative difference between the actual and expected number of incidents, $\Sigma(x - wT)$ and enter it below the chart.
7. Find the maximum and minimum values of the weighted cusum for scaling purposes.
8. Calculate the standard deviation, s, the formula is:

$$s = \sqrt{\frac{\bar{w}}{(n-1)} \sum \frac{(x - w\bar{x})^2}{w}}$$

where:
$\bar{w}$ = the average number of hours per month = 13.924/35 = 0.398,
n = 35 months,
$\bar{x}$ = average number of incidents per million hours = 60/13.924 = 4.31.
(i.e. the same as T in this case).

The calculations for $\dfrac{(x - w\bar{x})^2}{w}$ are given below the chart, and the total is 224.

Therefore

$$s = \sqrt{\frac{0.398}{(35-1)} \times 224} = 1.62$$

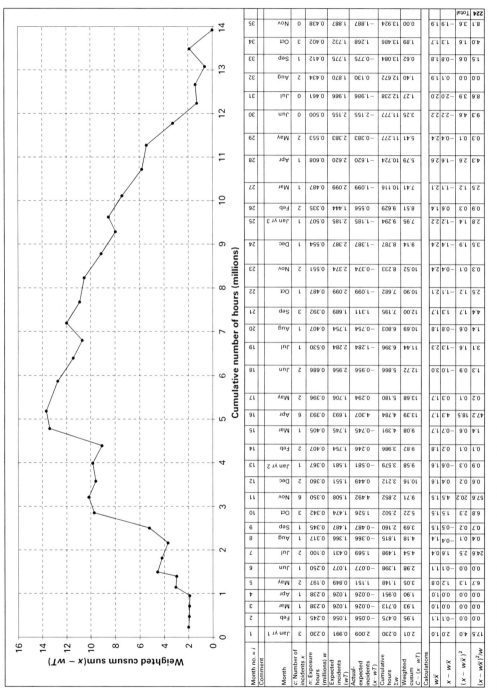

Chart 24.10 Weighted cusum chart for incidents

9. Scale the chart so that 2s units on the vertical axis corresponds to one unit on the horizontal axis. In this case, $2s = 3.2$ should cover approximately the same distance on the vertical axis as 1 million exposure hours on the horizontal axis.

10. Plot the data.

11. Interpret the chart in the same way as for the non-weighted cusum.

There is a major change in slope at around months 16 and 17. A mask or decision lines could be constructed to confirm this, or the corresponding u chart can be reviewed.

The formula for calculating the process average for a span of data between any two points is still straightforward, but a little more complicated than before. To determine the process average between two points, i and j, $\bar{x}_{i+1,j}$ The formula for doing so is:

$$\bar{x}_{i+1,j} = \frac{C_j - C_i}{\Sigma W_i - \Sigma W_i} + T$$

where C_i is the cusum value for month i. ΣW_i is the cumulative exposure hours for week i. For example, the average number of incidents per million exposure hours from month $i + 1 = 5$ to $j = 17$ is given by:

$$\bar{x}_{5,17} = \frac{13.68 - 1.90}{5.18 - 0.95} + 4.31 = 7.1.$$

To check this, the total number of incidents during this period was 30 and the total number of exposure hours was 4.229 resulting in 7.1 incidents per million hours exposure.

Summary

In this chapter we provided an introduction to cumulative (cusum) charts. We saw that:

- The cusum sum chart is a very powerful chart for identifying changes in the process average.
- They monitor the cumulative difference between the recorded values and a target value. The target is usually chosen to be the process average.
- Cusums can be used for both variables and attributes data.
- Cusum charts complement, and are usually used in conjunction with other control charts. The usual Schewhart charts are better at identifying all process signals except small changes in process average whilst the cusum chart is particularly good at identifying small sustained changes in average.
- The chart is more difficult to draw and interpret, and interpretation is by analysing changes in slope.
- To aid interpretation either masks can be constructed or decision lines can be drawn on the chart.

25 Issues for the more advanced SPC users

Introduction

In this chapter we investigate some important issues of interest to those who have a grounding in statistical process control.

In section *the number of observations required to identify a process change*, we see how to estimate how many observations it will take for a control chart to identify a process change of a certain magnitude. Where it is possible to select sample sizes or frequency of measurement, we can balance the cost of collecting more data with the benefit of identifying process changes more quickly.

The theory of variables control charts is developed around the assumption of data being normally distributed. In section *identifying and dealing with non-normally distributed data*, we introduce methods of identifying and dealing with non-normally distributed data.

In many statistical analyses it is also tacitly assumed that data are independent of one another. For control charts this means that successive observations are no more related to each other than two randomly selected observations. In section *identifying and dealing with auto-correlation*, we investigate how to identify the existence of auto-correlation.

The issue of how to deal with rare events such as natural disasters has been tackled in some of the case studies. In section *dealing with rare events data*, we summarise the problems and offer several techniques for coping with rare event data.

In some situations, particularly but not only manufacturing, we need to know how to subgroup data. The method of subgrouping is important where there are several sources of variation each of which need to be understood. This is addressed in section *analysing data in groups and subgroups*.

The number of observations required to identify a process change (average run length)

Introduction

One of the many advantages of using control charts is that we can estimate the average number of observations (known as the average run length or ARL) required before a process change of a certain size will be identified. This is useful to us for at least two reasons:

1. It gives us some idea of how long after an event occurs we are likely to have to wait before a control chart signals that the event has occurred.
2. If we are able to select the frequency and/or sample size we can choose it to give us a particular ARL.

The bad news is that the ARL depends on:

- what rules are being used to interpret the chart,
- how large the process change is,
- the data following a specific distribution.

Some theory

If you want to know a little of the statistical theory behind the ARL, read on, otherwise skip to the next section.

If data are normally distributed with mean ($\bar{x}$) and standard deviation (s) (Figure 25.1), then we know (amongst other things) that for a process in a state of control:

- the chance of getting a result above the upper action limit (UAL) of $\bar{x} + 3s$ is about 1.5 in 1000;
- the chance of getting a result above the upper warning limit of $\bar{x} + 2s$ is about 2.5 in 100 (i.e. 1 in 40, marked A in Figure 25.1).

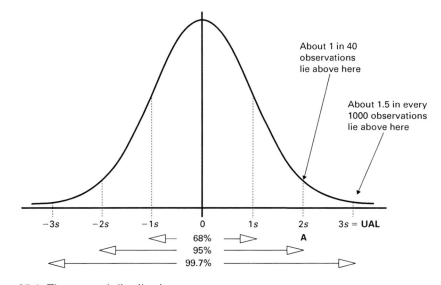

Figure 25.1 The normal distribution

If the process changes in such a way that the average increases by the equivalent of one standard deviation (Figure 25.2), the chance of getting a result above the old UAL, now only 2s above the new average, is 1 in 40. Therefore on average we need to wait for 40 observations before we receive a result above the original UAL.

Using this methodology (and statistical tables of the normal distribution) it is possible to calculate the ARL to an observation above the UAL after a process shift of any size. Similarly, of course, we can calculate the ARL for a reduction in process average or any size.

We can also determine the probability of getting a signal from runs above/below the average and any of the other rules that we use when analysing control charts.

Statistical theory similarly allows us to determine ARLs for X/MR, $\bar{X}$, R and other charts including cusum charts.

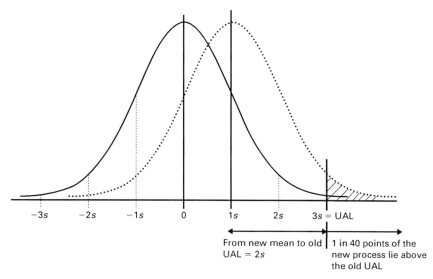

-3s   -2s   -1s   0   1s   2s   3s = UAL

From new mean to old | 1 in 40 points of the
UAL = 2s | new process lie above
the old UAL

Figure 25.2 Process shift

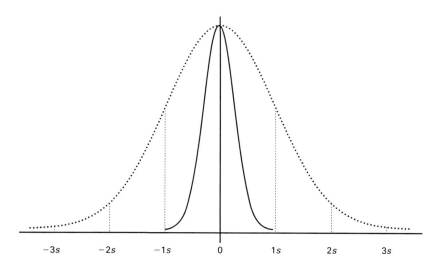

-3s   -2s   -1s   0   1s   2s   3s

Figure 25.3 Distribution $\bar{X}$ with sample size 9, standard deviation $= s/3$

We can also investigate the benefits of increasing sample sizes with the intention of reducing the ARL. In the above example of an X chart we were considering individual observations. If we were to sample four items at a time the mean would be the same, $\bar{x}$, but the standard deviation would be $s/\sqrt{n}$. So, if we take samples of size 4, the standard deviation of the means of these samples is $s/2$, and if we take sample sizes of 16, the standard deviation is $s/4$.

Considering the case of sample size $= 9$ and standard deviation $= s/3$, the UAL will be at $\bar{x} + 3 \times (s/3) = \bar{x} + 1s$ (Figure 25.3). If the mean now increases by an amount equal to s, then half the values will now be above the old UAL, and so the ARL will be 2 (Figure 25.4).

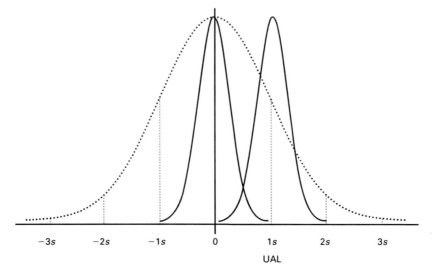

Figure 25.4 Process shift by 1s

In practice

In practice we seldom want to know the exact ARL, but it is useful to have some idea of what it is. Fortunately there are charts and tables available. Figure 25.5 is the ARL chart for identifying changes in the average when we only use the action limits for detecting changes. The chart consists of a number of curves, each for a specific value of n, the sample size. The horizontal axis is the size of the change measured in standard deviations and the vertical axis is the ARL. As an example of how to use the chart, suppose we have sample size of 4 and we want to estimate the ARL for a change of 0.75 standard deviations:

1. Select 0.75 on the horizontal axis.
2. Read vertically up to the curve labelled $n = 4$.
3. Read across to the ARL axis (the value is approximately 14).

Therefore, the average number of samples required to identify a change of 0.75 standard deviations with a sample size of 4 is 14. There are also charts, called operating characteristic (OC) charts, for calculating the probability of 1 observation detecting a change when one has occurred (Figure 25.6). The horizontal axis is the size of the change measured in standard deviations and the vertical axis is the probability of not detecting a change. As an example of how to use the chart, suppose we have sample sizes of 4 and we want to estimate the ARL for a change of 0.75 standard deviations:

1. Select 0.75 on the horizontal axis.
2. Read vertically upto the curve labelled $n = 4$.
3. Read across to the axis – the value is approximately 0.93. This is the probability of NOT detecting a change.

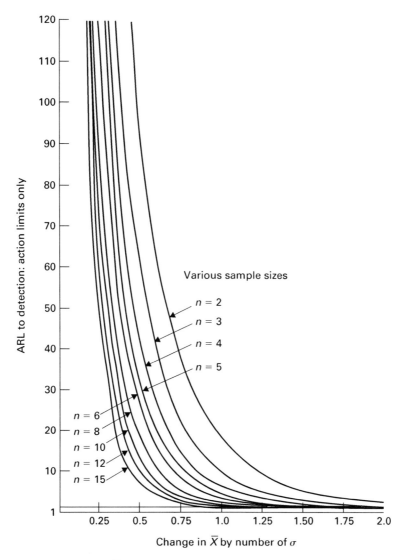

Figure 25.5 ARL curves for $\bar{X}$ charts. Oakland, J. Reproduced with permission from *Statistical Process Control*, 5th edn. Butterworth-Heinemann, p. 425

4. Calculate the probability if finding a change, by subtracting the value in 3 from 1.0 (i.e. $1 - 0.93 = 0.07$).

The values obtained from the two charts are related. If the probability of finding a change is 0.07, then the ARL can be calculated as $1/0.07 = 14$, the same value as read off the ARL chart.

There are also tables and OC curves for cusum charts which can be found in BS5703.

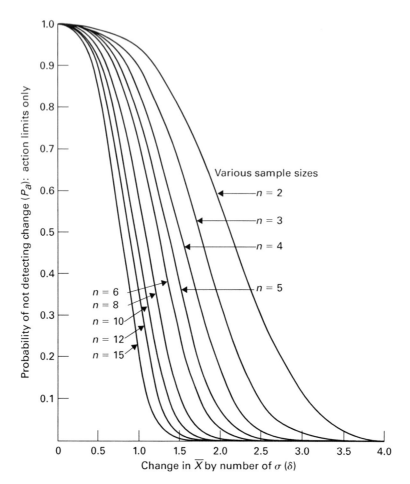

Figure 25.6 OC curve for means. Oakland, J. Reproduced with permission from *Statistical Process Control*, 5th edn. p. 424

Identifying and dealing with non-normally distributed data

Variables data are assumed to be normally distributed
The underlying assumption of the variables charts (e.g. X/MR, $\bar{X}$/range and standard deviation charts) is that the data are randomly scattered about the average and normally distributed, as described in Chapter 1. The implicit analysis carried out when analysing these charts is to test for non-normality.

Attributes data are not usually normally distributed
Attributes charts do not generally follow the normal distribution, which is why they use different formula. However, in certain circumstances, as explained earlier, the distributions come close to the normal distribution and this is why the X/MR chart can sometimes be used in place of the attributes charts.

Variables data are not always normally distributed

There will be occasions where the data do not follow the normal distribution. Typical examples include the amount of impurity in a substance; the cost or time to carry out distinct tasks (such as drill a well, complete a construction project or maintenance task) or the time to recover from surgical intervention. In these situations there may be a technical lower limit, but no upper limit. Most of the time the process operates reasonably near the technical limit, but as problems crop up so the time and/or costs increase, and there is no upper limit. In these cases the distribution will be skewed.

It is important to be aware of non-normality

It is important to be aware of any non-normality inherent in the data and there are a number of methods of testing for non-normality. The simplest is to draw a histogram of the data along with the control chart, and there are several case studies where the histogram is drawn next to the chart (see e.g. Chapter 7). If we are concerned about the potential non-normality of a set of data there are a variety of tests that can be carried out; they are outside the scope of this book, but may be found in many statistical texts.

What do we do with non-normal data?

When the data are non-normally distributed, it is useful to identify the cause. If the non-normality is due to data from two or more streams being mixed before measuring (e.g. several machine outputs; results from different laboratories, different operatives, etc.) then the usual method is to stratify the data into its individual streams and analyse it separately. Chapter 7 is a simple example of the results from two analysts being combined before charting.

If the data are naturally non-normal (e.g. the level of contamination may not be normally distributed), another approach is to determine statistically what distribution the data follow and determine the limits empirically from the data. It is also possible to transform the data so that it is normally distributed and plot the transformed data as usual. Both these methods are beyond the scope of this book, and the advice of a statistician should be sought.

Non-normality is mainly of concern with the X/MR chart, though it gives reasonable results even with non-normal data. We are far more protected with $\bar{X}$ charts because of fact that whatever the underlying distribution, where individual values are averaged, the distribution of the average tends to normality. Put loosely, the effect of extreme values is reduced. For this reason, if data are thought to be non-normally distributed, it is better to use an $\bar{X}$/range chart rather than an X/MR chart, where there is a choice.

Finally, the above brief discussion serves to illustrate a key aspect of control charts: they are a tool to be used to help understand what information there may be in a set of data. They are a starting point to generate theories, not an end point. Sometimes we may not know how best to calculate control limits, or how to cope with skewed or missing data. In addition we have seen in several case studies that even if we use the "wrong" chart (i.e. one for which the distribution does not follow the theoretical distribution) we frequently find that the results are still useful.

Identifying and dealing with auto-correlation

If two variables are related in such a way that a change in one variable is reflected by a change in the other variable, the two variables are said to be correlated. Examples of correlated variables are given in the case study in Chapter 16.

Data are auto-correlated if each value is correlated to the previous value. Typical examples of auto-correlation include:

- Number of cars passing a specific point measured every 5 minutes throughout a day.
- Outside temperature recorded every hour.
- Weekly weight of pregnant women.
- Output of any process that drifts, where several readings are taken during the drift.

Auto-correlation is a problem because control charts, and indeed most statistical analyses, assume that the data are not auto-correlated.

A quick and simple method of checking for auto-correlation is to draw a scatter diagram of each value, x_i, against the previous value, x_{i-1}. Details of how to draw and interpret a scatter diagram are given in Chapter 26, and the method is exactly the same for auto-correlated variables. Chart 25.1 shows a scatter diagram of an auto-correlated variable.

The strength of the relationship between x_i and x_{i-1} can be determined by calculating the correlation coefficient, which is explained in a little more detail in Chapter 26.

If data are found to be correlated there are a number of steps we can take, depending on the cause of the correlation:

- Sample less frequently, checking that there is a minimum of auto-correlation. This is a typical solution when sampling from batch processes where the output changes slowly.
- If the process is being sampled after the mingling of alternate outputs of several processes, moving upstream of the mingling may solve the problem.
- If there are technical reasons why the data should be auto-correlated it may be necessary to consult a statistician or technical expert in the process before proceeding.

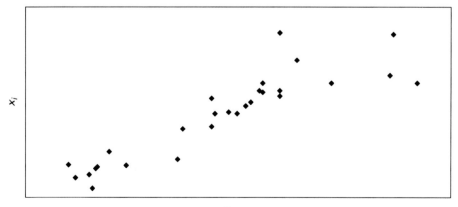

Chart 25.1 Scatter diagram showing auto-correlation

The discussion above relates to auto-correlation between consecutive values, that is, of lag 1. It is also possible to calculate the auto-correlation of lag 2 (i.e. between values two observations apart) or any number of observations apart. The correlation coefficient can be calculated for lags 1, 2, 3, etc. and it should steadily decrease. For some value n, the correlation coefficient will not be significant and this gives us the minimum time that should be allowed between sampling. The theory and practice behind this technique is beyond the scope of this book.

Dealing with rare events data

When monitoring the occurrence of events it will sometimes happen that much of the time we are recording zero and only occasionally non-zero values. A typical example could be the number of aeroplane crashes per month, and there are several specific examples in this book, see especially the case study in Chapter 8.

In this section we discuss what a rare event is and provide some techniques for overcoming the difficulties of monitoring them.

In control charting a rare event is not just something that does not occur very often. We may reject only one in a million items, but if we inspect 10 million every day, we would still be recording an average of 10 rejects a day and would monitor rejects using an np chart (or a p chart if sample sizes vary). Often we can compensate in these situations simply by increasing the sample size or area of opportunity that we look at.

In control chart terms, a rare event is recognised when we record many zeros and only the occasional non-zero value. Typical examples would be the number of injurious incidents per million hours worked in an organisation. Such incidents may occur only once or twice a year, or even less. The resulting c chart (or u chart if the hours worked varied) of incidents per month, or even per quarter would contain many zeros. In these situations we could convert the count data into measurement data by monitoring the number of hours between incidents and converting it into a number of incidents per million (or 10 million, etc.) man hours. For example, if 625,000 work hours had been worked since the last incident, the value charted would be $(1,000,000/625,000) = 1.6$. Points are plotted on an X/MR chart whenever an incident occurs. Control limits are calculated and the chart interpreted in the usual way.

In terms of SPC there is no universally accepted definition of a rare event, but some useful guidelines are:

- if the process average is less than 1 or
- the lower action limit (LAL) is less than 0.

Like most things, these are not a strict cut-off criteria but indicators that the usefulness of the chart is being compromised.

It is useful to turn the question round and ask when do we NOT need to worry about rare events. Two useful guidelines are:

- when the LAL is greater than 0,
- when the process average is greater than 4.

These guidelines are different and a chart may meet one criteria and not the other. This highlights the fact that there is no hard cut-off for determining when rare events are occurring.

The problem of rare events is well illustrated by Chart 8.1 in Chapter 8. In this chart the average is 0.21 incidents per month and the UAL is between 1 and 2. The incident frequency could increase from 2 incidents per year, in two separate months, to one incident per month for 6 months without triggering a signal that the process had changed. If the UAL were less than 1, which would happen if the average number of incidents per month were less than 0.09, then every time there was even one accident in a month it would signal a special cause of variation. Clearly in these and similar situations monitoring the number of incidents does not help us.

Three possible solutions to rare events are as follows:

1. Increase the sample size.
2. Combine groups (see the case study in Chapter 10).
3. Measure the time between events (see the case study in Chapter 8).

Increasing the sample size is probably the most appropriate when we are sampling data and we have control over the sample size.

Example 1

A factory produces widgets in batches. From each batch 100 are randomly sampled and tested. The reject rate, $n\bar{p}$, is 1 per 100 (i.e. $\bar{p} = 0.01$). In this case many of the samples will have no defects, some will have 1 and a few more than 1.

The simple solution would be to increase the sample size, but to what? We would like the average plotted value, $n\bar{p}$ to be greater than or equal to 4; that is for the sample size, $n > 4/\bar{p} = 4/0.01 = 400$, we would need to sample 400 widgets in order that the average number of rejects per batch is 4.

If the sample size is a percentage of the batch size (e.g. 1% and the batch size is varying), the same technique and formula can be used for determining an appropriate average sample size.

In non-manufacturing organisations, n could be the number of late train/bus arrivals, number of patients and number of people surveyed giving a particular response.

Example 2

An organisation records the number of events (e.g. fires, resignations, litigations, serious injuries) per month. The average number of events per month is 1.6, and in many months a 0 is recorded. To raise the average to over 4 we could combine $4/1.6 = 2.5$ months, or more easily 3 months, and report quarterly rather than monthly.

Example 3

As an alternative to the organisation recording an average of 1.6 incidents per month, it would be possible to count the number of days between events and turn this into a rate per year (or 100 days or 1000 days, etc.). For example, if there are 25 days between events this equates to $100 \times 1/25 = 4$ events per 100 days, 40 per 1000 days or $365 \times 1/25 = 14.6$ per year.

Analysing data in groups and subgroups

Introduction

Many of the case studies in this book use examples where outputs from different subgroups of data have been combined. They show that breaking down the data into its subgroups and analysing the subgroups can lead to important discoveries about the data.

This is an important phenomenon and in this section we explore a little more the concepts of grouping data. Consider, for example, a facility (e.g. hospital, factory, processing plant) with several units working in parallel (e.g. wards or machines) as illustrated in Figure 25.7.

As an example, consider four casualty departments monitoring the number of daily admissions. A table similar to that presented in Table 25.1 below could be drawn up to summarise the data over any complete number of weeks.

Table 25.1 is designed for four casualty departments. The last row gives the average for each day (e.g. AMon is the average for Monday), and the last column gives the average for each casualty (e.g. AC1 is the average for casualty 1). The average number of admissions per department per day is given as AVE.

With a set of data in this format there are several sources of variation that can be investigated:

- Weekday-to-weekday variation (e.g. differences between Mondays at casualty 1).
- Time trends (e.g. seasonality, steady increase/decrease in admissions).
- Between days of the week (e.g. differences between Fridays and Mondays).
- Between casualty departments (e.g. one department has a higher daily admissions rate than another).

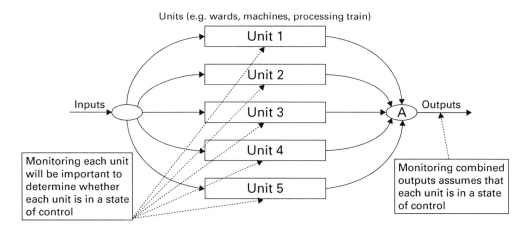

Figure 25.7 A typical process with units

Table 25.1 Number of casualty admission over

Casualty	Week							Average
	Mon	Tue	Wed	Thu	Fri	Sat	Sun	
C1								AC1
C2								AC2
C3								AC3
C4								AC4
Average	AMon	ATue	AWed	AThu	AFri	ASat	ASun	AVE

And because we have data from more than one casualty department, we can carry out analyses on the first three items for each casualty department separately and, if appropriate, combined.

Analysis introduction

This section focuses on understanding concepts of subgrouping. In order to help keep other information to a minimum, the charts presented are run charts rather than control charts and are not rigorously analysed. In practice a complete analysis of the appropriate control charts would be required.

The analyses carried out for any particular set of data will depend on what is discovered as the analysis proceeds and on an understanding of how the process operates. What is illustrated here is a typical analysis approach which illustrates the different methods of investigating this form of data set.

The data were collected from four casualty departments on a daily basis for 5 weeks, that is, a total of 4 departments × 7 days per week × 5 weeks = 140 data values.

Analysis

The first chart we would usually draw in this type of situation is of the daily admissions. As an example, a run chart for casualty 1 is given in Chart 25.2. The chart gives the data for 5 weeks and suggests the following:

- There is no general trend over time but
- Saturdays are always higher than other days of the week and
- Sundays are also high, but not as high as Saturday
- Monday to Friday appear to be stable.

Note that this one chart gives us some pictorial information on two sources of variation: trends over time and variation within the week.

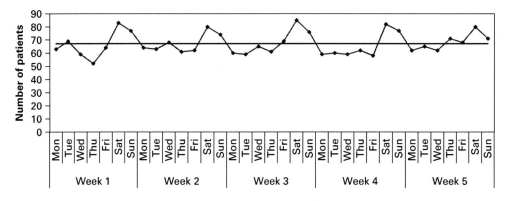

Chart 25.2 Run chart of daily admissions over a 5-week period for casualty 1

Whilst we have noticed the differences between the days of the week in Chart 25.2, it may not always be so obvious. We could further check for variability between days within a casualty department by drawing an $\bar{X}/R$ *or* $\bar{X}/s$ chart of the days over the 5 weeks for which data are available. As an example, Chart 25.3 gives the run chart of the averages, $\bar{X}$, for casualty 1. This highlights the increase in admission rates for Saturdays and Sundays.

Charts similar to Charts 25.2 and 25.3 would also be drawn for each of the four casualty departments. One way of presenting the average admissions for each casualty (i.e. the Chart 25.3) is to put all them on one chart, as shown in Chart 25.4.

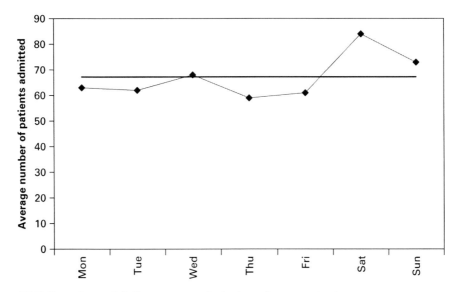

Chart 25.3 Run chart of daily average admissions for casualty 1

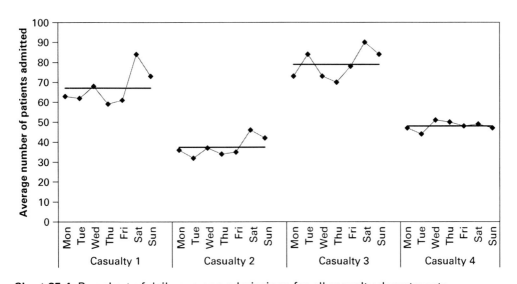

Chart 25.4 Run chart of daily average admissions for all casualty departments

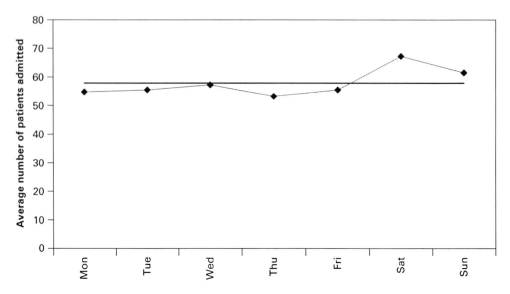

Chart 25.5 Run chart of daily average admissions for all casualty departments

If we find the same pattern occurring in all the casualty departments, we could combine the results from all departments to draw a chart of all casualties combined. Chart 25.2 would then become an $\bar{X}/R$ *or* $\bar{X}/s$ chart with 35 daily values, each value being the average of the four casualty figures for the day. This would tell us how the total admissions over the four casualty departments are behaving. As we have analysed the casualty departments separately, and found no trends, we would know that any trends found in the chart are not due to changes over time in any of the individual casualties.

In a similar way Chart 25.3 would become an $\bar{X}/R$ or $\bar{X}/s$ chart with 7 points, each being the average of 4 casualty departments × 5 weeks = 20 values (Chart 25.5). It confirms the general pattern that there is no difference between the days Monday to Friday, but that Saturday (in particular) and Sunday are high. Note, however, that casualty department 4 does not appear to follow this pattern (see Chart 25.4).

As we have already drawn the above charts for all casualty departments (i.e. we would have a version of Chart 25.2 for all casualties and/or Chart 25.4) we will have some idea of the variability between them. In this instance, the difference in admission rates shown in Chart 25.4 seems large and clear, but it may not always be so. To investigate these differences we draw an $\bar{X}/R$ or $\bar{X}/s$ for which the $\bar{X}$ run chart is shown in Chart 25.6. In this chart each point is the average of 35 days admissions, and the average line is the average number of admissions per department. Once again we note that casualty 4 does not follow the trend of having high values at the weekend and we could modify the analysis to take this into account.

We have noted that the weekends have higher admission rates compared to weekdays for departments 1, 2 and 3 but not for the department 4, and there are several ways of

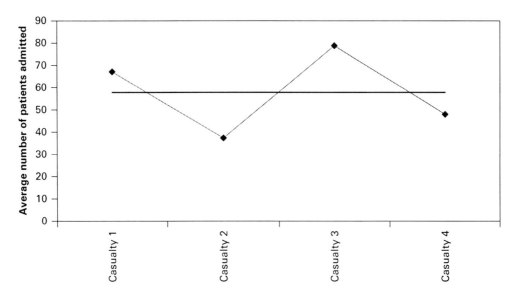

Chart 25.6 Run chart of average admissions by casualty

proceeding with the analyses once this has been discovered:

- If the pattern was the same for all departments the data could be included in the analyses as is done here;
- If the pattern is not the same for all departments, we should first investigate why there is a difference. We could then divide the departments into those showing similar patterns and/or split the data into two or three groups (weekdays, Saturdays, Sundays or weekdays and weekends) and analyse them separately.

Several of the charts in this discussion do not have the suggested 20 values for drawing a control chart. However, as illustrated in this example, useful information can be gained from charts with fewer points.

Further applications

This type of analysis has been illustrated for the health care industry, but the issues of sources of variation and methods of analysis are similar for many different situations.

For example, in manufacturing the units could be different suppliers, machines or production lines. Data could be collected every shift or hour, and the measured value could be lengths, weights, number of rejects or failures, etc.

In the service industry, the units could be people or groups of people, data could be collected every job or time period, and the values could be time to resolve customer queries.

Summary

In this chapter we have discussed some important issues related to control charting.

The ARL

- It is possible to calculate the ARL (average number of observations) to identify a change in process average of a specified magnitude.
- It is possible to calculate the probability of receiving a point outside the control limit if the process average has changed by a specified amount.

Normality of data

- Variables charts assume that the underlying distribution of the data is normal.
- Attributes charts usually follow non-normal distributions, which is why the formulae used in these charts are different.
- We should be aware of the underlying distribution of the data as this may give clues as to what is happening in the process.
- Histograms are a simple and useful tool for investigating data distributions.
- The X/MR chart is somewhat robust to non-normality; the $\bar{X}$ chart much more so.

Auto-correlation

- If the value of successive observations tend to be closer to each other than observations further apart, then the data are said to be auto-correlated.
- All control charts assume that data are not auto-correlated.
- An easy method of checking for auto-correlation is to draw a scatter diagram of each value vs. the previous value (x_i vs. x_{i-1}).
- There are several techniques of coping with auto-correlated data, the simplest of which is to take observations less frequently.

Rare events

- From a control charting viewpoint, rare events occur if the process average falls below 1 or the LAL is 0.
- Charting of rare events may be avoided by increasing the sample size or combining groups.
- Rare events are charted by monitoring the time between events.

Subgrouping

- Data can often be categorised into subgroups.
- Analysing subgroups with a control chart can lead to important discoveries about the process being monitored.

26 Data analysis tools

Introduction

In addition to the control chart there are a number of other data analysis tools that have been mentioned in this book. In this chapter we explain what these tools are, their applications and how to interpret them. The tools covered are:

1. Histograms
2. Run charts
3. Bar charts
4. Ranked bar charts and Pareto charts
5. Check sheet
6. Scatter diagrams.

These can readily be drawn using standard spreadsheet packages.

For the reader wishing to embark on process management, analysis and improvement, there are a number of other tools which are useful, chief amongst them is the flow chart. These are beyond the scope of this book, and details may be found in a wide variety of quality, business improvement and related books.

Histograms

A histogram is a graphic summary of the distribution of a set of data.

The rationale behind the histogram is:

- Almost any set of data will show variation.
- This variation will exhibit some pattern.
- The pattern can give us clues as to what is happening in the process from which the data were obtained.
- It is difficult to see this pattern in a table of data.
- The histogram presents the variation in a way that helps us understand what may be happening in the process.

Table 26.1 shows a set of viscosity measurements along with the date on which the measurements were taken. It is difficult to glean any information from the data. With some study it is possible to identify the maximum and minimum values, but the average, variability and existence of any trends or jumps are more difficult to appreciate. The histogram of the data in Figure 26.1, however, gives a quick indication of the average and spread of the data.

Table 26.1 Viscosity measurements

Viscosity	Date	Viscosity	Date	Viscosity	Date
14.18	11 Jan	14.71	23 Aug	14.21	18 Oct
14.21	31 Jan	14.57	26 Aug	14.09	20 Oct
14.22	12 Feb	14.55	02 Sep	14.61	21 Oct
14.31	19 Feb	14.41	06 Sep	14.44	23 Oct
14.14	04 Mar	14.57	06 Sep	14.38	28 Oct
14.24	22 Jul	14.34	08 Sep	14.63	11 Nov
14.36	26 Jul	14.29	15 Sep	14.61	15 Nov
14.21	05 Aug	14.64	15 Sep	14.49	16 Nov
14.39	06 Aug	14.25	19 Sep	14.31	21 Nov
14.14	08 Aug	14.31	21 Sep	13.96	29 Nov
14.33	11 Aug	14.38	23 Sep	14.36	29 Nov
14.38	12 Aug	14.72	27 Sep	14.44	02 Dec
14.75	17 Aug	15.01	29 Sep	14.98	06 Dec
14.41	17 Aug	14.52	03 Oct	13.96	08 Dec
14.72	18 Aug	14.72	06 Oct	14.29	12 Dec
14.74	18 Aug	14.67	08 Oct	14.33	20 Dec
14.66	19 Aug	14.47	13 Oct	14.06	22 Dec
14.65	22 Aug	14.19	18 Oct	14.83	28 Dec

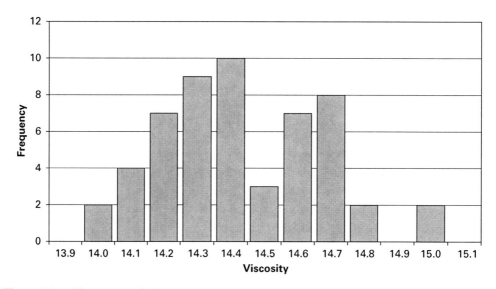

Figure 26.1 Histogram of viscosity

To plot the histogram the data are split into bands of 0.1, starting at 13.85 (rounded to 13.9). The number of readings in each band is plotted on the vertical axis. The main features of the histogram are as follows:

- The minimum is around 14 and the maximum just over 15.
- There are two peaks to the data, one at 14.4 and the other at 14.7.

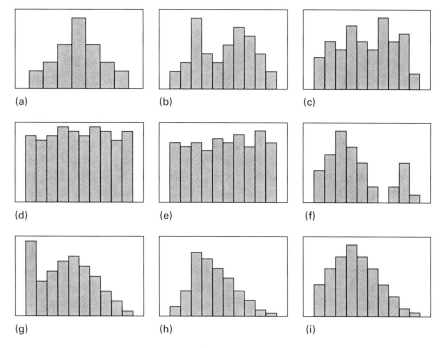

Figure 26.2 Typical histogram shapes (a) Bell-shaped distribution, (b) double-peaked distribution, (c) multi-peaked distribution, (d) flat distribution, (e) toothed distribution, (f) separated distribution, (g) edged distribution, (h) skewed distribution and (i) truncated distribution

- We can hazard a subjective guess at the limits within which the data are likely to lie. It seems unlikely that we will receive values less than about 13.9 or above about 15.1.
- We can also guess that the average is around 14.4.

The shape of the histogram also tells us something. Figure 26.2 gives examples of typical histogram shapes. Figure 26.2(a) is a histogram with a "bell" shaped or, as it is usually called a "normal" distribution. This is a very common shape of a well-behaved "in-control" process: the data are evenly scattered around the average, with fewer points the further away from the average we go. Whilst other shapes (especially skewed) are common with some types of data, variations from the normal distribution should usually be investigated.

Figure 26.2(b) has a double peak. A very common explanation for this distribution is that two process have been combined upstream of the measurement point. For example, results of two shifts, machines, suppliers may be mixed before charting. If there are many peaks, or if the distribution is flat (Figure 26.2(c) and (d)), it may indicate that the outputs of many processes have been combined upstream of the measurement point.

The "toothed" distribution – alternate high/low values – Figure 26.2(e), may be due to measurement error or an unfortunate choice of grouping when constructing the histogram, or a rounding error. Sometimes it is also a due to a special case of the flat distribution. Data collection and grouping should be reviewed before investigating the process-related causes.

Figure 26.2(f) is an example of a separated distribution. This is frequently an extreme example of the double-peaked distribution where the distributions averages are very different. If one of the two distributions has fewer values, it is likely to come from a little used process, for example, overtime, a spare machine, a little used supplier.

The edged distribution, Figure 26.2(g), is frequently seen where there is a specification limit or a target. People do not want to report values below the specification limit and so may falsely report values just above, resulting in an unexpectedly high number of observations at this value. If the specification limit is at higher levels, then the "edge" will be at the right of the chart. An extreme example of the edged distribution could occur, for example, when reporting equipment downtime. Some people not wanting to admit to having downtime report zero, whilst others report honestly. In this case there would be a single peak at zero, with a skewed distribution starting somewhere above zero.

Figure 26.2(h) is an example of a skewed distribution, where the mode (most frequently occurring value) is not central, and there is one long "tail" and one short "tail". This is a very common distribution in counts data where the average is low. For example, the number of accidents, failures or rejects typically follow this type of distribution. This is because although there is a practical minimum (of zero) there is no maximum. The skew can be in either direction, but that shown is by far the most common. Skewed data with the "hump" on the right should be investigated.

Finally, Figure 26.2(i) is a truncated distribution. This shape may occur, where for example, inspection has removed the items with lower values.

The power of the histogram: railway supplies

A railway stores area takes key measurements of critical equipment as it is delivered. A histogram of one such key measurement is shown in Figure 26.3(a). The histogram

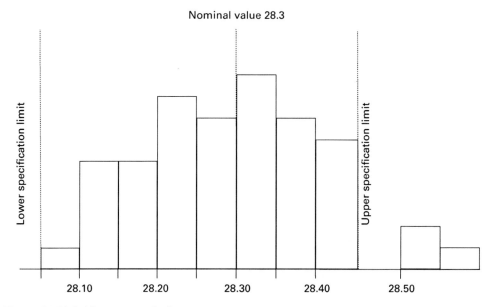

Figure 26.3(a) Histogram of a key measurement

includes specification limits and the nominal value. The histogram does not look like the bell-shaped normal distribution that we would expect from a measured value, and there are some values above the upper specification limit.

The chart suggests that there may be more than one supplier for this item, and on investigation it was found that there were three suppliers. The histograms for all three suppliers and the overall histogram are reproduced in Figure 26.3(b).

The reader might like to interpret these histograms. If the railway is intending to move to a single supplier, which one would be the first to talk with?

To help interpretation, proposed normal curves have been added to the histograms as shown in Figure 26.3(c).

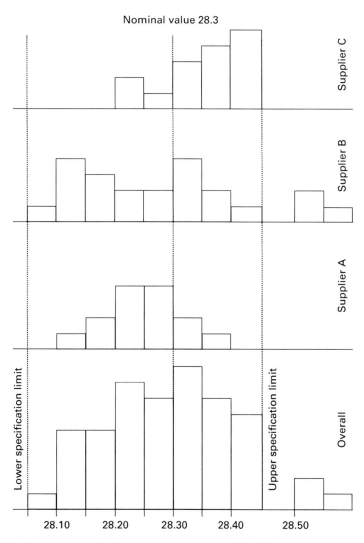

Figure 26.3(b) Histogram of a key measurement with all three suppliers and overall histogram

Analyses of these charts suggests:

- Supplier A's process is not properly centred (i.e. the nominal value does not coincide with the peak of the histogram). However, all the units delivered are well within specification, and have a variability of six units (i.e. the width of the histogram is six units).
- Supplier B appears to have three processes – perhaps three machines – each operating at different average values, none of which are centred on the nominal, and the one with the highest values, all above the upper specification limit, is only used occasionally – perhaps this is a standby machine? If the processes were centred, the variability would be five units (i.e. less than supplier A).
- Supplier C appears to be regularly producing units above the upper specification limit and carrying out inspection on site to reject those above the upper specification limit. The variability of the process is likely to be at least eight (assuming that we have reached the peak of the normal curve).

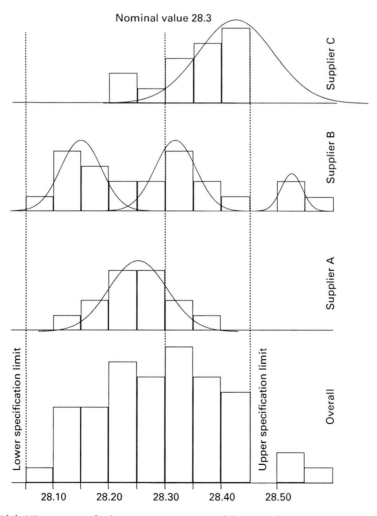

Figure 26.3(c) Histogram of a key measurement with normal curve superimposed

Since changing the nominal value of a production process is normally much easier than reducing variability, supplier B, with the smallest variability would be approached first. The railway should discuss the theories outlined above and seek their cooperation in re-centring the process and aiming to keep them centred by monitoring. If the analysis does reflect what is happening, and the supplier agrees to re-centre and monitor the process, we expect to see future units centred on the nominal with a span of five units.

The main (but not only) role of histograms within SPC is to tell us about the distribution of the data being charted, and it is useful to draw and check the histogram to ensure that the distribution is as expected.

Run charts

Helpful though the histogram is, it does not tell us anything about the behaviour of the process over time. For example, it could be that all the low values were recorded first and the high values last, or that after every low value there was a high value. The tool for revealing trends or patterns in the data is the run chart.

A run chart is a plot of a process characteristic, usually over time. Like the histogram, it is able to help us identify the maximum and minimum values and the average. Unlike the histogram, it can also help us to identify non-random patterns including trends, seasonality and process jumps.

Chart 26.1 is a run chart of the viscosity data discussed above. We see immediately that the process is chaotic. The first 12 points or so are reasonably stable, lying between 14.1 and 14.4, possibly with an increasing trend. The process then jumps to 14.8 and proceeds to trend downwards to around 14.3. Thereafter, the variability in the process increases dramatically and the process appears behaving wildly.

The control chart is a natural extension of the run chart as it is simply a run chart with control limits.

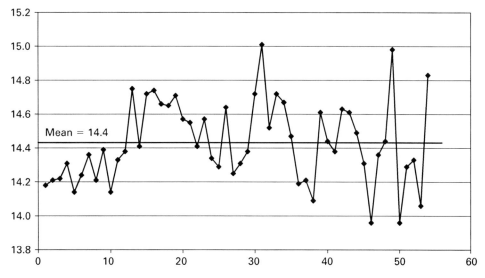

Chart 26.1 Run chart: viscosity

Bar charts

A bar chart displays the relationship between two variables one of which is numeric, the other of which is a category.

Figure 26.4(a) is a bar chart depicting the reasons for emergency hospital admissions. The reasons for admission are given along the bottom of the chart, and the number of admissions for each reason is given on the vertical axis.

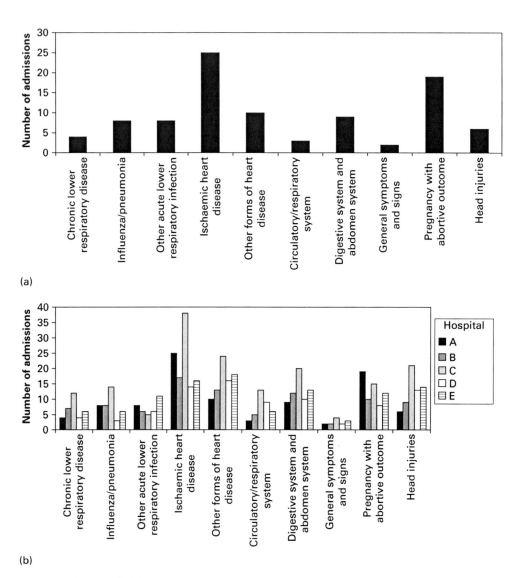

Figure 26.4(a and b) (a) Bar chart showing the reason for emergency admission in hospital A and (b) grouped bar chart for five hospitals

The chart shows that the most common reason for admission, for the period over which the data were collected, was heart disease.

It is also possible to show the numbers of emergency admissions at each of several hospitals using a grouped bar chart, as in Figure 26.4(b). Each group of bars represents the data from one particular reason from each of the five hospitals. This facilitates the comparison of admission reasons between different hospitals. For example:

- General symptoms is the least common reason for emergency admissions at all hospitals, and there is not a great difference in admissions between the hospitals.
- Ischaemic heart disease is the most common reason overall, but there is a large variability between hospitals, with hospital B having the largest number of admissions.
- Hospital C has the highest number of admissions overall, but has fewest admissions due to "other acute lower respiratory infection".

With further inspection, other comments could be added.

Another development of the basic bar chart is the stacked bar chart. Using the same data as in Figures 26.4(b), 26.4(c) compares the total admissions for each hospital. Each hospital has its own bar with strips representing a different "reason for admission" stacked on top of each other. This style of chart allows easy comparison of total admissions, and it is easy to see that hospital C has the highest admissions of around 170, with the other four hospitals having between about 85 and 105. With so many items in the stack, analysis of the stacks is difficult. If the number of categories is kept to less than about four, it is also possible to compare each element of the stack. However, as an example, a cursory glance will show that "pregnancy with abortive outcome" (the second strip from the top) is higher for hospital A than other hospitals, including hospital C which has a higher overall admission rate.

It would be possible to stack the bars by cause of admission as shown in Figure 26.4(d), allowing comparison of the major causes of admission among the five hospitals.

Returning to Figure 26.4(c), if we want to compare the percentage of different reasons for admissions rather than the number, we could re-scale the element of each stack not as the number, but as the percentage of admissions for each reason, as in Figure 26.4(e). Now it is much easier to compare the reasons for admissions in each hospital. For example:

- "Pregnancy with abortive outcome" (second highest strip) forms a higher percentage of admissions for hospital A than for other hospitals.
- "Head injuries" (top strip) is lower for hospital A than for other hospitals.

Again we note that with so many items in the stack it becomes difficult to differentiate between them.

With this set of data there are eight charts that we could draw:

1. Number of admissions by reason for admission (one chart for each hospital) (Figure 26.4(a)).
2. Number of admissions by hospital (one chart for each reason for admission) (not drawn).
3. Number of admissions by hospital grouped by reason for admission (Figure 26.4(b)).
4. Number of admissions by reason for admission grouped by hospital (not drawn).
5. Number of admissions stacked by reason by hospital (Figure 26.4(c)).

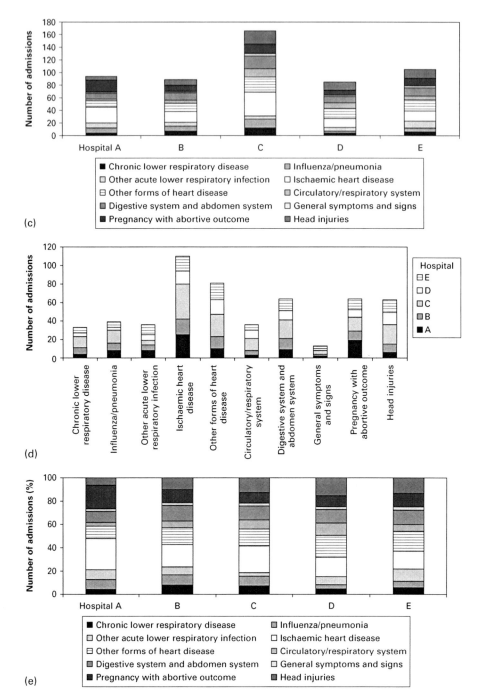

Figure 26.4(c–e) Stacked bar chart of (c) the five hospitals showing the reason for emergency admission, (d) the reason for emergency admission for five hospitals, (e) five hospitals showing by percentage the reason for emergency admission

6. Number of admissions stacked by hospital by reason (Figure 26.4(d)).
7. As Figure 26.4(c) but reported by percentage (proportion) (Figure 26.4(e)).
8. As Figure 26.4(d) but reported by percentage (proportion) (not drawn).

With so many charts even for a simple situation, it is easy to get swamped with charts and lost in the analysis.

These charts have a variety of purposes two of which are:

- To provide different views of the data and enable important features to be identified.
- To aid in communicating to others the conclusions reached during the analysis. It is important to select the format of the bar chart that focuses on the conclusion you want to draw (i.e. grouped or stacked, reported by number or proportion). The question to ask is: "what is the best way to illustrate this conclusion with a chart?"

Ranked bar charts and Pareto charts

Whilst a bar chart is better at conveying the information in data than a table, it can frequently be improved further by simply ordering the bars in decreasing frequency. This is illustrated by Figure 26.5(a), which is a ranked bar chart of Figure 26.4(a). We see immediately the major and minor reasons for admission. In addition, it is easy to see that the top two reasons are much higher than the remaining eight.

When used for process-improvement purposes (e.g. if our bar chart instead depicted reasons for delays in busses, and we wished to reduce the number of busses running late) we would focus on the top two causes. Dr J. M. Juran (Juran Institute) would call these the "vital few", as they are likely to give us a higher return on investment that focusing on the other "useful many" causes.

The ranked bar chart is the first step towards creating a Pareto chart (Figure 26.5(b)). The Pareto chart is a chart specifically designed to separate which are the "vital few"

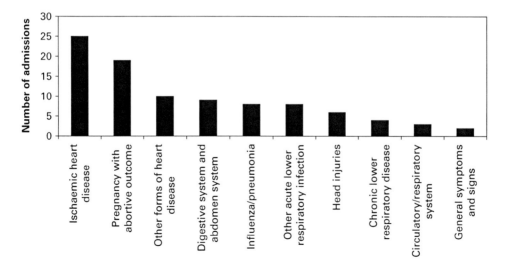

Figure 26.5(a) Ranked bar chart of the reason for emergency admission

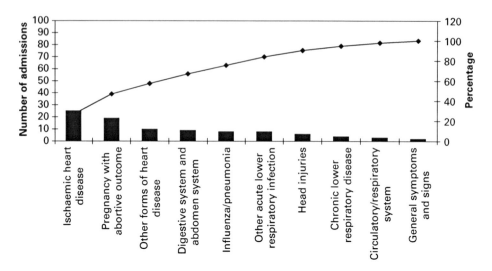

Figure 26.5(b) Pareto chart of the reason for emergency admission

from the useful many. In addition to the ranked bar chart, we calculate and plot the cumulative percentage of the categories. For example, in Figure 26.5(b) there are a total of 94 emergency admissions. The most frequent is due to ischaemic heart disease with 25 admissions. This equates to $100 \times (25/94) = 26.6\%$ of all admissions. The next most frequent is pregnancy with abortive outcome with 19 admissions. The top two reasons account for $25 + 19 = 44$ admissions, which equates to $100 \times (44/94) = 46.8\%$ of all admissions. After all the causes are added, the total of 100% is reached, as indicated on the right-hand side axis. Often, though not so clearly in this case, the line representing the cumulative totals is very steep to begin with and then flattens. It is the items to the left of where the line flattens that are deemed to be the vital few.

The concept behind the Pareto chart is important. Without the concept we may be tempted to try to reduce all the causes of breakdown, accidents, failures, rejects, etc. What the Pareto chart and the principal behind it encourage us to do is to recognise that there is unlikely to be one single problem or one single solution. It encourages us to break down the problem and look for and work on the vital few causes of the problem. It is far more likely that we will be able to solve these vital few cause one at a time rather than trying to find one solution to solve all problems.

Check sheets

The check sheet, also known as a tally chart, is a simple tool which can be used when we want to collect information on how often an event occurs. Figure 26.6 shows a check sheet used for reasons for emergency hospital admissions.

Turning a check sheet on its side gives a good indication of what the corresponding bar chart (or histograms) will look like.

Reason for emergency admission	Tick for each entry				
Chronic lower respiratory disease	1 1 1 1				
Influenza/pneumonia	1 1 1 1 1	1 1 1			
Other acute lower respiratory infection	1 1 1 1 1	1 1 1			
Ischaemic heart disease	1 1 1 1 1	1 1 1 1 1	1 1 1 1 1	1 1 1 1 1	1 1 1 1 1
Other forms of heart disease	1 1 1 1 1	1 1 1 1 1			
Symptoms and signs involving circulatory/respiratory system	1 1 1				
Symptoms and signs involving digestive system and abdomen system	1 1 1 1 1	1 1 1 1			
General symptoms and signs	1 1				
Pregnancy with abortive outcome	1 1 1 1 1	1 1 1 1 1	1 1 1 1 1	1 1 1 1	
Head injuries	1 1 1 1 1	1			

Figure 26.6 Check sheet for emergency hospital admissions

Scatter diagrams

A scatter diagram is a graphic representation of the relationship between two variables. The rationale behind the scatter diagram is as follows:

- Almost any set of data will show variation.
- This variation may be related to another variable.
- Relationships are easier to see in a scatter diagram than in a table.

Figure 26.7(a) shows the relationship between the weight of active ingredient in a certain medication and the time since the medication was manufactured. Figure 26.7(a) suggests:

- As time increases, the weight of active ingredient decreases.
- As time increases, the variability in the weight of active ingredient increases.

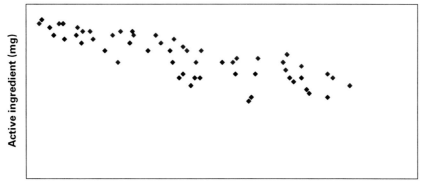

Figure 26.7(a) Scatter diagram of active ingredient vs. time since manufacture

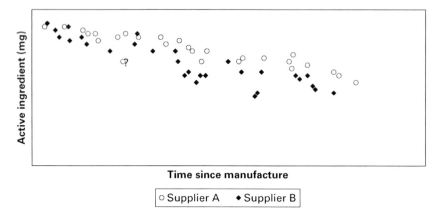

Figure 26.7(b) Scatter diagram of active ingredient vs. time since manufacture with two suppliers

In this example, it was known that there were two suppliers of the medication, and so the relationship between the active ingredient and time was drawn separately for each supplier as shown in Figure 26.7(b). For each supplier the amount of active ingredient decreases with age, but the amount of active ingredient appears to be higher for supplier A than supplier B. We also see that the point labelled "?" appears very low for supplier A compared with the rest of A's data. This could be a genuine outlier, but it is always wise to verify outlying data points. If the point is correct, can we find an explanation for it? If we can, it may help us understand the process of degradation or, if this is a manufacturing issue, identify a problem in manufacturing.

The fact that the variability increases with time is not unusual in some types of data.

Other common patterns that are likely to occur are shown in Figure 26.8 and include:

- *No relationship.* There is no obvious relationship between the variables plotted, the points are scattered.
- *Positive relationship.* As X increases so Y increases.
- *Negative relationship.* As X increases so Y decreases.
- *Complex relationship.* As X varies Y varies, but not in a straight line.

It is important to realise that just because two variables are related this does not imply that one "causes" the other. For example, there may be a strong relationship showing an increase in journeys taken on public transport and increases in the price charged for the journey, but this does not mean that increasing the cost of a journey causes an increase in the number of journeys, or vice versa. However, both may be caused by a third factor.

There are various methods of determining the strength of relationship between two variables. The most common is the correlation coefficient. The correlation coefficient is a value lying between $+1$ and -1 which measures the linear relationship between the variables (i.e. whether there is a straight line relationship). A correlation coefficient of zero indicates no relationship, of 1 a perfect positive relationship, that is X and Y

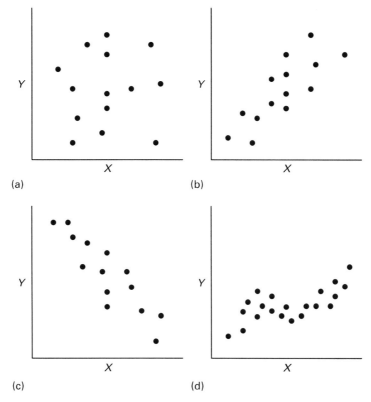

Figure 26.8 Patterns of relationship between two variables (a) No relationship, (b) positive relationship, (c) negative relationship and (d) complex relationship

increase together, and -1 a perfect negative relationship, where as X increases Y decreases. Details of calculating and interpreting the correlation coefficient may be found in basic statistical textbooks.

Summary

In this chapter we have introduced some common useful tool for analysing data. They, along with some less numerical tools, such as flow charting, cause–effect diagrams and brainstorming for the core of tools commonly used by many process-improvement teams.

Further information on process improvement can be found in many books that are available on the topic. A brief introduction to the more advanced methods of Six Sigma is provided in Part 6 of this book.

27 Setting up a processing monitoring system

Introduction

One of the main aims of this book is to encourage managers to use statistical process control (SPC) routinely in their organisation. This chapter focuses on how to begin effective control charting whether organisation wide, within a department or as an individual.

There are many ways of implementing SPC, or for that matter Total Quality Management (TQM), benchmarking, Six Sigma and the many other management tools and methodologies. Many books have been written on these subjects; books have also been written explaining why organisations fail to implement these tools and methodologies effectively. One of the advantages of implementing the ideas presented in this book is that whilst it is possible, with management commitment, to integrate SPC into the whole organisation, it is equally possible for individuals to begin using these ideas on their own and in isolation. Due to the expansive literature on implementing these business tools, we confine ourselves to the key stages of using control charts.

There is also a much simpler reason for cutting straight to the nitty-gritty of measurement: there is nothing like experience! A committed individual will succeed to some degree in using control charts to understand what is happening in their part of the organisation and improve performance. For this reason I suggest that you just start, try something. If that does not work think about why, learn from the experience and try something else. Pick something that is relatively simple, something of interest and for which you have data. Draw a few charts, and treat the exercise as an opportunity to learn. To begin with you do not need a control chart package (though you probably soon will). You can draw run charts in standard spreadsheet packages and add control limits yourself; the formulae are provided in this and many other books. Do read and learn from other SPC-related books.

Once you have a little experience in control charting, or if your organisation decides to integrate SPC with its way of working, you may wish to adapt a more formal process for using SPC. In this chapter we discuss an effective process for charting which is summarised in Figure 27.1.

The steps to effective use of control charts are:

1. deciding what to chart;
2. creating a framework for measurement;
3. collecting, charting, analysing and deciding on appropriate action.

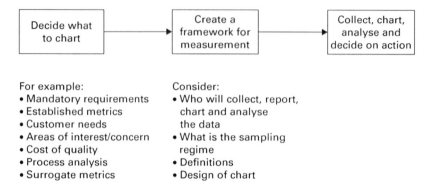

For example:
- Mandatory requirements
- Established metrics
- Customer needs
- Areas of interest/concern
- Cost of quality
- Process analysis
- Surrogate metrics

Consider:
- Who will collect, report, chart and analyse the data
- What is the sampling regime
- Definitions
- Design of chart

Figure 27.1 Implementing control charting

Deciding what to chart

In many situations the metrics of interest will be well known in the organisation. However, there may be occasions where it may be worthwhile reviewing the metrics that are being used in the organisation. There are a variety of methods of selecting what to measure. They include:

- Mandatory requirements
- Established metrics
- Customer needs
- Areas of interest/concern
- Cost of quality
- Process analysis
- Surrogate metrics.

Mandatory requirements
In all organisations there will certain metrics that need to be recorded for legal reasons. For all organisations these will include financial and safety statistics, other requirements will very depending on, for example, the industry and country.

Established metrics
All organisations will collect and record a variety of metrics. Some of these will be of practical value and some will be collected just because they have always been collected. These metrics may be published on noticeboards and/or reported to management, or just filed in computers. An excellent place to start when considering what to chart is to review these metrics and select some of the useful/interesting/appropriate ones.

At the same time, it is worthwhile reviewing what metrics are being collected because in all likelihood much of it is not being used. Frank Price in his book *Right First Time* developed the golden rules:

- No data collection without recording
- No recording without analysis
- No analysis without action.

and the three possible process actions are:

- Investigate because we have an unexpected out-of-control condition.
- Improve the process, because although the process is in a state of control, it is not operating at the level we would like it to.
- Do nothing because the process is in a state of control and we can get a better return on our effort by improving a different process.

Many organisations suffer from "data constipation". They have computers, desks, reports and noticeboards stuffed full of data which nobody wants and nobody uses. If data are not being used to make decisions, perhaps it is time to stop collecting it.

> One day I was talking to one particular manager who missed the deadline for issuing his monthly report. Illness and computer problems had delayed its production from one day to the next, and in the end the manager decided not to produce it. Since no one had commented on its absence, the following month he produced the report, but did not to issue it. Again there was no comment about its absence. That was the last time he produced the report. Each month since he had taken over management of the department he had issued 26 copies of the monthly report and it had taken the equivalent of about half a person per month to produce it.

Many organisations have a developed *Balanced Scorecard* approach to managing. This approach, pioneered by Kaplan and Norton is an excellent source of metrics ripe for charting. Similarly, many organisations have developed *Critical Success Factors* which are the ideal candidates for charting.

Customer needs

Another excellent source of metrics is to analyse customer needs and measure against them. For example, a hospital wanting to discover patients' needs may carry out a survey. Some needs are likely to be easy to measure against, such as "correct diagnosis", for which the metric could be the percentage of incorrect diagnoses. In some cases it may be difficult to measure the need. For example, to measure an item such as "friendly staff", the hospital may have to carry out regular patient surveys asking them to score friendliness of staff. In other cases we may have to "translate" the need into something measurable. For example, "good advice" could well be a requirement of the medical staff, but what does that mean? It may be necessary to attempt to specify correct advice for certain situations and then monitor, perhaps by sampling, how often the correct advice was given.

In most situations there will be many customers of the process outputs. Dr Juran refers to them as a "cast of customers". For example, the check-in system at an airport has at least two customers:

- The traveller who has a variety of needs including: fast check in, short queues, choice of seat, adequately tagged baggage.
- Boarding staff: who need accurate information about checked in passengers and luggage.

In many organisations (external) customer satisfaction is the key area to get right for success and growth, but departments also need to provide a good service to their internal customers. The information technology (IT) department may have no external customers, but it is vital to the success of most organisations that it provides a good service to all its internal customers.

Monitoring customer feedback is an extremely effective way of gauging overall performance levels of products and services. Typical metrics include returns, warrantee claims, complaints, customer retention and, though more difficult, loss of good will.

Areas of interest/concern

There may be areas of concern or interest that are under scrutiny, perhaps because of complaints or planned business growth, and these may provide opportunities for measurement.

Cost of poor quality studies

Some organisations carry out "cost of poor quality" studies to identify where they are wasting money, effort, losing customers or incurring some other loss. These and similar studies can be a useful source of metrics for monitoring.

Process analysis

There has been much talk, and many books written about process flow charting over the last 15 years or so. One of the many benefits of charting a process is that it provides clues as to which metrics to monitor (see Figure 27.2). For each check in the process,

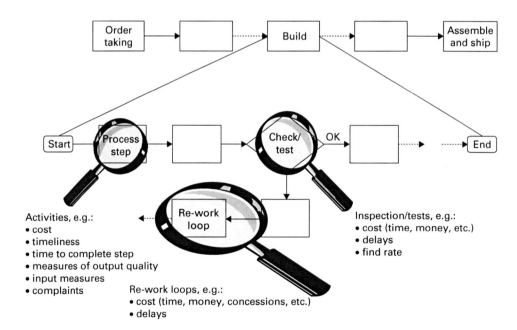

Figure 27.2 A flow chart is a good source of ideas for what to measure

represented by a diamond in the flow chart, we could consider measuring failure rates, cost, effort (time) and delays. If a check potentially creates scrap, we could include the cost of disposal, the cost and time invested in producing the scrapped item, and the costs resulting from any knock-on effects (such as late delivery). For re-work loops we could consider measuring the costs, effort and delays incurred by the re-work loop. For activities, represented by rectangles, we could measure the physical attributes of the inputs and outputs as well as costs, effort and elapsed time to complete the task. This type of analysis lends itself to both manufacturing and service industries. For example, the "inspection" may be the checking of forms, building application decisions or patient diagnosis. Activity costs and times could include the time taken to process an airline booking, phone order or job application.

One organisation charted one of their key business processes, which fell naturally into 15 major steps. Each step was charted as a deployment flowchart, typically with 50 elements. For each of these 15 major steps they selected 5–10 key metrics that would tell them how well that step was being carried out.

Surrogate metrics

It may be that it is not possible, or too expensive or difficult, to measure what you would really like to measure. In this situation it may be possible to measure a surrogate, or indicator variable which is correlated to the metric of interest. Consider, for example, health care where we might measure blood pressure to learn about some aspect of a patient's health. To take an example from an industrial process, we might record the downtime of a piece of equipment as an indicator of the quality of maintenance.

Creating a framework for measurement

Having identified what we want to measure the next step is to set up a framework for measuring it.

In many cases this will be obvious. The data may already be collected and reported in a consistent useable manner. However, if they are not there are some simple steps we can follow to help ensure that we create a good measurement framework:

● Review metrics
● Define exact requirements
● Define the method of collection and reporting
● Select and set up the control chart.

Reviewing metrics

It is worthwhile reviewing the list of metrics we have collected together and asking if there are too many or too few. Will we be swamped with data and charts? There is no answer as to how many we need, and if we choose the wrong number, or indeed the wrong metric entirely, we can easily revise what we are doing at any time. Remember that the purpose of the charts is to help us improve the way we run the organisation.

If the metric is not helping to do this, we can drop it, and if we are not monitoring something important, we can start to do so. In any event, organisational needs will change over time, and so the metrics we run the organisation on will also change.

Define exact requirements

It may be that at this stage the metric is not well defined. For example, we may want to measure the quality of water (or other liquid, gas or solid). One method of doing this is to measure the weight of different pollutants in the water sample and weight each according to its impact. The metric could then be calculated and reported as a *Pollution Index*. It is important that the exact item or event to be measured or counted is clearly defined. The definition of a clean room or table may depend on whether the room is to be used for a training course, or for surgery. How many people are in a conference room? Do we include delegates only, or staff as well? What about those who have left the room to take/make a phone call? Do we include those who registered but couldn't come? Presenters? And so on. If necessary ensure consistency in data recording by training those who will be responsible for data collection.

Define the method of collection and reporting

It is also necessary to determine how the data will be collected. For example:

- what equipment will be used;
- if appropriate, what is the sampling regime;
- when and how often will data be collected;
- who is responsible for collecting, recording and charting the data;
- who is responsible for analysing the chart and decision-making;
- if appropriate, develop a data collection form (possibly computerised). Include on it definitions, data collection procedure, etc. as appropriate.

It is beneficial to involve the people who will be measuring and recording the data in the development of the data collection process.

It is also useful to identify who measured/recorded the data, what equipment was used, operational comments and any other information which might help when checking recorded values, investigating out-of-control conditions or improving the process.

Much has been written on the topic of definitions. It is not the purpose of this book to do more than highlight the importance of, and need for, clear, concise and documented definitions not only, for collecting data, but also for other aspects of running an organisation effectively. A few examples should suffice to demonstrate why we need precise and workable definitions. What do we mean by a "clean" room? Does it mean the room has been vacuumed? Washed? It depends, of course, on the use of the room. For an ordinary office it will mean one thing, for a computer room another, and for an operating theatre something else yet again. Therefore, stating that an operating theatre shall be "clean" leaves room for interpretation. What about "late"? Again the meaning depends on the context. Civil engineers completing the construction of a 20-kilometre tunnel, a train operator, and a hairdresser with an appointments book will all interpret "late" differently.

In some cases we may need advice from experts in the field being monitored and/or from a statistician, but in many cases this will not be necessary.

A particular concern is over the frequency of measurement. There is a danger, for example, in continuous processes where changes occur slowly, that data may be auto correlated. Similarly, if we do not measure often enough we may miss important signals.

Select and set up the control chart

One of the tasks that many people find difficult when they first begin charting is selecting the appropriate chart to use. The examples and explanations in this book give advice on this topic, and as a case study shows, even though we select the "wrong" chart, the conclusions may still be reasonable!

Frequently the data are collected and recorded straight onto the control chart. For this reason it is convenient to have a hard copy of the control chart on the wall or the desk of the person collecting the data. In other situations it may be more convenient to use a control chart package, a general statistical package or a spreadsheet.

As previously suggested, it may be helpful to gather other information, such as the time, date, person recording the data, calculations, etc. When designing the control charts always consider who will use the chart, and what types of investigation/improvement activities are likely to follow and then decide what information should be recorded on the chart.

Collecting, charting, analysing and deciding on appropriate action

Deciding what to chart and setting up a framework are activities which are generally only done once, though they may occasionally be revised. The next stage (collecting, charting, analysing and acting) is a continuous loop.

When we are ready to begin recording data and entering them onto a control chart there are normally three distinct sub-processes that we need to go through:

1. *Establish the average and control limits*: This needs to be done for each new chart OR when a process change has been verified.
2. *Monitoring*: Once the average and control limits are established we enter the regular phase of plotting points on the chart and analysing the chart to determine whether the process is showing any out-of-control symptoms. If it is, we investigate the cause, if it is not we can choose whether to improve the process or not.
3. *Investigating the cause of an out-of-control signal*: If the chart exhibits out-of-control signals we need to investigate and identify the cause, so that we can take appropriate action to eliminate the cause or mitigate its effects.

Process improvement is not considered as one of the steps because it is not normally a continuous process. Only a small number of processes will usually be targeted for improvement at any one time. The impact on charting is only likely to occur when the process has been modified and the chart will hopefully identify the change as an out-of-control condition.

These three sub-processes are discussed in more detail below.

Establish the average and control limits

As the reader will be aware, process outputs vary and we use the data collected to esti-mate the true process average and standard deviation. Whilst we can estimate the average from only one data value (the average would be estimated as equal to that data value), it would be unwise to place much credence in the estimate.

As a simple, if somewhat crude experiment, this can be illustrated by throwing a stand-ard six-sided unbiased dice. The first throw may be a 2, and we would estimate the process average as 2. With only one data point we are not yet able to calculate the stand-ard deviation. The next throw may be a 3 yielding an estimated process average of $(2 + 3)/2 = 2.5$. The third throw may be a 6 yielding an estimated process average of $(2 + 3 + 6)/3 = 3.67$, and so on. As we increase the sample size so the estimate of the average should get nearer to the true average, which is 3.5. Exactly the same happens with the standard deviation: with more values our estimate of the standard deviation approaches the true standard deviation, and hence the estimate of the control limits approaches the true values for the control limits. The same concept applies to all meas-urements: incident rates, times, costs, etc.

When we are establishing the average and limits (see Figure 27.3) for the control chart we would not normally calculate them until we have about 10 data values (many authors suggest 20 or 30 data values). Whilst it is possible to work with fewer data values, we need to be extremely cautious when interpreting the results. Once we have plotted the average and limits we can begin interpreting the chart. As we collects val-ues we enter them on the chart, but we would recalculate the average and the limits either every few points, or when we record an unusually high/low value that we think will significantly change the average and limits. There is no formula as to exactly when this should be done, but with experience the analyst will develop his own understanding of when it is necessary. Certainly by the time we have 20 data values there should be relatively small changes to the calculated values, as each data value has only a limited influence on the data set. It is at this time that we would consider extending the average and control limits across the whole chart. For an example of this, see the *Rods Experiment*. After this time we may still decide to recalculate the average and limits occasionally, perhaps up to 30 or even 40 data values, but it should be unnecessary.

Of course, there are exceptions to the above guidelines. For example, if data are being collected every minute, we may decide to wait half-an-hour and calculate the average and control limits once. Similarly, if we are using a software package, the lim-its will probably be recalculated every time we plot a new value.

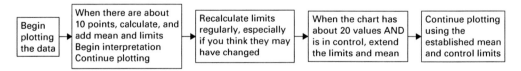

Figure 27.3 Establishing control chart average and limits

Monitoring

Once we have established the mean and control limits we enter the monitoring phase. During this phase the effort to maintain the control chart is minimal. Plotting a data value takes a few seconds and interpreting the chart normally just a few more. It is only when we suspect an out-of-control condition that we begin to look more closely. A typical monitoring process is given in Figure 27.4.

Having plotted the data point, we check to confirm that the process is in a state of control. If it is, we determine whether we wish to improve the process (normally, this would be only when there are resources available for embarking on an improvement project and we are considering which process to improve).

If the process does exhibit an out-of-control condition, or if we are just suspicious, it is possible that we have plotted the wrong value. We would check that it is indeed recorded and plotted correctly.

If the data value is correct, there are two further checks we can do. The order in which we do them depends on the cost, time delay and easy of carrying them out:

1. We can check the measuring equipment, or look for other sources of error in measuring the recorded value. If there is an error, re-sample and re-plot the point.
2. If appropriate/possible take another measurement to confirm the first reading.

In passing, it is worth noting that if the data are correct and the process is close to exhibiting an out-of-control condition, we may choose to collect more data to gather more information on whether the process is or is not exhibiting out-of-control conditions. For example, if we are monitoring the percentage of respondents to a survey we give a particular reply, an extra survey could be carried out or the sample size could be increased. Alternatively, where appropriate it may be decided to sample more frequently.

If we believe that the process has produced an out-of-control condition we need to investigate to find out what has happened (see below). There are two broad conclusions we can draw:

1. It was a process change (e.g. new machinery, new supplier, change in conditions). In this case we need to establish the new average and control limits (in addition to any action we might choose to take on the process).
2. It was an aberration (i.e. a one off-event that is unlikely to occur often: e.g. an act of god, new equipment installed that had teething problems, etc.). In these cases we normally:
   – include the point on the chart and annotate the chart so that we can see what occurred for future reference but
   – exclude the value from the calculations because it biases the average and inflates the standard deviation from the true normal operating values. Removing the point gives a more accurate estimate of the underlying average and standard deviation of the process.

In ALL cases whenever we investigate or find out something about the process that may be of use in future analysis we annotate the chart. In this way the chart becomes a history of the process.

362

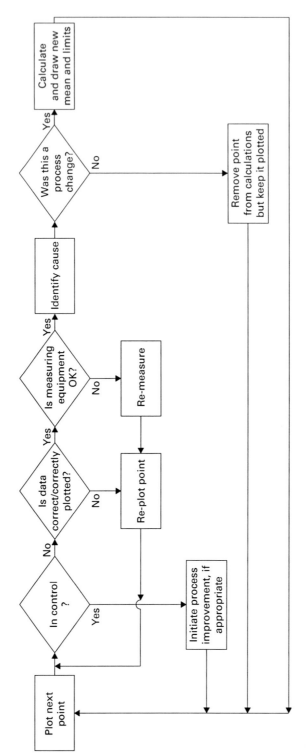

Figure 27.4 A typical monitoring process

Investigating an out-of-control signal

Having identified a genuine out-of-control condition the question that should be upper-most in our mind is "Why did we get this figure/series of figures?" and the type of signal will give us hints as to where the answer may lie.

There are typically three short-term actions that are taken when a process is found to be out of control (Figure 27.5):

- act on the item/event which is causing the out-of-control condition;
- consider what action to take on any items/events which have occurred since the previous in-control measurement;
- consider what immediate action to take to mitigate the effects of a repeat item/event.

For example, if this is a manufactured product, it may be repaired, downgraded or scrapped. It may also be appropriate to take immediate temporary action to minimise the chance of another such item getting through the system (e.g. by implementing 100% inspection). It may also be necessary to inspect all items manufactured since the previous inspection.

If the problem is due to faulty monitoring or testing equipment, for example in a hospital, it may be necessary to re-test all patients since the results were last known to be correct.

Note that these actions do not solve the problem in any way. The hope is that we may be able to continue working until the cause has been identified and resolved. In addition to the above we need to take long-term action.

The course of the investigation is dependent on the nature of the process, but there are some general points for consideration:

- What type of event could have caused this signal (see the section *interpreting a control charts* in Chapter 21 for some hints)? For example, a drifting process may suggest wear and tear; cycling may suggest two shifts or processes being mingled.
- Are there any comments on the chart that may give a clue as to the cause?
- If more than one chart is being plotted (e.g. X/MR chart) check both charts to see if there is a signal in both at the same time.

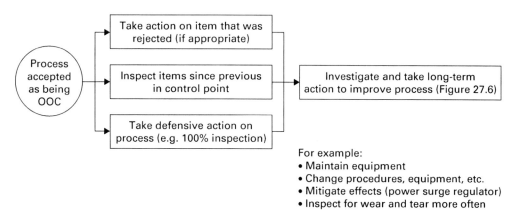

Figure 27.5 Out-of-control (OOC) signal: action on the outputs

- Draw and interpret a histogram of the data since the last process change.
- Have there been any changes to personnel, equipment, procedures, supplier, or similar that could have caused the observed value?

By asking the above, and other questions, we generate theories as to what happened to cause the unexpected result. Before taking action it is necessary to demonstrate that the theory is correct, and that we are treating the real cause and not the symptom. Re-starting a machine that continually trips is just as much treating with the symptom (tripping) as blaming staff for failures in business processes (the quality gurus all agree that most problems are outside the control of the individual). The problem is usually deeper: Why does the machine keep tripping? and Why does the process keep failing? Similarly, it is easy to confuse a theory or coincidence with a cause. Just because the computer network is offline more often since the new IT manager joined does not mean that the increased downtime has anything to do with him. It is necessary to find out and demonstrate that what he is doing has caused the increase. This is important as so many managers take a painkiller to stop the headache, but when the painkiller wears off the headache is still there!

A word on process improvement

There are three ways to "improve" the performance of a process:

1. *Edit the data* (i.e. mis-report the truth): This is a simple and effective method of making the process appear acceptable.
2. *Manipulate the system*: Sometimes, falsifying the data is not possible, so people manipulate the system. Simple examples include changing definitions (a delivery is only late if it does not arrive in the agreed day/week/month), putting injured people on "light duties" so that they are not absent from work, and hence do not appear on certain safety statistics.

   Manipulating the system is more difficult than editing the data, but often more difficult to detect.

3. *Improve the system*: It often seems to be a case that improving the system is the last resort when all else has failed. It is surprising that organisations will invest heavily on new plant, new people, new departments, buildings or on downsizing, but seldom on process improvement! Improving the process is usually difficult. The tools and techniques of process improvement are well understood and published, and their success undeniable.

> One creative production manager I knew who would personally collect goods that he had issued to stores and place them back on the production line so that they would be counted twice, thus apparently increasing production and possibly also his bonus. His actions were only discovered after many months of investigating, during which time a new security system was installed and barbed wire erected to thwart what was believed to be theft.

If a process is in a state of control but not operating at the desired level, sooner or later a decision will be made to improve it. There are many methodologies on how to do this. Unfortunately, perhaps, the most common is to go from symptom to solution without analysis.

There are many well-documented effective improvement processes, methodologies and philosophies, one of which is outlined in Figure 27.6. In this book we discuss the Juran

One purchasing department of a large organisation I worked with regularly "punished" suppliers whenever deliveries were wrong, late, lacked certification, etc. The first step was to write a letter of complaint. If there was another problem with the supplier within a few weeks they were suspended from the approved supplier list. By the warehouse manager's own experience deliveries did not seems to improve, but he did not know what else to do. On investigation (i.e. data collection, analysis, theorising as to causes and then testing to see which were the actual causes) it was found that majority of problems lay with the organisation's purchasing department. They would, for example, telephone through urgent orders, failing to ask for certification. The supplier would rush through the verbal order and deliver it before receiving the purchase order that requested certification. The warehouse would check the purchase order (with its certification requirement) against the delivery and quarantine the goods because there was no certification.

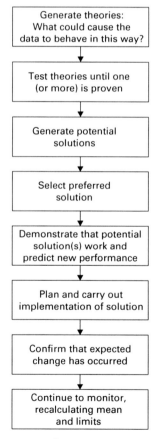

Figure 27.6 Typical process for process improvement

Institute's approach to Six Sigma. Other improvement tools include benchmarking, business process re-engineering, quality improvement, ISO 9000, supply chain management and quality circles amongst others. They are not discussed here, but the reader intent on improvement should become familiar with them.

A final word

In the early days, interpreting the chart may seem a daunting task, and we will often be uncertain as to what, if anything, is happening. This will improve with time. There are many case studies in this book, and many other books on SPC which can help gain experience and knowledge. Sometimes wrong decisions may be made, but probably far less often than they were being made without the use of control charts.

Summary

There are a variety of methods of deciding what to chart including:

- Mandatory requirements
- Established metrics
- Customer needs
- Areas of interest/concern
- Cost of quality
- Process analysis
- Surrogate metrics.

Having decided what to measure it is important to design a framework for measurement which will help ensure that the data collected meets the data requirements, that is, is collected routinely, is accurate, with required supporting information and traceability so that it can be used with confidence during monitoring, investigation and improvement activities.

When setting up a control chart it is possible to begin estimating averages and control limits with as few as 10 or so values. However, the averages and limits should be recalculated regularly until we are sure we have a stable process, usually with about 30 values. At this point we have established to base performance of the process.

When interpreting a control chart there are only three types of process decision that can be made:

- Investigate because we believe we have a special cause of variation.
- Improve the process because it is in a state of control, but not operating at the level we want.
- Do nothing because the process is in a state of control and we can get a better return on investment by improving other processes.

There are standard well-documented process improvement methodologies and techniques, of which Six Sigma is discussed in Part 6.

28 Potential process performance metrics

Introduction

In the previous chapter we discussed methods of identifying possible metrics. The purpose of this chapter is to provide generic lists of possible metrics. Before doing so, however, we review some of the issues that need to be taken into account when selecting metrics.

First, we briefly review the power of different chart and data types. This is important because having decided that we want to report, for example, the late deliveries, this may influence whether we report the percentage of deliveries which are delivered late or the difference between planned and actual delivery dates.

Secondly, we discuss the normalisation of data. Sometimes it is not appropriate to compare data directly. For example, comparing the number of traffic accidents between two cities is not appropriate unless we normalise the data in some way to take into account, for example, differing amounts of traffic.

Next, we briefly review some of the issues regarding sample size and frequency.

To create a complete list of possible metrics would be an impossible task and run to many pages. The lists provided aim is to illustrate the huge scope of metrics that are amenable to analysis by control chart, and provide a starting point for those wanting to develop metrics in their own organisation.

The power of different chart types

Some charts are more powerful than others. This means that for the same number of values they are more likely to correctly identify out-of-control signals. This phenomenon is discussed in several places in this book, but see particularly the Average run length section of see Chapters 22, 23 and 25. The conclusion of those discussions is that where we have a choice of what data to collect and how, we should take the "power" of the different charts into account. Variables charts are more powerful than the attributes charts, and of the variables charts the $\bar{X}/R$ and $\bar{X}/s$ charts are more powerful than the X/MR chart. Of the attributes charts the c and the u charts are more powerful than the np and p charts. The cusum chart is not included in this ranking because it is used in addition to any of these charts.

Another way of considering the power of the chart is to consider the data being collected. The type of data we collect will influence the amount of information we can glean from it. For example, if we monitor the percentage of late deliveries per week we will learn far less, and it will take longer to identify a process change than if we monitor the difference between expected and actual delivery dates.

Example

Suppose there is a goal that all trains should arrive within 10 minutes of the scheduled arrival time, we could collect the following data:

1. The percentage of trains arriving more than 10 minutes late.
2. The difference in minutes between scheduled and arrival time.

The first option does not tell us anything about outliers, 11 minutes late is as "bad" as 100 minutes late. It also gives no ideas as to what is possible; maybe the vast majority of trains arrive within 2 minutes. However, the advantage of reporting the percentage of trains late is that it is easier to record and analyse the data. This begs the question: What will be the data used for? If they are only being used to measure against a target of the percentage of trains arriving late, then recording the percentage late is all we need to do. If, however, we plan to use the information to help reduce the number of late trains, we would want to collect the difference in minutes, and probably for each train subsidiary information such as its route, time of day, cause of delay, etc.

A word on normalisation

In some cases it is not possible to compare data directly and we need to normalise data before comparing it. Typical examples of normalisations are:

- *Manufacturing*: In a factory, the cost or hours per unit produced (e.g. cost per car, hours per car).
- *Batch processing*: Cost or hours per unit produced.
- *Health care*: The survival rate after surgical operations depends on a variety of factors including the type of operation, the age and sex of patient. We could either develop a "risk" factor to take these and other factors into account, and report a "risk-adjusted" survival figure, or we could categorise data into groups such as males between the age of 30 and 40 years undergoing open heart surgery. Whilst the risk factor is more elegant and results in a larger quantity of data that can be directly compared, the development of such a factor may be difficult. Categorising the data is easier, but the more categories we create, the less data we have for analyses purposes in each category.
- *Education*: To compare schools, a possible metric is the number of students progressing from school to university. However, we need to categorise the schools to take into account, for example, the background of the students.

In all cases the process owner must agree with, and preferably be involved in, developing the measurement system or the results may not be accepted.

Sample size/frequency of measurement

While it is often possible to select sample sizes and sampling frequency, there are a number of issues to consider:

- The more frequently we measure the process the sooner we will identify process changes. However, if we measure too often, the data may be auto-correlated. The problem, identification and methods of coping with auto-correlation are discussed in the previous chapter.
- In general the smaller the average, the less powerful are the attributes charts. The problems associated with very small averages, rare events, are discussed in the previous chapter.
- For $\bar{X}$/R and $\bar{X}$/s charts, the larger the sample size, the more sensitive the charts are, as explained in the Average run length section of the previous chapter.

Where we are able to choose sample sizes for the $\bar{X}$/R and $\bar{X}$/s charts we need to balance the cost of extra measurements against the benefit of identifying process changes more quickly. Whilst we could carry out a cost–benefit analysis, this is probably only appropriate if the costs are high. Generally a reasonable and common starting point is to take sample size of $n = 4$.

Outcome vs. process measures

An outcome of a process is something that the process delivers. The benefit of measuring outcomes is that they are the final word on process performance. The disadvantages include that we do not know how well we are performing until the end of the process, and they cannot be managed. Whilst this may be acceptable when producing low-value mass produced products or services that can be discarded, it is not acceptable for high-value one-off products and services. In these situations we can measure process variables that will indicate how well the process is performing and we can then make adjustments as necessary. One way of viewing the difference between process and outcome measures is that a process measure is an early indicator of an outcome or result. Process metrics can often be identified by asking what might affect the outcome or result. Both process and outcome measures are important.

Examples

- Most financial measures, such as profit, are the outcomes. The number of orders is a process variable because they can be used, with other metrics, to predict, for example, profit.
- The number of road accidents are outcomes, the speed that people drive is a process variable.
- A train's arrival time is an output, its speed and breakdowns are process variables. However, the number of breakdowns are themselves output variables for which the amount and timeliness of maintenance are process variables.

Generic metrics applicable to a wide variety of organisations and sectors

The number (or frequency) of incidents

These may be reported simply as a number, or as a rate. For example, the number of accidents or the number of accidents per million hours worked.

Examples: Accidents; errors; fires; spillages; flaws in an item; customers/clients/users, grievances/complaints/returns/rejects; suggestions; late deliveries; late flight/train/bus arrivals; incorrectly completed forms; incorrect diagnoses; number of people leaving/ joining the organisation; job vacancies; spares; attendance at meetings/presentations/ conferences.

Time between events

Often we will monitor incidents, as described above, per unit time (e.g. complaints per month or complaints per 1000 clients per month). However, sometimes events occur very rarely, for example, plane crashes. In these cases most months we would record 0 and occasionally a 1. In these situations the usual approach is to measure the time between incidents and calculate that as an incident rate per year or per million units produced, etc.

For example, if there are 304 days between two train derailments, this would be calculated as 365/304 = 1.2 derailments per year. If the next derailment occurs 243 days later, the next value plotted would be 365/243 = 1.5 incidents per year.

Typical examples are included in above, but it is not applicable to all the examples listed there.

Times and costs to complete tasks

A task can vary from one small step in a process, for example time taken to take a telephone order measured in minutes or even seconds, to the time taken to complete whole processes or sub-processes such as the design of a new aircraft.

Times can be measured in absolute values (weeks, minutes, etc.) or against a target, requirement, plan or historic figure. To measure against a target, for example, we may use the following formula:

$$\text{monitored value} = \frac{\text{planned time} - \text{actual time}}{\text{planned time}}.$$

Examples: Time to: complete a task, project, order or analysis; complete checks; dispose of rejects/waste/inspection failures. Difference between planned/promised/agreed delivery/completion and against actual time.

Times can also be measured as a percentage of total time. For example, in a hospital, nursing time as a percentage of total time worked by all staff.

In addition to or as well as measuring times to complete tasks, it is possible to measure costs. In particular, costs of inspections/tests and re-work/failure/warrantee/scrap are commonly used to monitor non-added value costs.

Activity-specific metrics

In addition to the generic metrics, there are a large number of metrics that pertain to specific departments, industries or other groups.

Sales/retail

EXAMPLES:
There has been a lot of work done on identifying metrics for retail outlets, and they are well known in the industry. For the sake of completion we list a few here.

Sales by volume, value, number of sales, maybe per visitor, site, employee or sales person. Transactions per teller, check-out person, check-in person (e.g. airports). Shrinkage, sales per square metre, sales per employee, inventory turns.

Downtime

Downtime/unavailability of equipment, information technology (IT) systems/telecommunications equipment, plant, services and associated costs. Lost production/throughput, etc. in terms of time, cost or delays due to unavailability of equipment/tools, etc.

Physical qualities

Typical and traditional areas for measurement, especially in manufacturing are physical qualities.

EXAMPLES:
Length, weight, volume, porosity, chemical concentration, contamination levels, density, permeability, brightness.

Industry-specific metrics

Health care

Whilst many of the appropriate metrics are implicit in the above discussions, there are others which are specific to health care. We include a section on health care here in recognition of its importance to society. How some of these are measured is an ongoing debate and best decided by those actively involved in the area. Some of those identified below will be applicable to non-health care applications (e.g. the legal profession).

Examples

- Accessibility to care, pain management.
- Number or percentage of medical errors/mis-diagnoses/medication errors, unplanned re-admissions.
- Patient safety incidents (e.g. falls).
- Mortality rates for different medical conditions or surgery.
- Complaints from staff or patients, stratified (e.g. by department, type of complaint).
- Times from patient initial contact through each step of the process until discharge or beyond.
- Times for laboratory/X-ray, etc. turnaround stratified by type of test (e.g. urine, blood).
- Times between request for and completion of test/surgery, etc. (e.g. between an abnormal mammogram and obtaining a definitive biopsy).
- Spare beds as a percentage of capacity.
- % patients re-admitted for the same diagnosis, perhaps within a certain time period.
- % returns to operating theatre.
- % patients with complications after a certain treatment.
- % vaginal (or c-section) births.
- % patients picking up illness whilst in hospital.
- Post-operative length of stay.
- Results from questionnaires (e.g. "bedside manner").
- Condition monitoring: blood sugar, cholesterol, blood pressure, temperature.

Oil and gas/process/chemical industries

Many of the metrics in these industries are covered by those already discussed. However, there are others peculiar to these industries, a few of which are given below. Similarly, some of those below are also appropriate in other industries.

Examples

- Viscosity, impurities/contaminants and other characteristics of produced fluids, gases, waste products/discharges/flares/vents.
- Values of environment monitoring concentrations, etc.
- Number of process upsets; false alarms; results from condition/vibration/corrosion, etc.; monitoring; number of off-specification/rejected, etc.; blends; flow rates; extrusion rates; etc.
- Yield, chemical/film/paint thickness.
- Corrosion/wear rates.

Business (general)

Though we do not discuss specific metrics, the following areas are often monitored using a wide variety of metrics:

- Business plan performance (actual against plan)
- Sales performance (actual against plan)
- Production (actual against plan)

- Availability of materials/parts/sub-assemblies, etc.
- Plant/equipment utilisation (both actual and plan)
- Inventory management including obsolete inventory
- Work in progress management
- Bill of materials' accuracy
- Delivery performance
- Incoming materials/assemblies, etc.; timeliness/defects/problems
- Receipt/inspection/warehousing metrics.

Education

- Number of courses offered
- Pass or failure rates
- Drop out rates for courses
- School leavers destinations (e.g. percentage to further education)
- Truancy rates
- Class sizes, pupil to teacher ratio.

See also Chapters 6 and 11.

Transport systems

- Delays and cancellations
- Service frequency
- Road usage
- Traffic flow rates
- Numbers of accidents (e.g. on a particular stretch or road, at specific times)
- Number of speeding and other traffic offences
- Journey times.

PART 5

Developing SPC Skills: Organisational Review Questions, Workshops and Exercises

This part of the book is aimed at helping develop SPC understanding and skills.

The Rods Experiment (Chapter 29) is a practical experiment that can be used to teach control charting and is written as an experiment that SPC tutors can use with course delegates. However, it can also be treated as a case study and it is possible to carry out the charting and calculations in parallel as you read through it. It is particularly valuable for those new to SPC as it shows how a wide variety of charts can be used to analyse a set of data in real time.

The second chapter (Chapter 30) in this section is a series of tools and exercises, which can be used in two ways:

- The first set of questions are aimed at helping gather information on the monitoring, analysis and reporting methods in an organisation. This information can be used to appraise how effective the systems are and identify where improvements are needed.
- The remaining exercises are aimed at developing understanding of SPC.

Many of the workshops are short case studies and can be treated as an extension of Part 3 of the book. The data are presented for analysis while suggested analyses and conclusions are provided in Chapter 31.

29 The Rods Experiment
A practical case study that can be used for training

Introduction

The Rods Experiment is a practical experiment that can be used to teach the process of control charting. Whilst it is written initially for statistical process control (SPC) tutors, anyone can benefit from reading through the chapter as it shows the relationship between the following tools and how they contribute to monitoring process performance:

- Table of data.
- Check sheet.
- Histogram.
- Run chart.
- X/MR chart with a known change of average in the rod length, and demonstrates how to maintain a control chart in "real time".
- $\bar{X}$/range chart.
- $\bar{X}$/s chart.
- Cusum chart.
- Difference chart.
- Z chart.

Materials

The experiment is explained on the assumption that each person will draw an X/MR chart, and $\bar{X}$/range chart, a cusum chart and two histograms. The experiment can equally be run using charting software.

If you are intending to run this experiment you will need:

- One set of rods of various lengths cut to an average of 6 inches with a standard deviation of 0.2 inches, cut to the nearest 0.1 inch. I suggest a minimum of 50 rods, but any number can be made up. For ease of description these will be called the red rods (Figure 29.1).
- One set of rods as above, but set to an average of 6.2 inches. These rods must be marked in some way to distinguish them from the red rods. For the sake of explanation, they will be called the yellow rods.
- One ruler at least 7 inches long marked off in units of 0.1 inch.
- One box to keep the rods in.
- One data collection sheet per person.
- One check sheet per person.
- One blank X/MR chart per person.

- One blank $\overline{X}$/range chart per person.
- One blank cusum chart per person.
- One difference chart per person.
- One Z chart per person.

It is not necessary to run the experiment for ALL the charts. For an introduction to SPC the last four charts are usually omitted as they are for more advanced delegates. If you want to discuss the other charts, but do not want to take the time for the delegates to draw them, you could issue pre-drawn charts as examples, but of the course the data for these charts is unlikely to be the same as the data collected by the delegates. With a little experience the experiment can be varied to meet specific delegate needs and available time.

Setup

The experiment as described here has the following steps:

1. Collect the data.
2. Fill in a check sheet.
3. Draw a histogram.
4. Create a run chart on the control chart.
5. Complete the X/MR control chart.
6. Introduce the process change and monitor data collection in real time.
7. Draw and interpret an $\overline{X}$/range chart.
8. Draw and interpret a cusum chart.

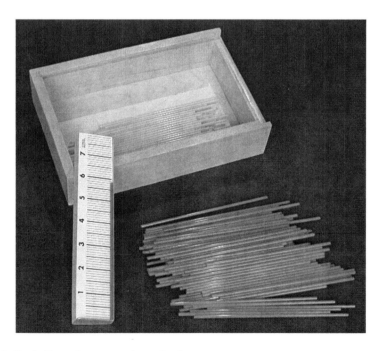

Figure 29.1 Rods Experiment: rods and ruler

9. Draw and interpret a difference chart.
10. Draw and interpret Z chart.

Remove the yellow rods from the box, hand out the data collection sheet.

The setup is explained from the viewpoint of a manufacturing company making and selling rods to a customer. There are various scenarios that the delegates can be asked to assume; the scenario used here is that the delegates are production team leaders in the company and we are about to sample their output.

Note: The advantage of adopting a scenario related to manufacturing is that it is easy for delegates to translate what is happening in the classroom to a real situation with which they can identify. When working within your own organisation you may wish to adapt the method explained here to meet your own needs.

Data generation and collection (Charts 29.1 and 29.2)

The data are collected in the following way:

1. Each person in turn will mix the (red) rods in the box, select a rod from the box and measure it calling out the length of the rod.
2. When the length of the rod is called out, the data collection sheet is filled in with the value (see Chart 29.1).
3. The rod is returned to the box, and the box passed to the next delegate.
4. Once the box has been round all delegates, it is returned to the first delegate and the process begins again (round 2). If you are intending to also draw an $\bar{X}$/range chart, ensure that the delegates enter the data from the second time round in the second row.
5. At least four rounds of the delegates will be required to draw an $\bar{X}$/range chart. Otherwise between 20 and 30 measurements in total will be sufficient.

A completed sheet for eight delegates each selecting four rods is given in Chart 29.2.

	Rod lengths															
Round 1																
Round 2																
Round 3																
Round 4																
Round 5																
Round 6																
Round 7																
Round 8																
Round 9																
Round 10																
																Grand total
Total																
																Grand average ($\bar{\bar{x}}$)
Average ($\bar{x}$)																
																Average range ($\bar{R}$)
Range (R)																

Chart 29.1 Rods Experiment data recording sheet

Rod lengths

Round 1	5.9	6.1	6.2	5.9	6.0	6.0	5.8	6.4				
Round 2	6.1	5.9	6.1	5.9	5.6	5.8	6.2	6.1				
Round 3	5.8	5.9	6.0	6.0	5.8	6.5	6.1	5.9				
Round 4	6.2	5.8	5.7	5.9	6.0	6.1	5.8	6.0				
Round 5												
Round 6												
Round 7												
Round 8												
Round 9												
Round 10												

Total _____   Grand total _____

Average ($\bar{x}$) _____   Grand average ($\bar{\bar{x}}$) _____

Range (R) _____   Average range ($\bar{R}$) _____

Chart 29.2 Rods Experiment data recording sheet (completed)

As you become familiar with the experiment you can begin to comment on each rod measured. For example, once a rod of less than 6 inches is drawn you can explain that the customer will not be happy. They require rods to be at least 6 inches long and any excess they cut off. If the following rod happens to be longer, you can explain that this "demonstrates" that telling the workforce what is required has a positive effect. On the other hand, if the following rod is still less than 6 inches, you can ask if the delegate understands the requirement; as further short rods are drawn you can feign anger, offer to pay for good performance, etc. Whenever a rod of exactly 6 inches is drawn the delegate can be declared the "employee of the month" and congratulated. Should you happen to catch a delegate lying (i.e. pretending to have a 6 inch rod), ask if this is what happens in their own organisation when management pay for good performance (or blame individuals when things go wrong).

Check sheet (Chart 29.3)

This step is optional. It demonstrates how to complete a check sheet and is used in the next step to show that a check sheet can be used as a histogram.

Delegates transfer the data from the data collection sheet onto the check sheet. You can use this time to introduce the check sheet and its uses, making the comment that if we want to use a check sheet it would have been faster to record the data straight onto it. However, data entered directly on a check sheet cannot be used to draw a control chart because we will not have a record of the order in which the data were. The completed check sheet for our example data is given in Chart 29.3.

Histogram (Charts 29.4 and 29.5)

Delegates create a histogram from the data collection sheet. The easy way to do this is to read off each value and blank out the corresponding square on the histogram.

Class interval	Midpoint			
5.05–5.25	5.15			
5.25–5.45	5.35			
5.45–5.65	5.55	I		
5.65–5.85	5.75	IIIII	II	
5.85–6.05	5.95	IIIII	IIIII	III
6.05–6.25	6.15	IIIII	IIIII	
6.25–6.45	6.35	I		
6.45–6.65	6.55	I		
6.65–6.85	6.75			
6.85–7.05	6.95			

Chart 29.3 Rods Experiment check sheet

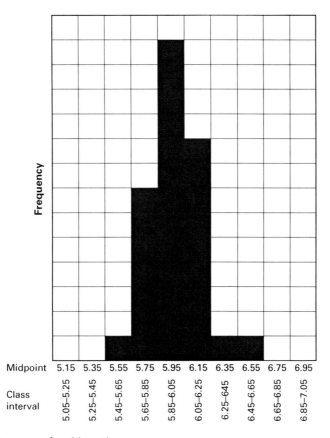

Chart 29.4 Histogram of rod lengths

Chart 29.4 shows the completed histogram. You can use this time to introduce the histogram and its uses.

Ask the delegates to interpret the histogram. If the sampling is unbiased we expect to see a "normal" distribution.

Point out to the delegates that turning the check sheet through 90 degrees results in a histogram.

Creating a run chart (Chart 29.5)

Ask the delegates to fill in the rod length, and plot the values on the x part of the control chart, thus plotting a run chart (a plot of a process characteristic over time), Chart 29.5. Ask them to interpret the chart. To prompt comments ask:

- Are the rod lengths getting longer or shorter or staying the same?
- What is the mean?
- What are the maximum and minimum rod lengths?
- If we carry on measuring, what will future rod lengths be? What will the maximum and minimum rod lengths be?

Hopefully you will get a variety of opinions. It could be useful to record these, for example on a flip chart, and ask the same questions after drawing a control chart, by which time the answers should be much more consistent.

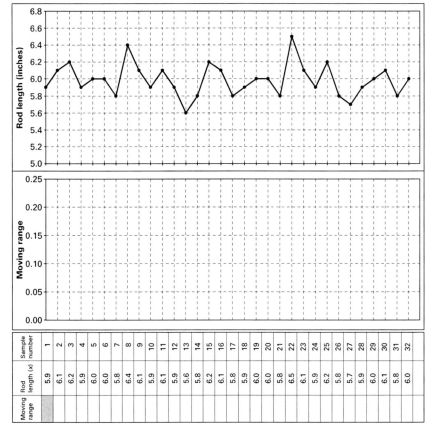

Chart 29.5 Rods run chart

Completing the X/MR control chart (Chart 29.6)

Ask the delegates to:

- calculate and plot the moving ranges,
- add lines for the mean and control limits.

Once this is done, ask the delegates to interpret the chart again, prompting with the same questions as before. See Chart 29.6:

- The average, $\bar{x}$, is calculated in the usual way = 5.98.
- The average moving range, $\bar{R}$ = 0.23.
- S_{mr}, the standard deviation calculated from the moving range = 0.23/1.128 = 0.20.
- The upper and lower warning limits are calculated in the usual way as $\bar{x} \pm 2S_{mr}$ and the action limits are $\bar{x} \pm 3S_{mr}$. The values are printed on the chart.
- For the MR chart, the UAL = $3.27\bar{R}$ = 0.75.

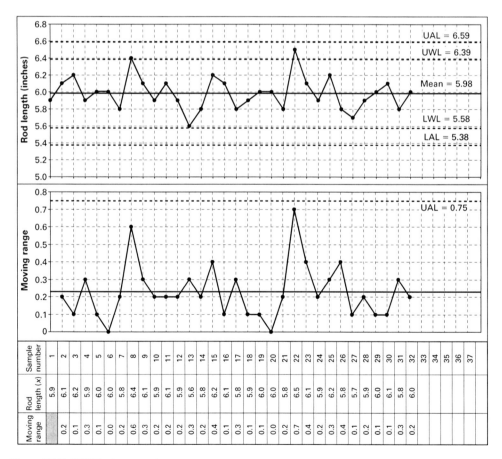

Chart 29.6 X/MR chart: rods

The interpretation should be that the process is in a state of control. The average rod length should be 6.0 inches (since rods are measured to the nearest 0.1 inch). The maximum rod length must be 6.6 inches or less as the longest rod is 6.6 inches and similarly the minimum rod length must be 5.4 inches or more.

In our example the process is in a state of control, and we can extend the average, warning and control limits beyond the data as these represent the expected process performance into the future. The MR chart can be completed if required.

Introducing the process change. Monitoring and analysing in real time (Chart 29.7 and 29.8)

Remove the red rods from the box and replace them with the yellow rods (average length 6.2 inches, standard deviation 0.2 inches).

Brief the delegates on the process change

- Tell the delegates that the customer is not happy with the number of short rods and has complained. Since the process is in a state of control, it is possible to analyse the process and attempt to improve it. (Or preferably discuss the problem and encourage the delegates to come to this conclusion.)
- Tell the delegates that whilst they were on holiday a team has been working on improving the process and it has now been changed. To distinguish the rods from the changed process, they are coloured yellow and we will continue sampling.

Brief the delegates on data collection

From now on we are going to monitor the process in real time, just as would be done manually in an organisation. Sampling will continue as before, however, once a rod has been measured, the value will be written in the rod length row of the chart, the moving range calculated and written in the moving range row and the two values plotted on the respective charts. After each point is plotted ask the delegates if there is evidence that the process has changed (i.e. an out-of-control condition).

Chart 29.7 gives a sample completed chart. In this case, samples 36 and 37 are both near the warning limits, and this may be a cause of concern. Certainly at sample 39 we have had two out of the last four points virtually on the upper warning line and one near it, and the alarm bells should be ringing! However, many delegates, being new to control charting would probably wait for a value which is above the upper action limit (UAL) before signalling an out-of-control condition. Note that there are no signals in the moving range. This suggests that only the average has changed, not the variability – which is correct.

It is now possible to ask delegates to recalculate the average and control limits from the known process change sample (i.e. sample 33 in this case) and the resulting chart is given in Chart 29.8. There are 14 samples with the yellow rods and the average length is 6.24, and the action limits are at 6.92 and 5.79. In fact, the yellow rods have an average of 6.2 inches, the longest is 7.0 inches and the shortest is 5.6 inches.

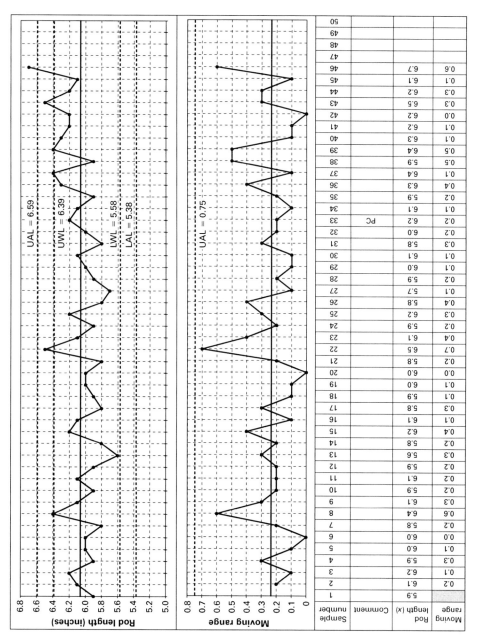

Sample number	Comment	Rod length (x)	Moving range
1		5.9	
2		6.1	0.2
3		6.2	0.1
4		5.9	0.3
5		6.0	0.1
6		6.0	0.0
7		5.8	0.2
8		6.4	0.6
9		6.1	0.3
10		5.9	0.2
11		6.1	0.2
12		5.9	0.2
13		5.6	0.3
14		5.8	0.2
15		6.2	0.4
16		6.1	0.1
17		5.8	0.3
18		5.9	0.1
19		6.0	0.1
20		6.0	0.0
21		5.8	0.2
22		6.5	0.7
23		6.1	0.4
24		5.9	0.2
25		6.2	0.3
26		5.8	0.4
27		5.7	0.1
28		5.9	0.2
29		6.0	0.1
30		6.1	0.1
31		5.8	0.3
32		6.0	0.2
33	PC	6.2	0.2
34		6.1	0.1
35		5.9	0.2
36		6.3	0.4
37		6.4	0.1
38		5.9	0.5
39		6.4	0.5
40		6.3	0.1
41		6.2	0.1
42		6.2	0.0
43		6.5	0.3
44		6.2	0.3
45		6.1	0.1
46		6.7	0.6
47			
48			
49			
50			

Chart 29.7 X/MR chart: rods after process change; PC: process change

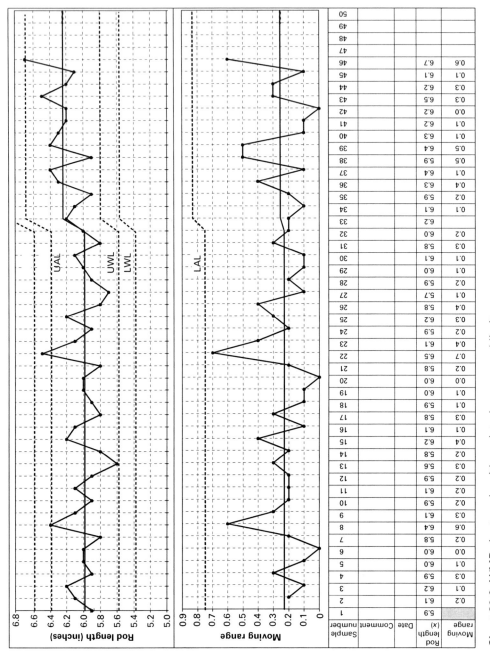

Chart 29.8 X/MR chart: rods with updated means and limits

In this situation we know when the process was changed, and it is easy to identify the onset of the process change and recalculate the limits. If you wish, you can discuss with the delegates:

1. How they would react if they were not expecting a process change.
   - Would they investigate what happened to cause an out-of-control point?
   - Would they look back to see if the process change occurred earlier? If so how?
   In our example, the high value at sample 43, and the three out of four high values between samples 36 and 39, plus the run of points above the average since sample 39 all point to the fact that sample 46 is just the latest in a whole string of suspect samples.
2. How long does it take to identify a process change (i.e. the average run length)?

Drawing and interpreting an $\bar{X}$/range chart (Charts 29.9 and 29.10)

The experiment can be extended to produce an $\bar{X}$/range chart in a number of ways, depending on what the aim of the session is. Potential options are:

- Compare the performance of the delegates by averaging each delegate's scores. The chart should initially be drawn for the red rods only, since the yellow rods do not come from the same process. The chart will have one average and one range for each delegate. Many of the calculations are shown in Chart 29.9. Having calculated the limits and average for the red rods, the delegates average rod lengths for the yellow rods can be added.

   One of the advantages of using this option is that it gives the opportunity to show how control charts can be used to chart data in ways other than time related. The disadvantage is that there is now no time relationship between this chart and the X/MR charts.

Rod lengths (inches)

Delegate Round	1	2	3	4	5	6	7	8					
1	5.9	6.1	6.2	5.9	6.0	6.0	5.8	6.4					
2	6.1	5.9	6.1	5.9	5.6	5.8	6.2	6.1					
3	5.8	5.9	6.0	6.0	5.8	6.5	6.1	5.9					
4	6.2	5.8	5.7	5.9	6.0	6.1	5.8	6.0					
5													
6													
7													
8													
9													
10													

									Grand total
Total	24.0	23.7	24.0	23.7	23.4	24.4	23.9	24.4	191.5

									Grand average ($\bar{x}$)
Average ($\bar{x}$)	6.00	5.93	6.00	5.93	5.85	6.10	5.98	6.10	5.98

									Average range (R)
Range (R)	0.4	0.3	0.5	0.1	0.4	0.7	0.4	0.5	0.41

Chart 29.9 Rods Experiment data recording sheet

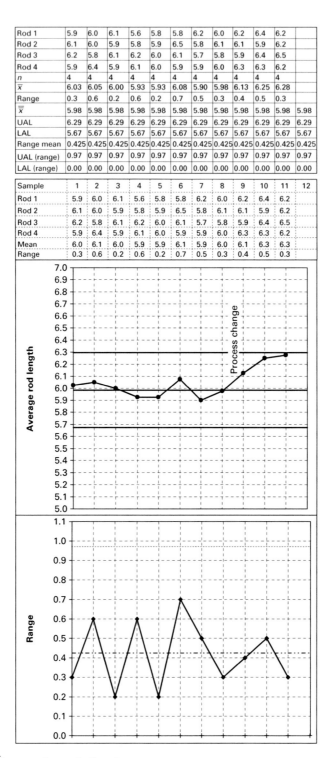

Rod 1	5.9	6.0	6.1	5.6	5.8	5.8	6.2	6.0	6.2	6.4	6.2	
Rod 2	6.1	6.0	5.9	5.8	5.9	6.5	5.8	6.1	6.1	5.9	6.2	
Rod 3	6.2	5.8	6.1	6.2	6.0	6.1	5.7	5.8	5.9	6.4	6.5	
Rod 4	5.9	6.4	5.9	6.1	6.0	5.9	5.9	6.0	6.3	6.3	6.2	
n	4	4	4	4	4	4	4	4	4	4	4	
$\bar{x}$	6.03	6.05	6.00	5.93	5.93	6.08	5.90	5.98	6.13	6.25	6.28	
Range	0.3	0.6	0.2	0.6	0.2	0.7	0.5	0.3	0.4	0.5	0.3	
$\bar{\bar{x}}$	5.98	5.98	5.98	5.98	5.98	5.98	5.98	5.98	5.98	5.98	5.98	5.98
UAL	6.29	6.29	6.29	6.29	6.29	6.29	6.29	6.29	6.29	6.29	6.29	6.29
LAL	5.67	5.67	5.67	5.67	5.67	5.67	5.67	5.67	5.67	5.67	5.67	5.67
Range mean	0.425	0.425	0.425	0.425	0.425	0.425	0.425	0.425	0.425	0.425	0.425	0.425
UAL (range)	0.97	0.97	0.97	0.97	0.97	0.97	0.97	0.97	0.97	0.97	0.97	0.97
LAL (range)	0.00	0.00	0.00	0.00	0.00	0.00	0.00	0.00	0.00	0.00	0.00	0.00

Sample	1	2	3	4	5	6	7	8	9	10	11	12
Rod 1	5.9	6.0	6.1	5.6	5.8	5.8	6.2	6.0	6.2	6.4	6.2	
Rod 2	6.1	6.0	5.9	5.8	5.9	6.5	5.8	6.1	6.1	5.9	6.2	
Rod 3	6.2	5.8	6.1	6.2	6.0	6.1	5.7	5.8	5.9	6.4	6.5	
Rod 4	5.9	6.4	5.9	6.1	6.0	5.9	5.9	6.0	6.3	6.3	6.2	
Mean	6.0	6.1	6.0	5.9	5.9	6.1	5.9	6.0	6.1	6.3	6.3	
Range	0.3	0.6	0.2	0.6	0.2	0.7	0.5	0.3	0.4	0.5	0.3	

Chart 29.10 $\bar{X}$/range chart of all rods

- Change the scenario to view the data as samples of four observations at a time (e.g. equating four samples to a shift). In this case there are eight samples of four red rods followed by three samples of four yellow rods. The corresponding $\overline{X}$/range chart is given in Chart 29.10. Note that the limits and average are based on the red rod data only. The advantage of this scenario is that the rods are being used on the chart in the order in which they were drawn and so the chart can be compared directly with the X/MR charts. The disadvantage is that it is necessary to change the scenario part way through the experiment.
- If you think that the standard deviation has changed, draw an $\overline{X}$/R chart of normalised data (described below, see Z chart), if you think that only the mean has changed draw an $\overline{X}$/range chart of the difference data (described below, see difference chart).
- It is possible to check for a difference between rounds by plotting an $\overline{X}$/R chart consisting of eight samples from each of the four rounds using the red rod data. If you want to try this option, you really need more rounds and perhaps fewer people. With this set of data there would be only four rounds with eight measurements in each round.

The data collection sheet can be used to calculate means and averages, or the control chart itself can be used, or both. For the sake of illustration, both the data collection sheet (Chart 29.9) and the control chart (Chart 29.10) have been completed.

The data and calculations are given above the control chart. The limits and mean are based on the red rods data only: that is, the eight samples of four rods:

- The first four rows reproduce the data.
- Row 5 is the number of values in each sample, n, which is 4 in this case.
- Row 6 is the average, $\bar{x}$, of each sample of four rods.
- Row 7 is the range = maximum − minimum for each sample of four rods.
- Row 8 is the mean of all 32 red rods, $\bar{\bar{x}}$ = 5.98.
- Row 9 is the UAL = $\bar{\bar{x}} + A_2\overline{R}$ = 5.98 + (0.729 × 0.425) = 6.29.
- Row 10 is the lower action limit (LAL) = $\bar{\bar{x}} - A_2\overline{R}$ = 5.98 − (0.729 × 0.425) = 5.67.
- Row 11 is the average range, $\overline{R}$, for the first eight samples (the red rods) = 0.425.
- Row 12 is the UAL for the range chart = $D_4 \times \overline{R}$ = 2.282 × 0.425 = 0.97.
- Row 13 is the LAL for the range chart, all values are zero.

The $\overline{X}$/s chart can also be drawn in addition to or in place of the $\overline{X}$/R chart. If drawn in addition to the $\overline{X}$/R chart, it will be possible to directly compare the two charts.

Drawing and interpreting a cusum chart (Chart 29.11)

Note: If you are intending to use this data to draw a cusum chart, you will need to explain the purpose, development and interpretation of cusum charts.

It is possible to draw a cusum chart of the individual rods data; Chart 29.11. A target value of 6 has been used as this is near to the red rods average. Selected target values are usually either the required value, in which case the cusum monitors actual against plan, or the average, in which case the cusum monitors against overall average. In this

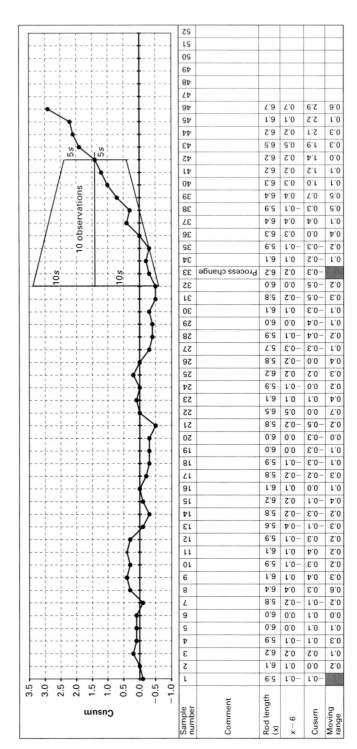

Chart 29.11 Cusum chart: rods

instance, we could have selected the overall average for red plus yellow rods, 6.063, but this is close to 6. You might like to try 6.063 as an alternative target. The cusum is calculated in the usual way as the cumulative differences from the target.

With cusum charts it is necessary to scale the chart carefully, and to do this we calculate the standard deviation in the normal way, by calculating the average moving range. For the red rods the average moving range is 0.229 and $s_r = 0.229/1.128 = 0.203$, and for the yellow rods $s_y = 0.254/1.128 = 0.225$.

For scaling it is convenient to use $2s = 0.4$ so the distance between successive observations should be equal to the same as 0.4 on the cusum axis.

The mask is created in the usual way, with $s = 0.2$, $5s = 1.0$ and $10s = 2.0$.

The mask is superimposed on the cusum chart at sample 42 and the cusum just cuts the mask, indicating a process change. If the mask is placed on sample 40, or on 41, the cusum just fails to cut the lower limb of the mask, whilst if placed on sample 43 the cusum clearly cuts the lower limb of the mask. The reader may want to construct a mask to check this.

Having identified that the process average has changed, by inspection of the chart we see that this probably occurred after the low point at samples 31 and 32 where the slope of the cusum changed. The cusum often gives an indication not only that the process has changed, but also approximately where it happened. When investigating for possible causes we would first look at what was happening at these times.

The difference chart (Chart 29.12)

Up to now we have had to split the red and yellow rods data. It could have been that the machine was set to 6 inches for the red rods and once the batch was completed, the machine re-set to 6.2 inches for the yellow rods. Similarly in our own organisations we may have a process that changes regularly. For example, on a production line, we may only take a few readings before re-setting the machine for a different product. In a hospital operating theatre we may have a few operations of one type followed by a different type. In situations where we are swapping between a limited number of outputs, for example red rods and yellow rods, we could keep a separate chart for each type of rod or output and these may be useful charts to keep. In extreme cases we may only have one reading before the process is changed, for example in project work, and therefore keeping charts for each project is not an option as each chart would only have one point. There are two methods of dealing with these situations that we discuss here:

1. The difference chart, used when the average changes but the variability does not.
2. The Z chart, used where both the average and the variability change.

For the difference chart, we simply subtract the nominal (or average or target, etc.) value from each reading. Thus for the red rods we subtract 6.0 inches (the target length for the red rods) from each reading and for the yellow rods we subtract 6.2 inches. The resulting data are plotted on the X/MR chart and the calculations carried out in the normal manner using the differences. Chart 29.12 is the resulting chart for the red and yellow rods.

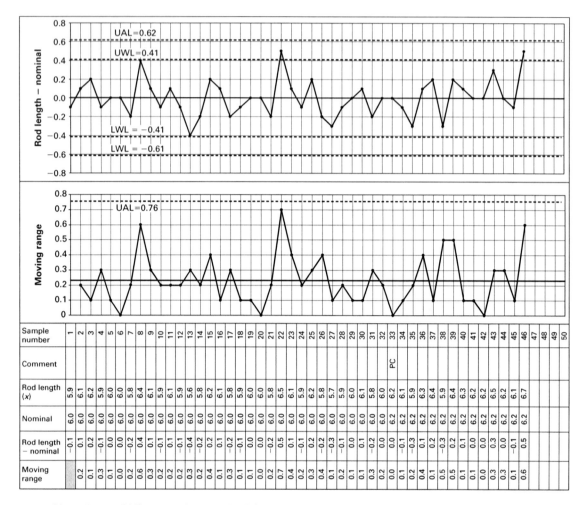

Sample number	1	2	3	4	5	6	7	8	9	10	11	12	13	14	15	16	17	18	19	20	21	22	23	24	25	26	27	28	29	30	31	32	33	34	35	36	37	38	39	40	41	42	43	44	45	46	47	48	49	50
Comment																																PC																		
Rod length (x)	5.9	6.1	6.2	5.9	6.0	6.0	5.8	6.4	6.1	5.9	6.1	5.9	5.6	5.8	6.2	6.1	5.8	5.9	6.0	6.0	5.8	6.5	6.1	5.9	6.2	5.8	5.7	5.9	6.0	6.1	5.8	6.0	6.2	6.1	5.9	6.3	6.4	5.9	6.4	6.3	6.2	6.5	6.2	6.1	6.7					
Nominal	6.0	6.0	6.0	6.0	6.0	6.0	6.0	6.0	6.0	6.0	6.0	6.0	6.0	6.0	6.0	6.0	6.0	6.0	6.0	6.0	6.0	6.0	6.0	6.0	6.0	6.0	6.0	6.0	6.0	6.0	6.0	6.2	6.2	6.2	6.2	6.2	6.2	6.2	6.2	6.2	6.2	6.2	6.2	6.2	6.2					
Rod length − nominal	−0.1	0.1	0.2	−0.1	0.0	0.0	−0.2	0.4	0.1	−0.1	0.1	−0.1	−0.4	−0.2	0.2	0.1	−0.2	−0.1	0.0	0.0	−0.2	0.5	0.1	−0.1	0.2	−0.2	−0.3	−0.1	0.0	0.1	−0.2	−0.2	0.0	−0.1	−0.3	0.1	0.2	−0.3	0.2	0.1	0.0	0.3	0.0	−0.1	0.5					
Moving range		0.2	0.1	0.3	0.1	0.0	0.2	0.6	0.3	0.2	0.2	0.2	0.3	0.2	0.4	0.1	0.3	0.1	0.1	0.0	0.2	0.7	0.4	0.2	0.3	0.4	0.1	0.2	0.1	0.1	0.3	0.2	0.0	0.1	0.2	0.4	0.1	0.5	0.5	0.1	0.1	0.0	0.3	0.3	0.1	0.6				

Chart 29.12 Difference chart: rods, PC: process change

The chart shape is, of course, the same as the equivalent X/MR chart except that the yellow rod values are effectively pulled down in line with the red rods. We have based the standard deviation on all red and yellow rods and the result is that $s = 0.208$ which is between the red rods value of 0.203 and the yellow rods values of 0.225, and is weighted towards the red rods as we have more red rods than yellow. The chart now represents a process in control, which is what we would wish to see if we were purposefully changing the mean and no other changes were occurring.

The reader might like to calculate the mean and limits as an exercise.

The Z chart (Chart 29.13)

The assumption behind the difference chart is that the standard deviation is the same for all the subsets being plotted. If this is not the case, we need to use the more complex Z chart (Chart 29.13).

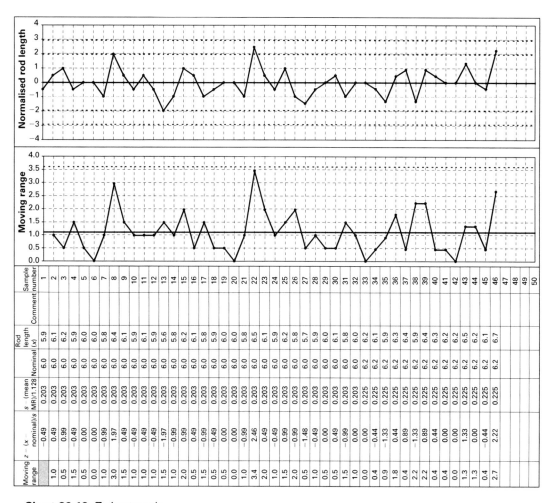

Chart 29.13 Z chart: rods

For each value, *x*, we calculate a transformed value as $(x - \text{nominal})/s$ where *s* is the standard deviation.

For the red rods data, the nominal value is 6.0 and $s = 0.203$ whilst for the yellow rods the nominal is 6.2 and $s = 0.225$.

The mean should be zero, or very close to it since we subtracted the average from each value. As we normalised the variability by dividing by *s*, the limits will be at ±3. It is interesting to note that if we use the usual formula for an X/MR chart, the mean is 0.004, the control limits are at +2.94 and −2.93, i.e. very close to the expected values.

The moving range chart for normalised data, is usually called a W chart. The average should be at 1.128 and the UAL at 3.686.

For further information on the difference and Z charts see Don Wheeler's book Short Run SPC.

Comments

- The experiment may be extended and/or varied by making other sets of rods with different variances or different distributions.
- After running this experiment many times I discovered that the average length of red rod was NOT 6 inches as would be expected, but somewhat longer. After watching delegates select rods I realised that many would tilt the box and run their fingers down the inside of the box until the end of a rod was found and select that one. Of course, this resulted in the shorter rods never being selected. This demonstrates the importance of ensuring that sampling methods (both in this experiment and in practice) are unbiased. You can ask delegates to sample by selecting the rod from the middle, thus avoiding the problem.
- After each rod is returned to the box the rods should be mixed. Failure to do so may result in biased sampling because there may be a tendency to select rods from the same place.
- It makes the case study more interesting and easier for the delegates to understand, if the "rounds" and rod lengths can be related to delegates' own work situations.

30 Organisational review questions, workshops and exercises

Most of these exercises can be done individually or in teams. They can also be set as exercises for training purposes.

Reviewing what is happening in your organisation today

1. Make a list of the reports produced by your organisation for internal use.
   Obtain samples of as many of these reports as you can.
   Review the reports. Consider:
   – What do the charts and tables tell you about performance?
   – Are things getting better or worse or staying the same?
   – What actions were taken based on the data?
   – Were these data knee-jerk reactions or did the data really suggest that the action be taken?
   – Where action was taken in the past, what effect has it had? How do you know?
   If there is enough data, try drawing a control chart. If you cannot manage a control chart, perhaps draw a run chart.
2. Find out who receives reports in your organisation. Visit some of them. Ask them:
   – Whether they actually receive the report (perhaps they do not know that they do!).
   – What they do with the report.
   – Are they making use of the information? How?
   – If not, why do they receive the report (…perhaps they like to be on the mailing list because their boss is, or because they want their name to be seen).
3. Walk round the building. Look at the notice boards.
   Where organisation performance is displayed, consider:
   – What do the charts and tables tell you about performance?
   – Are things getting better or worse or staying the same?
   – What actions were taken based on the data?
   – Were these data knee-jerk reactions or did the data really suggest that the action be taken?
   – Where action was taken in the past, what effect has it had? How do you know?
   If there is enough data, try drawing a control chart. If you cannot manage a control chart, perhaps draw a run chart.

> It was because I saw safety data being mis-reported on a notice board that I analysed it and demonstrated the mis-information in moving averages. It is that experience that forms the moving average case study.

4. Look for evidence of "tampering". Listen to the decision-making process. Do people go from symptom to cause (we were late with that delivery, we must speed up the next one).

   Keep a log of examples of tampering in your organisation: What are the effects on the process, the results and on people?

5. *Target setting*: Are targets within the organisation set on the basis of what can be achieved and evidence that it can be achieved, or are they randomly selected (e.g. 10% better than last time)? What evidence is there that the target is:
   – appropriate (not too low/high)?
   – achievable?
   Do people understand that if a process is in a state of (statistical) control, you will get what the process is delivering, and that to achieve a new target the process has to be analysed and improved?

6. Where targets (or similar) are set or where there are rewards or punitive actions taken against individuals, do people "massage" the data to ensure that targets are met? How do you know? If data are being massaged:
   – What is the effect?
   – Why?
   – Does it matter? – If not, why was the target set? If yes, what harm is set doing?

7. Do people predict future process performance? If so, how? Ask around and make a list.
   Are these predictions based on, for example, wishes, plans, targets, edicts from above, 10% less/more than last time? How often are they based on sound statistical analysis of what has happened in the past and hence what will happen next (e.g. control charts or something similar)?

Selecting performance indicators

1. Think of a process with which you are familiar (e.g. something you are involved with at work, or booking a holiday, going to hospital, or being involved with a training course, university or school):
   – make a list (brainstorm) of potential performance metrics;
   – review the list and select six that, if implemented, would monitor how well the process is performing.

2. Think of another process with which you are familiar:
   – Who are the customers of the process?
   – Select the three most important customers.
   – What do you think are the needs of these customers?
   – Select the three key needs for each customer.
   – Create the six most important metrics that would best monitor all these needs.

3. If you have a process flow chart, analyse the chart and identify potential metrics for each activity on the chart.

4. Make a list of six metrics that you are aware of in your organisation. If you do not have any, make a list of six that you think would be useful. Create a framework for measurement for each one (note that the "measurement" could be counting failures, or answers to questionnaires, etc.). You should include in your framework, as appropriate:
   – a definition of the metric;

- units of measure;
- any required calculations;
- specification of any equipment that may be needed to take the measurement;
- the sampling regime;
- how often will the measurement be taken;
- the name/job title of person who will take the measurement;
- how will the data be recorded (e.g. onto a control chart, into a computer, on a form, etc.; if necessary, draft out forms);
- who will be responsible for analysing the data;
- who will be responsible for deciding what action to take.

If you do not know the answer to some technical issues such as sampling regime, how would you find out?

Selecting the correct control chart

Which chart would you use to analyse the following metrics:

Metric	Chart
Flange diameter (maximum)	
Weight of product coming off a production line	
Weight of product coming off a production line (measurements taken in samples of 4 every 10 minutes)	
Weight of product coming off a production line (measurements taken in samples of 10 every 10 minutes)	
Days from order receipt to order fulfilment	
Number of enquiries per day	
Percentage of failures	
Percentage of time a facility (e.g. operating theatre) is idle	
Number of wrongly administered medications/ 1000 administrations	
Absenteeism rate	
(Actual value − planned value)/planned value	
Percentage of people who smoke on a daily basis	
At a conference, presentations are graded by delegates as Excellent (5 points), Good (4 points), Average (3 points), Below Average (2 points) and poor (1 point). What charts would you used to compare the quality of each presentation?	
Accidents per million man hours worked	

(*continued*)

Metric	Chart
Asthmatics use a peak exhalation flow meter to monitor their peak exhalation flow rate. What chart would be used to monitor the daily readings. (Readings can vary by more than a factor of 3, e.g. from under 100 to over 300.)	
Sometimes patients are put in restraints for their own safety. The time in minutes out of the restraint is recorded every shift. What chart would be used for monitoring this?	
Length of time spent on phone calls by help desk	
For a sales organisation, each sale either does or does not result in a complaint. What charts would be used to monitor the number of complaints per week: (1) if sales remain constant (2) if sales vary from one period to another (3) per 1000 sales	
An insurance company is reviewing claims forms and monitoring form errors. The number of forms arriving each week varies. What charts should be used in for monitoring: (1) The number of incorrect forms when 50 are sampled per week (2) The proportion of incorrect forms when 10% are selected at random each week (3) 30 forms are sampled each week, and the number of errors on the forms are monitored (4) 10% of forms are sampled each week and the number of errors on the forms are monitored	
Proportion of orders shipped on time	
Every week a health insurance company monitors the total number of days their members spent in hospital as a proportion of the total membership days	
A carpet manufacturer inspects the same area of carpet from every role produced and monitors the number of flaws per carpet. The data are recorded at the end of the day, and the number of carpets inspected every day is: (1) similar (within 25%) (2) varies widely	

Control chart interpretation

There are many un-interpreted charts in the book which the reader is encouraged to review.

Chart 30.1 is a control chart with many out-of-control signals. How many can you spot?

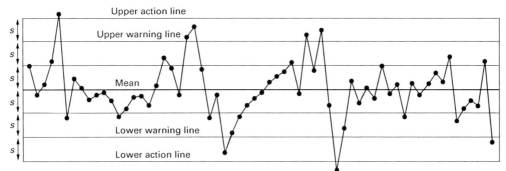

A process exhibits a lack of control if:

1. any one point is beyond the action lines
2. any two out of three consecutive points fall outside the warning limits on the same side of the chart
3. any eight consecutive points lie on the same side of the centreline
4. any eight consecutive points move upward or downward in value
5. any eight consecutive points oscillate between high and low values relative to each other
Note: The above rules do NOT hold for moving range charts. See notes on moving range for details.

Chart 30.1 Out-of-control conditions

Data workshops and case studies

In several of the case studies it is possible to start with the data presented in the tables or in the first chart and complete the analyses, checking your results with those in the book. In many cases the averages and control limits are either provided or can be read off the chart.

To complete these workshops you will either need hard copy blank charts or a standard spreadsheet package.

Red rods workshop

The Red Rods Experiment has a number of opportunities for drawing and interpreting different charts. These are described in the relevant chapter 29.

Downtime

Many organisations such as hospitals, oil companies, and the construction industry hire equipment on long term hires. In Table 30.1, the total number of hours each month when the equipment should have been available for use and the percentage downtime are given. Draw an appropriate control chart for analysing downtime and interpret it.

Repairs workshop

Units of equipment are hired out from various suppliers on a task-by-task basis. The number of repairs carried out on each of these equipment units (and incidentally other critical units of equipment) are monitored and logged. Several months after the data collection had begun a table was produced (Table 30.2). Which control chart should be used to monitor the number of repairs? Draw and interpret the chart.

Table 30.1

Month	Hours available	% downtime	Month	Hours available	% downtime
1	2360	9	17	3688	25
2	2998	27	18	6103	40
3	3159	18	19	4307	26
4	3704	12	20	2930	28
5	4278	24	21	3616	23
6	5056	32	22	4658	23
7	3198	21	23	5879	24
8	2317	25	24	5852	16
9	3399	29	25	5170	28
10	5389	29	26	6550	24
11	3877	17	27	5792	26
12	4090	17	28	4475	38
13	3504	20	29	4869	18
14	4335	21	30	6089	19
15	5704	25	31	5094	18
16	4027	17			

Table 30.2

Task ID	Number of hours usage (in '0000)	Number of repairs	Machine number	Supplier number
A	1.5140	12	1	1
B	1.5936	1	2	2
C	1.4529	9	3	1
D	1.0276	0	4	3
E	1.4535	21	3	1
F	0.6575	0	1	1
G	1.2050	0	4	3
H	1.5400	0	2	2
I	1.3180	19	5	1
J	1.3701	6	1	1

Days taken to raise invoices workshop

The number of days taken to raise invoices for 11 companies is given below. Between 4 and 6 invoices were taken for each company. (Actually, it should have been 5 for each company, but an error was made at data collection.) Draw a control chart to compare performance and interpret the chart.

Company	1	2	3	4	5	6	7	8	9	10	11
Invoice 1	12	6	12	11	12	14	15	19	28	31	10
Invoice 2	13	13	7	11	10	12	30	22	25	22	13
Invoice 3	14	7	7	7	7	9	47	13	20	20	13
Invoice 4	17	20	18	19	17	10	17	19	28	31	10
Invoice 5	13	18	19	6	6	11	17	11	29	29	
Invoice 6							19				

Blending workshop: Part 1

A lubricant centre produces 10,000 blends annually of over 700 different types, sold to customers in 2500 different grade/pack combinations. One of the important measures of plant performance is the proportion of blends that fail the first laboratory test.

Week	Number of failures	Week	Number of failures	Week	Number of failures
1	31	11	23	21	29
2	28	12	26	22	49
3	30	13	24	23	32
4	31	14	38	24	33
5	28	15	32	25	26
6	26	16	26		
7	34	17	28		
8	27	18	30		
9	28	19	23		
10	18	20	45		

In the 52 weeks prior to this data set, 10,113 blends were produced, of which 1547 failed at this test. Using an np chart:

1. Calculate the control limits based on last year's data.
2. Plot the data from the above table and interpret the chart as you do so.
3. What assumption was made in using an np chart?
4. What data do you need to check this assumption?

Blending workshop: Part 2

Following on from Part 1 above, the table below gives the number of blends tested each week, along with the number of failures.

Week	Number of blends	Number of failures	Week	Number of blends	Number of failures
1	222	31	14	210	38
2	176	28	15	190	32
3	214	30	16	207	26
4	197	31	17	215	28
5	180	28	18	218	30
6	216	26	19	183	23
7	229	34	20	220	45
8	199	27	21	213	29
9	187	28	22	200	49
10	64	18	23	194	32
11	142	23	24	209	33
12	197	26	25	171	26
13	209	24			

Total number of blends = 4862.
Total number of failures = 745.

1. Is the assumption for using the np chart valid? Why?
2. What chart should be used?
3. Draw and interpret the appropriate chart using data from this year only.

Rejected tenders

An organisation has 2000 tenders that they wish to analyse. Most of the tenders were accepted, but some were rejected. The tenders are in date order and have been split into 20 batches of 100. Each batch has been analysed and the causes of rejects categorised into one of five main reasons and "other". A tender may be rejected for one or more reasons. The resulting data table is reproduced below:

Batch	1	2	3	4	5	6	7	8	9	10	11	12	13	14	15	16	17	18	19	20
Reason for rejecting:																				
Price		1	3		2	1	1		1			3	1		2		2		2	1
Delivery		1	1	2	3	1	3	5	1	2				3	1	1	1		3	
Specification	1	1	2		1	1	2	1		1	1	2	1		1		1		1	3
Track record		1	2	1	3	1		2	1	1	1		2				2			1
Too remote			1		1				2	1			1	1	2	1		1	1	2
Other			2	2		1							1			1		1		1
Total	1	4	11	5	10	5	6	8	5	5	2	5	6	4	6	3	6	2	7	8

1. Draw and interpret the appropriate multivariate chart.
2. On the assumption that the chart is in a state of control, and based only on the information you have, where would you focus improvement attention? Why?

Washouts and twist-offs

1. Referring to the washouts and twist-offs case study (Chapter 12), what chart could you use to help identify the month in which process averages changed?
2. Draw a weighted cusum charts of washouts and/or twist-offs. What are your conclusions?

Chapter 18: workshop

Referring to the batch production process data (Chapter 18), draw and interpret a p chart for the number of off-specification batches. Compare the conclusions with the p chart for off-specification tons.

Hospitality

In the hospitality case study (Chapter 16) we discovered that the number of functions and the number of guests were related. Draw:

- Histogram of the number of guests (similar to Chart 16.8).
- $\bar{X}$/range chart of the number of functions (similar to Chart 16.9).

Do the interpretations of these charts agree with the corresponding interpretations in the case study?

Rare medical errors

Using the data in the medical errors case study (Chapter 10), draw p charts for each centre by:

1. Combining months to produce 24 bi-monthly data points.
2. Combining the half-yearly data to produce 8 half-yearly points.

Compare these charts with the monthly and quarterly charts given in the case study. What are your conclusions?

Surgical complications

Using the data in the surgical complications case study (Chapter 9), draw p charts for each centre by combining months to produce 16 quarterly data points. Compare these charts with the monthly and quarterly charts given in the case study. What are your conclusions?

Moving averages

Using the data in the cost per foot case study (Chapter 15), draw a moving average of span 6 and compare the results with the control chart. Does the moving average give more or less information than the control chart? Why?

Comparison of attributes and variables charts

There are a number of attributes data and charts in this book. Use the attributes data to draw X/MR charts and compare the results. What are the differences? What does this tell you about selecting the appropriate chart? Can attributes charts always be replaced by variables charts?

Discussions

"The central problem in management today is the failure to understand the nature and interpretation of variation" (Lloyd. S. Nelson, Nashua Corporation). What would the effect be if management understood the information in variation as explained by control charts?

"If you have a stable process there is no point setting a goal: you will get what the process will deliver".

What are the key blocks to implementing charting in an (your) organisation?
What could be done to overcome these blocks?
What would be the benefits in your organisation if control charts were used wherever appropriate?

Data Experiment

Both of the following are excellent for training courses and university course students. In addition to teaching the elements of control charting, much can be taught about management.

Buy or make a set of rods and repeat the Rods Experiment.

A similar experiment, using a c chart is explained in, for example, *Out of the Crisis* by Dr Edwards Deming (2000). Buy a set of beads and repeat the experiment.

31 Answers to exercises in Chapter 30

Selecting the right control chart

Possible answers are given in the table below:

Metric	Chart
Flange diameter (maximum)	X/MR
Weight of product coming off a production line	X/MR
Weight of product coming off a production line (measurements taken in samples of four every 10 minutes)	$\bar{X}$/range
Weight of product coming off a production line (measurements taken in samples of 10 every 10 minutes)	$\bar{X}$/s
Days from order receipt to order fulfilment	X/MR
Number of enquiries per day	c
Percentage failures	X/MR
Percentage time a facility (e.g. operating theatre) is idle	X/MR
Number of wrongly administered medications/1000 administrations	u
Absenteeism rate	X/MR
(Actual value − planned value)/planned value	X/MR
Percentage of people who smoke on a daily basis	X/MR
At a conference, presentations are graded by delegates as Excellent (5 points), Good (4 points), Average (3 points), Below Average (2 points) and Poor (1 point). What charts would you used to compare the quality of each presentation?	X/MR
Accidents per million man hours worked	u
Asthmatics use a peak exhalation flow meter to monitor their peak exhalation flow rate. What chart would be used to monitor the daily readings (readings can vary by more than a factor of 3, e.g. from under 100 to over 300)?	X/MR

(*continued*)

Metric	Chart
Sometimes patients are put in restraints for their own safety. The time in minutes out of the restraint is recorded every shift. What chart would be used for monitoring this?	X/MR
Length of time spent on phone calls by help desk	X/MR
For a sales organisation, each sale either does or does not result in a complaint. What charts would be used to monitor the number of complaints per week (1) If sales remain constant (2) If sales vary from one period to another (3) Per 1000 sales	(1) np (2) p (3) np (a p chart could be used for 1. and 3. as well, but the calculations for an np chart are simpler)
An insurance company is reviewing claims forms and monitoring form errors. The number of forms arriving each week varies. What charts should be used in for monitoring: (1) The number of incorrect forms when 50 are sampled per week (2) The proportion of incorrect forms when 10% are selected at random each week (3) 30 forms are sampled each week, and the number of errors on the forms are monitored (4) 10% of forms are sampled each week and the number of errors on the forms are monitored	(1) np (2) p (3) c (4) u The p chart may be used for (1) but the calculations for the np chart are simpler. Similarly, the u chart may be used for (3) but the calculations for the c chart are simpler. A multivariate chart would help with further analysis for (3) and (4)
Proportion of orders shipped on time	p
Every week a health insurance company monitors the total number of days their members spent in hospital as a proportion of the total membership days	p
A carpet manufacturer inspects the same area of carpet from every role produced and monitors the number of flaws per carpet. The data are recorded at the end of the day, and the number of carpets inspected every day is: (1) Similar (within 25%) (2) Varies widely	(1) c (2) u The u chart could be used for (1) but the calculations for the c chart are simpler. Could use a multivariate chart to record the type of flaw and this would help with further analysis

Control chart interpretation

Most of the out-of-control points are identified in Chart 31.1.

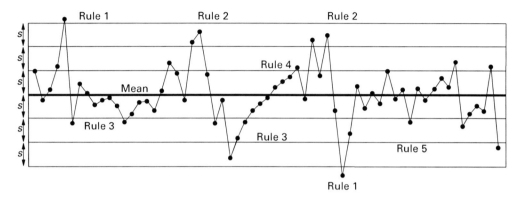

Chart 31.1 Out-of-control conditions

Downtime workshop (Chart 31.2)

The X/MR chart with calculations and histogram is given in Chart 31.2. The chart looks reasonably in control, though the histogram has a gap between the top bar at 37.5–42.6 and the rest of the data with the peak at the high end. When drawing a histogram, a useful guideline is that each bar, except perhaps for the two extreme ones, should be constructed to contain at least five observations. In this case, perhaps we do not quite have enough data, possibly the choice of category limits have contributed to the problem. It may be interesting to use different categories. However, we cannot escape the gap in the data between around 28 and 37. This probably is just a quirk of the data.

Repairs workshop (Chart 31.3)

The u chart is given with the calculations and shows that there are two out-of-control conditions. With such a small amount of data we can see by inspection that it is not always the same machine that breaks down, however, machines from supplier 1 accounts for all but one of the breakdowns. On querying this with the appropriate engineers, the explanation was that supplier 1 is the only supplier that does not supply spare machines and repairs are only logged when they result in a holdup. Where there is a spare that can be substituted no repair is logged unless the spare also fails resulting in a holdup.

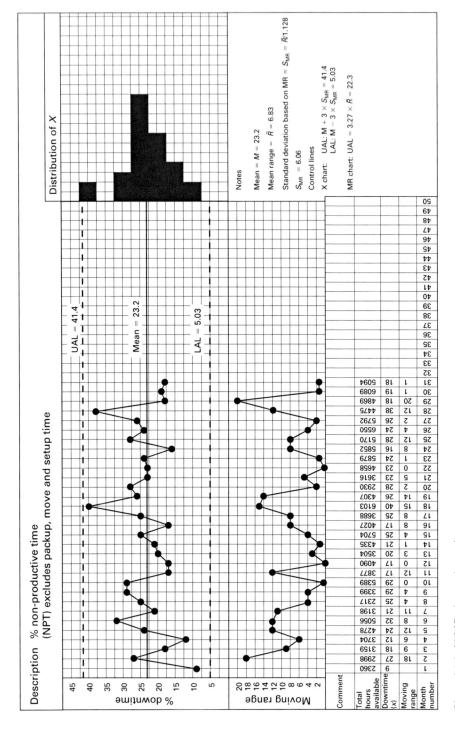

Chart 31.2 X/MR chart: downtime

Description

Equipment repairs per 10,000 hours usage by task
Tasks 5 and 9 are unexpectedly high
Machine numbers 1, 3, 5 appear to have high repair rates
Supplier 1 operates all machines with high repair rates

Calculations

n = number of hours usage
c = number repairs in the sample
$u = c/n$
$\bar{n}$ = total number of hours (10,000)/number of tasks = 13.1/10 = 1.31
$\bar{u}$ = total number of repairs/total hours usage (10,000) = 68/13.1 = 5.2
$s = \sqrt{(\bar{u}/n)} = \sqrt{5.2/1.3} = 2.0$

Limits

$\text{UAL} = \bar{u} + 3 \times s = 5.2 + (3 \times 2.0) = 11.2$
$\text{LAL} = \bar{u} - 3 \times s$

Comment	Supplier Number	Machine Number	Task	Number of hours usage in '0000 (n)	Number of repairs (c)	Repairs rate (u = c/n)	Sample number
	1	1	1	1.5140	12	8	1
	2	2	2	1.5936	1	1	2
	1	3	3	1.4529	9	6	3
	3	4	4	1.0276	0	0	4
	1	3	5	1.4535	21	14	5
	1	1	6	.6575	0	0	6
	3	4	7	1.2050	0	0	7
	2	2	8	1.5400	0	0	8
	1	5	9	1.3180	19	14	9
	1	1	10	1.3701	6	4	10

Repairs per 10,000 hours (u)

UAL = 13.6   UAL = 11.2   Mean = 5.2

Chart 31.3 u chart: repairs per 10,000 hours usage by task

Days taken to raise invoices workshop (Chart 31.4)

The $\bar{X}$/range chart suggests that there is a difference between companies. The raw data are provided above the chart, along with the results of the calculations.

The trend in the chart over the first five companies is chance, there is, of course no meaning in trends for this type of data as the companies have been ordered randomly.

The average value, $\bar{X}$, for each company is calculated as the average number of days the company takes to process invoices. For example, for company 11 this equals $(10 + 13 + 13 + 10)/4 = 11.5$ days.

The range is the difference between the maximum and minimum values. For company 11 this is $(13 - 10) = 3$ days.

The average for all the data, $\bar{\bar{X}}$, is calculated as:

$$\frac{\text{sum of all days to process all the invoices}}{\text{number of invoices}} = \frac{896 \text{ days}}{55 \text{ invoices}} = 16.29.$$

$$\text{the average range} = \frac{(5 + 14 + 12 + 13 + 11 + 5 + 32 + 11 + 9 + 11 + 3)}{11} = 11.5$$

The upper action limit (UAL) $= 16.29 + (0.577 \times 11.5) = 22.9$ for all companies with five observations.

The constant 0.577 comes from tables (explained in the procedure for drawing $\bar{X}$/range charts). The constant is 0.483 for company 7 which has six observations and 0.729 for company 11 which has four observations.

The lower action limit (LAL) is calculated in a similar way. For companies with five observations the calculation is:

$$\text{LAL} = 16.29 - (0.577 \times 11.5) = 9.68.$$

For the range chart the calculations are:

Mean range $= 11.5$ as above.
The UAL $= 2.114 \times$ mean range $= 2.114 \times 11.5 = 24.21$.

Again, the constant 2.114 comes from tables (reproduced in the procedure for drawing $\bar{X}$/range charts) and varies depending on the number of observations.

Blending workshop – Part 1 (Chart 31.5)

From the previous year's data the calculations are as follows:

p = proportion of blends of off-specification $= 1547/10113 = 0.153$.
n = average number of blends per week $= 10113/52 = 194.48$.
Average number of off-specification blends per week $= 1547/52 = 29.75$.

Company	1	2	3	4	5	6	7	8	9	10	11
Invoice 1	12	6	12	11	12	14	15	19	28	31	10
Invoice 2	13	13	7	11	10	12	30	22	25	22	13
Invoice 3	14	7	7	7	7	9	47	13	20	22	13
Invoice 4	17	20	18	19	17	10	17	19	28	31	10
Invoice 5	13	18	19	6	6	11	17	11	29	29	
Invoice 6							19				
Number of invoices	5	5	5	5	5	5	6	5	5	5	4
X̄	13.80	12.80	12.60	10.80	10.40	11.20	24.17	16.80	26.00	26.60	11.50
Range	5	14	12	13	11	5	32	11	9	11	3
X̄ mean	16.29	16.29	16.29	16.29	16.29	16.29	16.29	16.29	16.29	16.29	16.29
UAL	22.90	22.90	22.90	22.90	22.90	22.90	21.82	22.90	22.90	22.90	24.64
LAL	9.68	9.68	9.68	9.68	9.68	9.68	10.76	9.68	9.68	9.68	7.94
Range mean	11.45	11.45	11.45	11.45	11.45	11.45	11.45	11.45	11.45	11.45	11.45
UAL	24.21	24.21	24.21	24.21	24.21	24.21	22.95	24.21	24.21	24.21	26.14
LAL	.00	.00	.00	.00	.00	.00	.00	.00	.00	.00	.00
Number of invoices	5	5	5	5	5	5	6	5	5	5	4

Number	1	2	3	4	5	6	7	8	9	10	11
Company	A	B	C	D	E	G	H	I	J	K	L
Comment							OOC		OOC	OOC	
Mean	13.8	12.8	12.6	10.8	10.4	11.2	24.2	16.8	26.0	26.6	11.5
Range	5	14	12	13	11	5	32	11	9	11	3

Days taken to raise invoice — UAL; Mean = 16.3; LAL

Range — Mean = 11.5

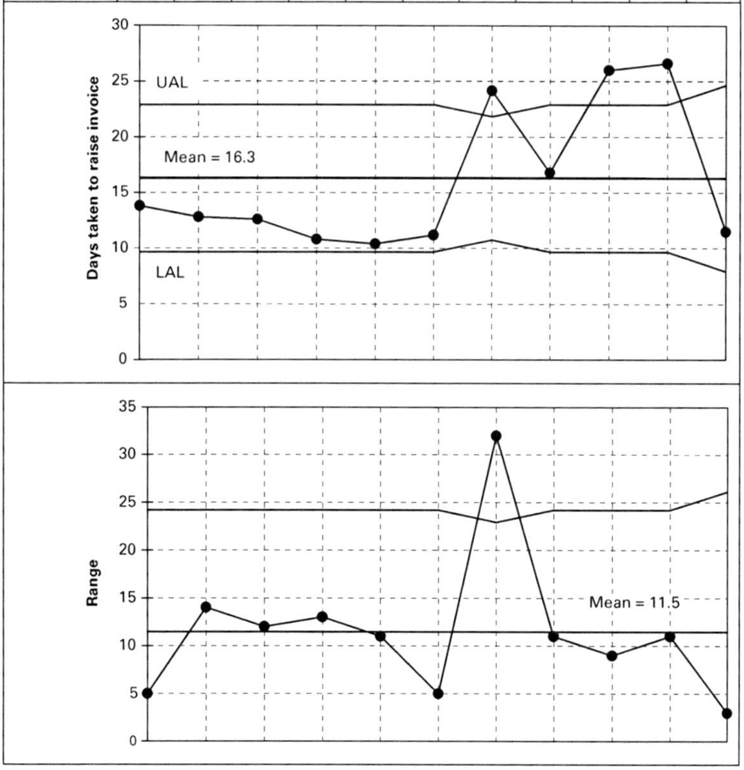

Chart 31.4 X̄/R chart: Days taken to raise invoices; OOC: out of control

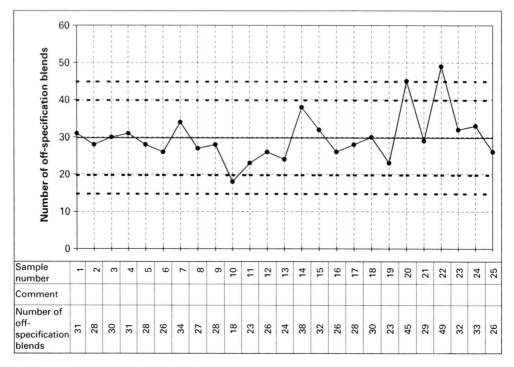

Chart 31.5 np chart: off-specification blends

$$s = \sqrt{194.48 \times 0.153(1 - 0.153)} = 5.02.$$

UAL $= 29.75 + (3 \times 5.02) = 44.81.$
LAL $= 29.75 - (3 \times 5.02) = 14.69.$

Interpretation:
Point 22 is out of control, and point 20 is on the UAL. Point 10 is low and there is a run of 6 points below the average.

Assumptions:

- We assume that last year's data is in a state of control (if not, then it is not appropriate to use these limits for any chart).
- We assume that no process changes have occurred at or near year end.
- We can easily check whether the failure rate per week is similar for last year and this year:
  - Failures per week last year $= 29.75$ (as calculated above).
  - Failures per week this year $= 745$ failures/25 weeks $= 29.8.$
    This does NOT automatically mean that it is acceptable to use last year's data to calculate the limits for this year, but it does give a small amount of confidence that it may be.
- The assumption for using an np chart is that the number of blends made, and hence tested, each week are approximately the same, that is within 25% of the average, which is between 194.48 ± 48.62. To check this assumption we need the number of blends tested per week.

Part 2 (Chart 31.6)

The assumption for using the np chart is NOT valid because the number of blends checked each week varies from 64 to 229. Therefore, we should use the p chart.

The p chart shows that the process is not in a state of control. Week 22 is above the control limit, and week 10 is very close to the limit.

Comment:

- Week 22 was the week after the Easter holiday and week 10 was the week after the Christmas holiday. An investigation should be carried out to confirm that failure rates increase after holiday periods, and if confirmed, the reason sought.
- Note that in the np chart week 10 was very low, but in the p chart it is very high. This illustrates that sometimes it is important to select the appropriate chart. (Note that in other case studies in this book, we have demonstrated that often selecting the wrong chart still leads to correct conclusions.)

Calculations:

$$\bar{p} = \frac{\text{total number of off-specification blends}}{\text{total number of blends}} = \frac{745}{4862} = 0.153.$$

The standard deviation can be calculated for each value, as shown on the chart using the formula:

$$s = \sqrt{\frac{\bar{p}(1 - \bar{p})}{n}}.$$

The control limits are then calculated as $\bar{p} \pm 3s$ and the warning limits as $\bar{p} \pm 2s$.

Alternatively, if the calculations are being carried out by hand we can use the following:

$$\bar{n} = \text{average number of blends each week} = \frac{4862}{25} = 194.48.$$

$$s = \sqrt{\frac{\bar{p}(1 - \bar{p})}{\bar{n}}} = \sqrt{\frac{0.153(1 - 0.153)}{194.48}} = 0.0258.$$

This value of s is only valid for weeks where the number of blends checked lies between (0.75×194) and (1.25×194), that is 146 and 243.

In these cases:

UAL = 0.153 + (3 × 0.0258) = 0.230.
LAL = 0.153 − (3 × 0.0258) = 0.076.

For other weeks s and the limits must be calculated separately. For example, week 10 had 64 checks and so:

$$s = \sqrt{\frac{0.153(1 - 0.153)}{64}} = 0.045.$$

UAL = 0.153 + (3 × 0.045) = 0.288.
LAL = 0.153 − (3 × 0.045) = 0.018.

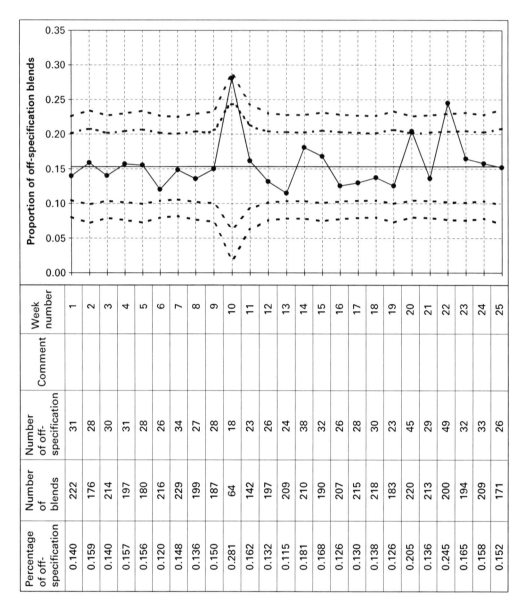

Chart 31.6 p chart: off-specification blends

Rejected tenders (Chart 31.7)

The chart is somewhat suspect. From batch 12 to 18 there appears to be some cycling. In general the reject rate appears to have reduced (note: e.g. the very high number of rejects in batches 3 and 5, and the lack of low reject rates in batches 2–10 compared to later batches).

If, however, the process were in a state of control, we could draw a Pareto chart to help show the key reject reasons. Delivery has the highest number rejects, and so we would focus attention on rejects. The first step would be to chart the number of rejected

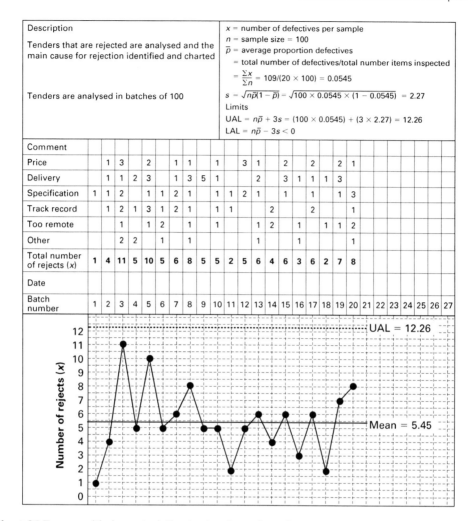

Description																												
Tenders that are rejected are analysed and the main cause for rejection identified and charted																												
Tenders are analysed in batches of 100																												

x = number of defectives per sample
n = sample size = 100
$\bar{p}$ = average proportion defectives
= total number of defectives/total number items inspected

$= \dfrac{\Sigma x}{\Sigma n} = 109/(20 \times 100) = 0.0545$

$s = \sqrt{n\bar{p}(1 - \bar{p})} = \sqrt{100 \times 0.0545 \times (1 - 0.0545)} = 2.27$

Limits
UAL = $n\bar{p} + 3s$ = $(100 \times 0.0545) + (3 \times 2.27) = 12.26$
LAL = $n\bar{p} - 3s < 0$

Comment																												
Price		1	3		2		1	1		1		3	1		2		2		2	1								
Delivery		1	1	2	3		1	3	5	1		2		3	1	1	1	3										
Specification	1	1	2		1	1	2	1		1	1	2	1		1		1			1	3							
Track record		1	2	1	3	1	2	1		1	1			2			2			1								
Too remote			1		1	2		1		1		1	2		1			1	1	2								
Other			2	2		1		1				1			1				1									
Total number of rejects (x)	1	4	11	5	10	5	6	8	5	5	2	5	6	4	6	3	6	2	7	8								
Date																												
Batch number	1	2	3	4	5	6	7	8	9	10	11	12	13	14	15	16	17	18	19	20	21	22	23	24	25	26	27

Chart 31.7 np multi-characteristic chart: rejected tenders

tenders due to delivery to determine whether it is in a state of control, and if it is, we can use all the data over this period to look for causes of delivery problems. An example of how this is done is given in the case study, Chapter 19.

The calculations are given on the chart.

Surgical complications (Charts 31.8–31.10)

Charts for each hospital (Charts 31.9(a)–31.9(d)) and all hospitals combined (Chart 31.8) are given, with some calculation details on Chart 31.8.

Chart 31.8, the results from all hospitals combined, is in a state of control, but after the first point shows little variation. However, the individual hospital charts show that they are not in a state of control:

● Chart 31.9(a), for hospital A, is the nearest to being in a state of control, but of the 11 points starting from Q2 year 1, 10 are below the average.

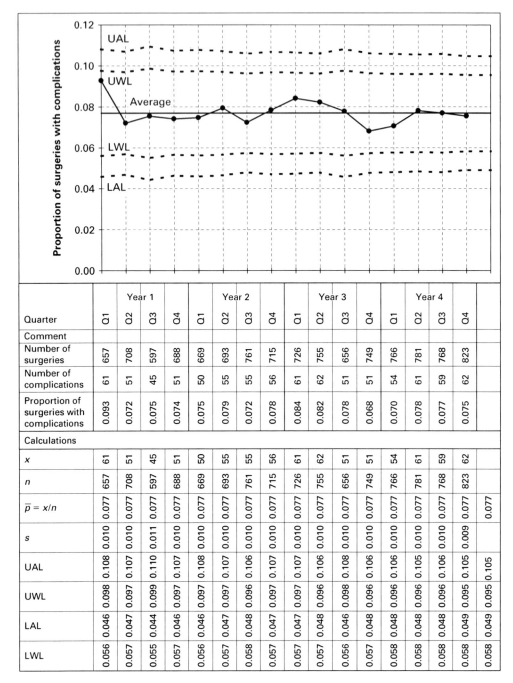

	Year 1				Year 2				Year 3				Year 4				
Quarter	Q1	Q2	Q3	Q4	Q1	Q2	Q3	Q4	Q1	Q2	Q3	Q4	Q1	Q2	Q3	Q4	
Comment																	
Number of surgeries	657	708	597	688	669	693	761	715	726	755	656	749	766	781	768	823	
Number of complications	61	51	45	51	50	55	55	56	61	62	51	51	54	61	59	62	
Proportion of surgeries with complications	0.093	0.072	0.075	0.074	0.075	0.079	0.072	0.078	0.084	0.082	0.078	0.068	0.070	0.078	0.077	0.075	
Calculations																	
x	61	51	45	51	50	55	55	56	61	62	51	51	54	61	59	62	
n	657	708	597	688	669	693	761	715	726	755	656	749	766	781	768	823	
$\bar{p} = x/n$	0.077	0.077	0.077	0.077	0.077	0.077	0.077	0.077	0.077	0.077	0.077	0.077	0.077	0.077	0.077	0.077	0.077
s	0.010	0.010	0.011	0.010	0.010	0.010	0.010	0.010	0.010	0.010	0.010	0.010	0.010	0.010	0.010	0.009	
UAL	0.108	0.107	0.110	0.107	0.108	0.107	0.106	0.107	0.107	0.106	0.108	0.106	0.106	0.105	0.106	0.105	0.105
UWL	0.098	0.097	0.099	0.097	0.097	0.097	0.096	0.097	0.097	0.096	0.098	0.096	0.096	0.096	0.096	0.095	0.095
LAL	0.046	0.047	0.044	0.046	0.046	0.047	0.048	0.047	0.047	0.048	0.046	0.048	0.048	0.048	0.048	0.049	0.049
LWL	0.056	0.057	0.055	0.057	0.056	0.057	0.058	0.057	0.057	0.057	0.056	0.057	0.058	0.058	0.058	0.058	0.058

Chart 31.8 p chart: complications during surgery – all hospitals; UWL: upper warning limit; LWL: lower warning limit

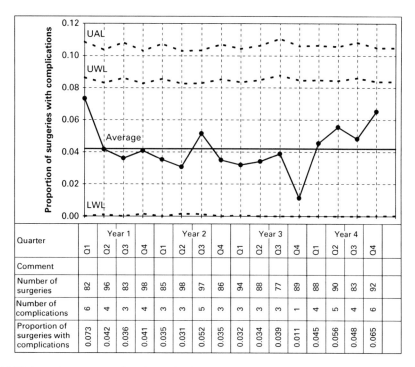

Chart 31.9(a) p chart: complications during surgery – hospital A; UWL: upper warning limit; LWL: lower warning limit

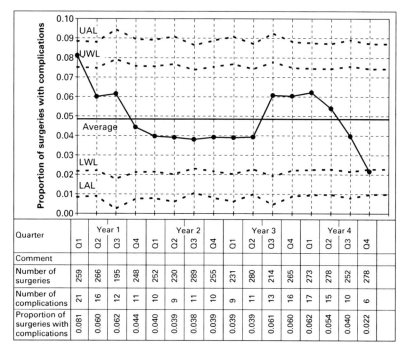

Chart 31.9(b) p chart: complications during surgery – hospital B; UWL: upper warning limit; LWL: lower warning limit

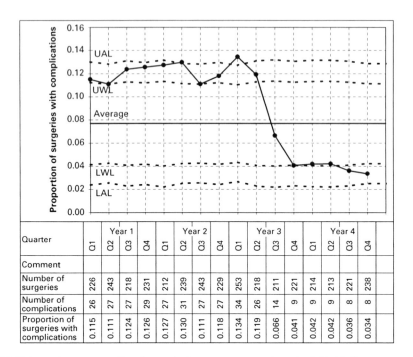

Chart 31.9(c) p chart: complications during surgery – hospital C; UWL: upper warning limit; LWL: lower warning limit

Quarter		Year 1				Year 2				Year 3				Year 4			
	Q1	Q2	Q3	Q4	Q1	Q2	Q3	Q4	Q1	Q2	Q3	Q4	Q1	Q2	Q3	Q4	
Comment																	
Number of surgeries	226	243	218	231	212	239	243	229	253	218	211	221	214	213	221	238	
Number of complications	26	27	27	29	27	31	27	27	34	26	14	9	9	9	8	8	
Proportion of surgeries with complications	0.115	0.111	0.124	0.126	0.127	0.130	0.111	0.118	0.134	0.119	0.066	0.041	0.042	0.042	0.036	0.034	

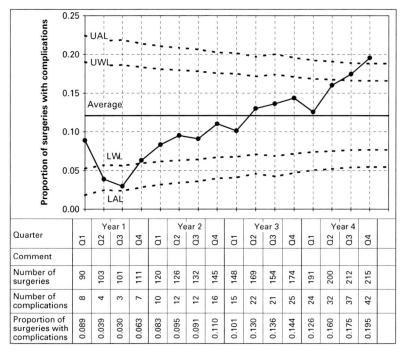

Chart 31.9(d) p chart: complications during surgery – hospital D UWL: upper warning limit; LWL: lower warning limit

Quarter		Year 1				Year 2				Year 3				Year 4			
	Q1	Q2	Q3	Q4	Q1	Q2	Q3	Q4	Q1	Q2	Q3	Q4	Q1	Q2	Q3	Q4	
Comment																	
Number of surgeries	90	103	101	111	120	126	132	145	148	169	154	174	191	200	212	215	
Number of complications	8	4	3	7	10	12	12	16	15	22	21	25	24	32	37	42	
Proportion of surgeries with complications	0.089	0.039	0.030	0.063	0.083	0.095	0.091	0.110	0.101	0.130	0.136	0.144	0.126	0.160	0.175	0.195	

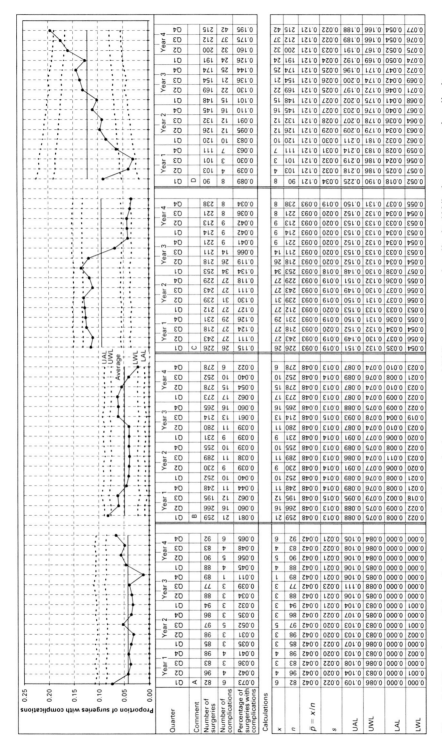

Chart 31.10 p chart: all hospitals: complications during surgery; UWL: upper warning limit; LWL: lower warning limit

- Chart 31.9(b), for hospital B, has a run of 7 points below the average, beginning Q4 year 1, the last six of which show very little variation.
- Chart 31.9(c), for hospital C, has a sharp drop in Q3 year 3, followed by another smaller drop in the following quarter.
- Chart 31.9(d), for hospital D, shows a steady increase in both the number of surgeries and the proportion with complications.

This demonstrates the importance of ensuring that subgroups of data are in a state of control before combining the data.

PART 6

An Introduction to Six Sigma

Luis Miguel Giménez

Six Sigma is the latest in a stream of successful management tools and philosophies aimed at improving organisational performance. The relationship between process improvement and SPC has already been mentioned and discussed in earlier parts of the book.

This Part, written by Luis Miguel Giménez of the Juran Institute España, SA, provides a more detailed introduction to Six Sigma and is supported by a case study to illustrate the approach in more detail.

32 An introduction to Six Sigma
Luis Miguel Giménez

Introduction

This chapter provides an introduction to Six Sigma and is based on the learning materials used by the specialists in Six Sigma, called Black Belts (BBs). It is aimed at the organisation's management, all those involved with Six Sigma projects as well as those interested in understanding the basics of Six Sigma.

Six Sigma was born in the 1980s in Motorola as a program of drastic reduction of defects in the products it manufactured. The success of this program made other companies adopt similar programs, and was originally geared towards and constrained to the reduction of variability and defects in manufactured goods.

However, Six Sigma did not become a phenomenon until the mid-1990s. That was when Jack Welch, CEO of General Electric, adopted Six Sigma as management philosophy, embodied the organisation with this philosophy, and turned "a monster into an efficient organisation". It was credited as multiplying by several times the stock value of the organisation.

What is Six Sigma?

Six Sigma provides companies with a series methodology and statistical tools that lead to breakthrough improvements in profitability and quantum gains in quality, whether a company's products are durable goods or services.

Sigma (σ) is a letter in the Greek alphabet used to denote the standard deviation of a process. (The difference between σ and s to denote standard deviation need not concern us at this stage.)

A process with "Six Sigma" capability means having 12 standard deviations of process output between the upper and lower specification limits. Essentially, process variation is reduced so that not more than 3.4 parts per million fall outside of the specification limits.

The "Six Sigma" term also refers to a philosophy, goal and/or methodology utilised to drive out waste and improve the quality, cost and time performance of any business. On average, one Six Sigma project will save an organisation between $150,000 and $200,000. BBs with 100% of their time allocated to projects can execute five or six projects during a 12-month period, potentially adding over $1M to annual profits.

Six Sigma implementation is achieved through a series of successful projects. Projects can be of different size and duration. We define a project as a structured and systematic approach to achieving Six Sigma levels of improvement. Depending on the scope of the project, they are categorised as:

- Transactional Business Process Project – an improvement of a transactional business process that extends across an organisation; such as order processing, inventory control and customer service.
- Traditional Quality Improvement Project – aimed at solving chronic problems crossing multiple functions of an organisation.
- Design for Six Sigma Project – a project aimed at incorporating the "voice of the customer" (i.e. customer needs) and Six Sigma level targets into the design of products, services or processes.

The basis of Six Sigma

Six Sigma is based on several key factors that allow an organisation to achieve important and sustained results. An organisation that adopts the "model" Six Sigma is transformed. These factors are:

- Focus on the customer. In Six Sigma the customer is the most important element.
- Focus on the process. Improvement efforts are focused on processes.
- Implement projects, establishing goals and expected benefits for each project.
- Rigorous and systematic use of cost of poor quality (COPQ) measurement on process output and the process effectiveness and efficiency.

The three key roles in Six Sigma: Management, Specialists and Staff

Six Sigma is a program in which the whole organisation participates, from top management down. There are three key roles in Six Sigma necessary to achieve success:

1. Management decides to adopt Six Sigma, selects and defines the projects, and commits to support cited projects to achieve results. The managers responsible for the projects usually are called Champions.
2. Specialists known as Black Belts, Green Belts, Yellow Belts, selected from among the best, are intensely trained so that they are able to apply the Six Sigma methodology in projects.
3. Staff of the management, who participate, with their knowledge and experience, collaborating with the specialists in the development of the projects.

The meaning of quality in Six Sigma

The concept of "quality" in Six Sigma implies giving the customer exactly what is wanted and without defects.

This approach sets up two methodologies within the Six Sigma programs:

- **Six Sigma (DMAIC)** consisting of improving current products or services and associated processes to levels of excellence.

- **Design for Six Sigma (DPSS/DMADV)** consists of designing new products or services that, from the beginning, satisfy the customer's key needs and are produced without defects.

Six Sigma programs generally start with projects geared towards the improvement of current processes. When the required degree of maturity is achieved, the organisation begins projects on designing new products and the processes that will create them.

The two key measures in Six Sigma

There are two key measures used in Six Sigma and they are different from the traditional way of measuring the functioning of the processes:

- Efficiency measures illustrate how well the process is operating and achieving its goal.
- Effectiveness measures refer not only to the number of defects, but also to the defects per opportunity.

Selecting improvement projects

To ensure the success of improvement projects, some technical and strategic requisites must be followed. The technical requirements are:

- **Chronic**, that is to say, that the problem to be solved happens frequently.
- **Manageable**, that is, it can be carried out in a short period of time, from 3 to 6 months.
- **Significant**, the expected result and benefits will have an important impact on the organisation's goals.
- **Measurable**, that there is quantifiable data, such as quantity, cost and time.

The strategic requirements are specific to the organisation, sometimes driven by the strategic plan, sometimes by the urgency of certain improvements. One also has to take into account the foreseeable resistance to change or risk possible failure.

The Six Sigma improvement methodology

To be able to understand the Six Sigma methodology it is first necessary to understand some concepts and terms that are used in projects.

CTQs or critical to qualities correspond to those characteristics of the product or service that are key for the customer and, therefore, for customer satisfaction.

Ys are the process outputs and the measures that are used to evaluate the functioning of the process and the degree to which it meets CTQs.

Xs are causes of the problem or variables of the process affecting the result.

Y = f(Xs) is the formula expressing that the final result of a process is a function of its inputs and variables.

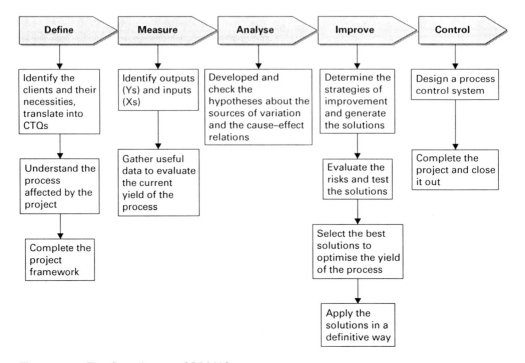

Figure 32.1 The five phases of DMAIC

Six Sigma improvement methodology is known as DMAIC, an acronym for the five phases Define, Measure, Analyse, Improve and Control. These phases and their steps are summarised in Figure 32.1.

The following describes in more detail the activities and tools that are used in each of the steps and the results that are expected when they are fulfilled.

Though there are many tools that can be used throughout the methodology, those shown are, according to our experience, those used mostly in projects. The BB must be able to select the most suitable one from the many, in each step and for each project.

Define

In the Define phase, potential Six Sigma projects are identified. Nominations can come from various sources, including customers, reports and employees.

The project problem and mission statements, as well as a team charter, are prepared and later confirmed by the management. Management selects the most appropriate team for the project and assigns the necessary priority.

The goal of this phase is to specify all the elements related to the project including the reason why it is being carried out, the problem to be solved, the goal to be achieved, and the estimated benefits and, finally, who will support the BB and the project team.

Activities

Some of the activities of this step will have been carried out by the organisation's management (the initial definition is made by the Champion) when selecting the project. At this point in the Six Sigma process, projects have been selected and roles and responsibilities have been defined. The deliverables of this phase are:

1. Identify the customers, understand their needs and prioritise those needs from the point of view of the customer satisfaction to determine the CTQs of the project.
2. Identify the process producing those CTQs. Develop a high-level process map to understand its current functioning and how well those CTQs are being met.
3. Complete the project charter, indicating the project statement, the scope, the goal to be achieved and its expected benefit, and select the team members to work on the project with the BB as well as the roles and responsibilities.

Tools

- Research methods for customer needs; i.e. interviews, focus groups, surveys, etc.
- Processes flow diagrams (process maps).
- Prioritisation matrices of needs (QFD).

Measure

The objective Measure is twofold. It consists of identifying and selecting process variables that determine the output and, evaluating the current process capability, making use of valid data and expressing its yield as a "sigma" value.

Activities

Activities of this phase include identifying process performance measures and setting their goals according to the customer information. The current process is then evaluated against the targets. The work in the Measure phase is considerable and can take many directions, depending on the nature of the project:

1. Identify the necessary measures (Ys) of the process related to the determined CTQs. Link measures to the goal process.
2. Identify, select and prioritise the variables (Xs) that may be causing the (Ys).
3. Develop a plan to gather data on Xs and Ys in such a way as to ensure the validity of the data obtained.
4. Evaluate the current yield of the process and express its result in sigma terms.

Tools

There is a broad range of data and process tools used in Measure, including:

- Brainstorming
- Process flow diagram
- Cause–effect diagram
- Matrix diagram

- Potential Failure Mode and Effects Analysis (FMEA)
- Measurement Systems Analysis (R&R)
- Process capacity studies
- General charts and graphs.

Analyse

The objective of this step is to identify those vital few variables (causes) that determine the functioning of the process and, consequently, its output. This step involves statistical methods.

Activities

During the Measure phase, the Six Sigma team accumulates significant amount of data in order to measure the current performance of the process and identify the process critical X's and Y's. In the Analyse phase, the team analyses the data. Key information questions formulated in the previous phase are answered during this analysis. A statistical approach to decision-making involves moving from data to information to knowledge.

1. Analyse graphically the data obtained to answer the questions about the cause-effect relationships in the process.
2. Structure or stratify the data and complete the graphic analysis verifying statistically the cause–effect relationships in the process.
3. Identify the few variables cause (Xs) that determines process output.

Tools

Decision-making must be based on objective facts and knowledge, and statistics is the science that brings clarity to what is otherwise insignificant data. Appropriate statistical tools and techniques are used:

- Scatter plots
- Normality proofs
- Proportion proofs, contingency tables
- Variance equality proof, t-proofs
- Variance analysis (ANOVA)
- Correlation and regression.

Improve

The goal of Improve is to introduce the necessary changes in the process that will solve the problem.

Activities

There should be clear evidence that solutions generated and integrated into redesigned processes are capable of closing the gaps between the current process and the customers CTQs requirements. The new process should also show direct financial impact, given that

Six Sigma is a business strategy to enhance bottom line revenues. The Improve phase may be the most motivating part of a Six Sigma project. The chance to develop and experiment innovative solutions is the payoff for all the hard work done in prior phases:

1. Define the strategy of improvement, generating and selecting the most effective solutions to optimise the process.
2. Evaluate the risks of the selected solutions and determine which improvement will be implemented.
3. Evaluate the cost of improvement and the expected benefit, carrying out a cost–benefit analysis of each.
4. Develop an implementation plan of the improvements and transfer it to the Champion or process owner for application.
5. Implement the improvements on behalf of the Champion or the process owner.

Tools
- Design of experiment (DOE)
- Work-out, creative thinking
- Benchmarking
- Risks evaluation
- Simulation pilot proofs
- Benefit–cost analysis.

Control

Establish process controls to ensure the results sustained.

Activities
Closing the loop on a project is essential to the ongoing success of the company-wide Six Sigma effort, and the BB is the most appropriate person to ensure this happens.

1. Create a control plan and define a system to carry it out in a systematic way. Include possible adjustments to the process. Transfer the control plan to the process owner, along with the implementation plan.
2. Standardise the improvements. Elaborate or revise the documentation of the process (procedures, methods). Identify opportunities for replication.
3. Create a final report of the project.
4. Present **the project** to Management and be ready to show its results and receive feedback and recognition.

Tools
- Self-control techniques
- Statistical process control
- Error-proofing
- Rules, procedures, auditing.

Case Study

Pre-operative prophylactic antibiotic administration

Introduction

The case study is presented in a way that would be typical of a brief summary report for Six Sigma project. It is intended to outline and summarise a typical project, and we omit detailed explanations of the medical and statistical terminology nor the use of the tools used in the project.

Define

Problem statement
Inappropriate or sub-optimal use of antimicrobial prophylaxis (ABX) process prior to surgery.

Project scope
Pre-surgical patients receiving antibiotic prophylactic therapy.

Project goals
- Proper timing of administration of antimicrobial agent prior to surgical incision.
- Appropriate choice of antimicrobial agents.

Consequences of not engaging this project
- Increased surgical infections, which are seen as potentially preventable adverse events.
- Increased medical errors which decrease patient safety and customer confidence.
- Death or loss of limb from an infection is classified as a Sentinel Event.
- Increased cost of poor quality.

Key deliverables
- To provide the pre-surgical antimicrobial prophylaxis within 0–60 minutes prior to surgical incision to assure that a bactericidal concentration of the drug is established in the serum.
- To select the correct antimicrobial that will reduce Gram organisms, such as *Staphylococcus* aureus and *Streptococcus* that often cause post-operative infections.

Table 32.1 summarises the key customer needs as expressed by the customer, the issue behind the need and the CTQ requirement.

The process map (Figure 32.2) shows how the current inpatient ABX administration process works.

Measure

Process baseline performance
The Coordinator collected baseline data to determine current status of the process and to determine whether an improvement was required. Data revealed a normal
(continued)

Table 32.1 Key customer needs

Frame	Probe and prioritise	Translate
Voice of the customer	*Key issue*	*CTQ*
Are we providing the ABX at the right time to obtain pre-operative prophylaxis?	The standard of the providing the correct ABX is 0–60 minutes before the incision	The correct time window
Are we providing the right drug to obtain pre-operative prophylaxis?	Incorrect ABX are being administered	The use of the correct ABX

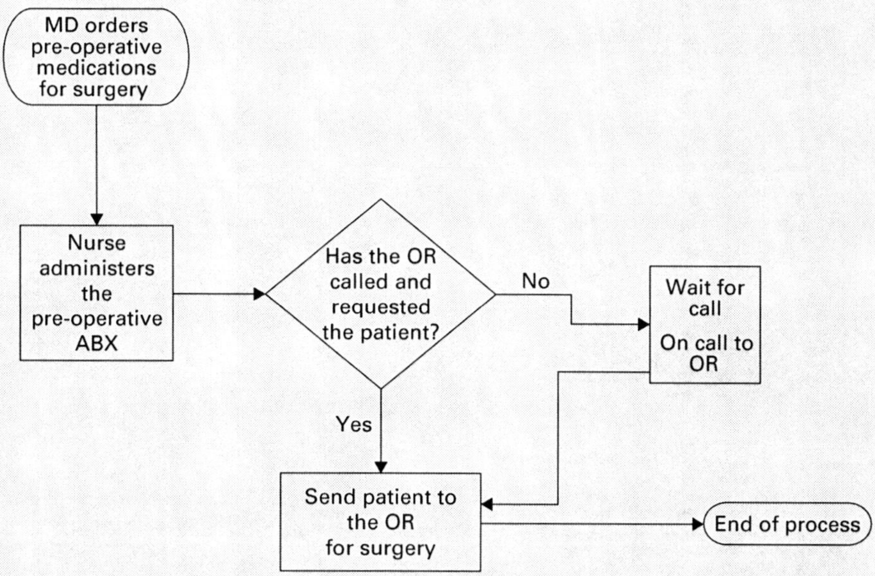

Figure 32.2 Process map inpatient ABX administration

process that was not operating with best practice specifications (see histogram and statistics for process capability analysis for time, Figure 32.3). The baseline data (removing cases where no ABX was given or given after the cut) indicates on average the patient receives the antibiotic 88.9 minutes prior to surgery and a median of 89 minutes. 80% of baseline cases exceeded the time window of 60 minutes. In addition, 17% of the patients did not receive any antibiotic and 12% of the patients received an incorrect antibiotic prior to the surgeon's incision.

Selecting variability factors
The current process of giving the antibiotic at a unit location does not permit a method to control and manage the antibiotic administration time within the window prior to surgery. The data and the current process indicate an obvious flaw in meeting the time requirement. Multi-voting among team members was used to select most

(continued)

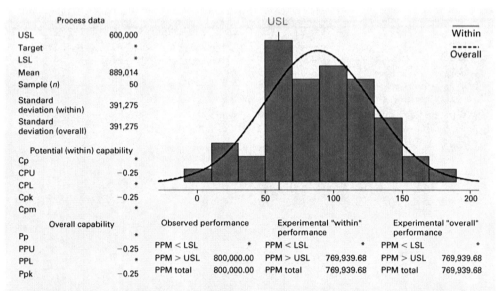

Process data			
USL	600,000		
Target	*		
LSL	*		
Mean	889,014		
Sample (n)	50		
Standard deviation (within)	391,275		
Standard deviation (overall)	391,275		

Potential (within) capability	
Cp	*
CPU	−0.25
CPL	*
Cpk	−0.25
Cpm	*

Overall capability	
Pp	*
PPU	−0.25
PPL	*
Ppk	−0.25

	Observed performance		Experimental "within" performance		Experimental "overall" performance	
PPM < LSL		*		*		*
PPM > USL		800,000.00		769,939.68		769,939.68
PPM total		800,000.00		769,939.68		769,939.68

Figure 32.3 Process capability analysis for TIME

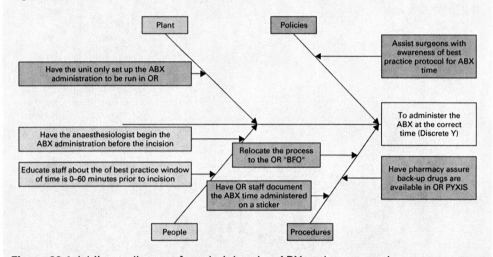

Figure 32.4 Ishikawa diagram for administering ABX at the correct time

critical factors form Ishikawa (cause–effect) diagram (Figures 32.4 and 32.5). An FMEA was conducted to explore and prioritise those factors.

Analyse

The individuals (X) control chart (Chart 32.1) demonstrate a mean of 88.9 minutes for the ABX administration, operating outside the 0–60 minutes. The data indicates an upper control limit of 202.5 minutes. Data is not operating within the specifications limits of 0–60 minutes.

The pie chart (Figure 32.6) indicates that at baseline a 71% compliance rate with 29% receiving the wrong or no antibiotic.

(continued)

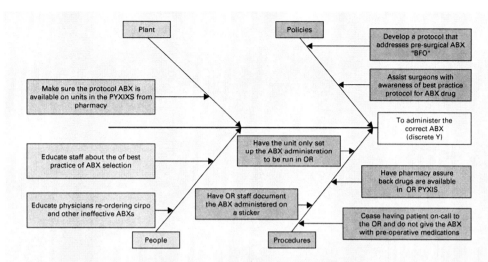

Figure 32.5 Ishikawa diagram for administering the correct ABX

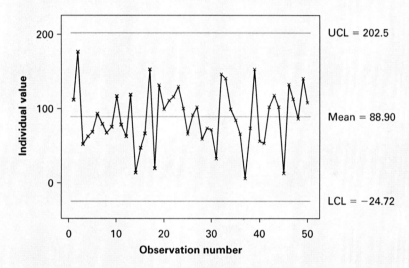

Chart 32.1 X chart of the number of minutes for the ABX administration

After several statistical analyses it was found that the "vital few" Xs are the:

- Hospital location of where the antibiotic administered.
- Protocol/standing order for selecting the correct antibiotic.

Improve

Potential solutions

- Team wanted the nurse on the floor to hang the ABX but not open the flow on the intravenous drip and assure the medication is sent with the patient.

(continued)

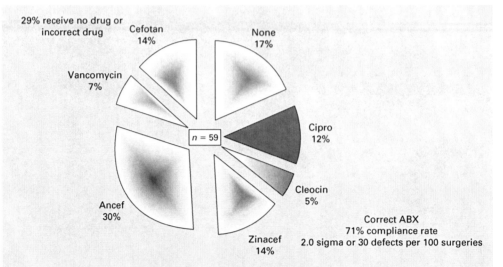

Figure 32.6 Pie chart of the percentage of ABX administered

Strength: ABX ready for infusion and no longer responsibility of unit nurse who cannot control the timing.

● The team selected to move the administration of the antibiotic to the operating room. Strength: able to judge infusion time and administer the ABX within the 0–60 minute target window.

The team used location change as the key driver in the improvement process. Also, the team realised a standing order was required to assure the physician he was selecting the correct antibiotic.

The team decided to relocate the process to the OR, developed and had approved the ABX standing order, and a sticker to collect data in a central location in the chart.

Refine solution

● The team decided to remove non-value added work from the nursing unit by not giving ABX with the pre-operative medications.
● The team moved the process to the operating room holding where capability exists to give ABX within the time window.
● The team decided to facilitate the correct drug selection by developing a pre-operative ABX protocol.

Control

Process control plan

The surgical services department has taken responsibility for monitoring this process and providing quarterly reports to Performance Improvement Council. Data collection, reporting and analyses procedures have been set up and there are procedures in place for following up in case of non-compliance with the requirement.

(*continued*)

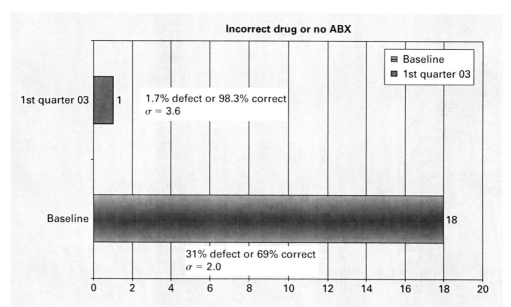

Figure 32.7 Frequency of incorrect drug or no ABX administered

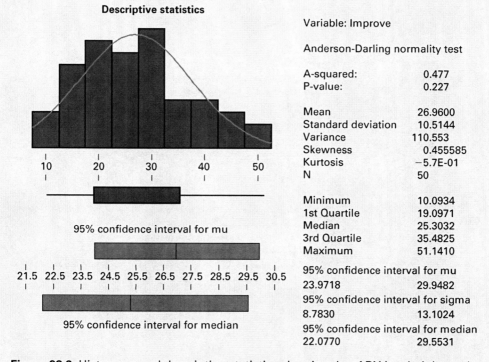

Figure 32.8 Histogram and descriptive statistics showing that ABX is administered within specifications limits

(continued)

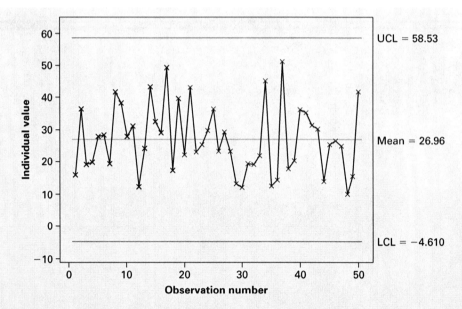

Chart 32.2 X chart of the number of minutes before incision that drug is administered

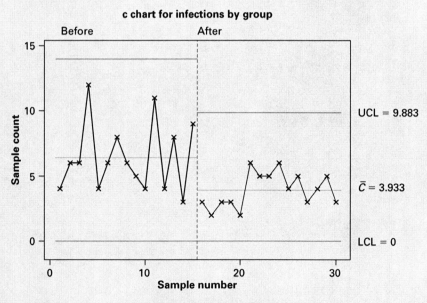

Chart 32.3 c chart showing the decrease in of the number of infections

(*continued*)

The Surgical Services department has also taken over responsibility for ongoing education and awareness of the requirements.

Results

Results for the 3 months after the process change are promising:

- The bar chart (Figure 32.7) shows that rate of administering no drug or the incorrect drug has fallen from 31% before the project to 1.7% in the first 3 months after the new process was implemented.
- The number of minutes prior to surgical incision that the drug was administered reduced from an average of 88.9 minutes to 27 minutes (see the histogram and descriptive statistics diagram, Figure 32.8). The I (individuals) chart (Chart 32.2) shows that the process is in a state of control with the UCL (=UAL) at 58.53 minutes, within the 60 minutes specification, and is in a state of control.
- The last chart (Chart 32.3) is a c chart of the number of infections per week. It shows that the infection rate decline from 6.3% to 3.9% in the 15 weeks after the new process was implemented.

Bibliography, references and other resources

Books

AIAG (1995) *Statistical Process Control* (Chrysler, Ford and GM).
This standard QS9000 text on SPC is condensed and relates to the manufacturing industry and has some excellent explanations on some of the SPC concepts. Most of the book is given over to the detailed drawing and interpretation of control charts. The maths in some places may seem a little daunting, but the explanations of subgrouping and common and special causes of variation are particularly clear.

Carey, R.G. (2002) *Improving Healthcare with Control Charts: Basic and Advanced SPC Methods and Case Studies.* American Society for Quality Control.
This easy-to-read book is aimed at the health care industry. However, the interpretation and methods of using control charts are applicable in other industries and makes a useful reference to anyone wanting examples of practical applications of SPC in non-manufacturing situations.

Caulcutt, R. and Porter, L.J. (1992) Control chart design – a review of standard practice. *Quality and Reliability Engineering International*, Vol. 8, pp. 113–122.
Like many books on SPC, when it comes to explaining how to draw and interpret control charts, this book presents simple, easy to understand guidelines that apply in very many situations. However, under certain circumstances, more advanced guidelines will be more robust and give a more accurate interpretation of actual process performance. This article discusses some of the shortcomings of the simple procedures and proposes more advanced methods. This is a suggested reading for the more advanced user.

Deming, W.E. (2000) *Out of the Crisis.* The MIT Press.
This highly respected book was a landmark in management thinking that is still relevant today. Do not be fooled by the easy style of the book, it provides food for deep thought about the way we manage our organisations. It is an excellent starting point for thinking about what we need to do to help ensure long-term success in our organisations. Deming gives several simple examples of control charts but focuses much more on the implication of variation within organisations. He also explains the Red Beads Experiment, an excellent tool for teaching many of the concepts and managerial implications of variation.

Hayes, B.E. (1997) *Measuring Customer Satisfaction: Survey Design, Use and Statistical Analysis Methods* (2nd edition). Quality Press, Milwaukee, WI.
This highly respected book explains the development and use of customer surveys.

Joiner, B.L. (1994) *Fourth Generation Management.* McGraw-Hill.
This excellent, if dated, book clearly explains many aspects of management from a systems-based approach and devotes 50 pages specifically to the understanding and management of variation. It includes a few pages on the Taguchi Loss Function. The main

thrust of the book is the Joiner Triangle of quality, teamwork and scientific approach to management.

Juran J.M.
Joseph Juran has written many ground breaking books on quality management over the past 40 years or so. The *Quality Control Handbook*, 1800 pages, is considered by many to be THE text on quality management and is an invaluable reference. Amongst the other excellent books, those most closely related to this book either authored or co-authored by Juran are The *Six Sigma Training Kit* and *Managerial Breakthrough: The Classic Book on Improving Management Performance*. In addition, he has written a variety of books on management, leadership, planning and healthcare.

Kaplan, R.S. and Norton, D.P. (1996) *The Balanced Scorecard: Translating Strategy into Action*. Harvard Business School Press.
   The Balanced Scorecard (BSC) is a management system that can help an organisation translate its mission and strategy into operational objectives and metrics which can be implemented at all levels in the organisation. This highly respected book quickly became a key for helping organisations determine, communicate and measure against key factors vital to success. If implemented correctly, the vision determines the metrics, and conversely, it should be possible to determine the vision by looking at the metrics. The authors demonstrate how senior executives in industries such as banking, oil and retailing are using the technique to evaluate current performance and target future performance based on financial and non-financial criteria such as customer satisfaction, internal processes and employee learning, and growth. If you are concerned that you may not be measuring the appropriate metrics, this book is worth reading.

Oakland, J.S. (2003) *Statistical Process Control* (5th edition). Butterworth-Heinemann.
If you are involved in statistical process control (SPC) you will need a standard SPC text. This down to earth book by a highly respected author is a good choice and covers most of the basics.

Wheeler, D.J. (1991) *Short Run SPC*. SPC Press.
This 60-page book does what is says in the title – explains how to analyse manufacturing data when the process is frequently being reset for different products. The ideas do apply outside manufacturing, but the book focuses on manufacturing.

Wheeler, D.J. (1992) *SPC at the Esquire Club*. SPC Press.
This short book is a case study of how waitresses in a nightclub were using SPC to improve the running of the club. If you think SPC only applies to large manufacturing organisations, you are in for a shock. It does not explain how and why to use the tools of SPC, but helps dispel the myth that SPC is only for manufacturing.

Wheeler, D.J. (1999) *Understanding Variation – the Key to Managing Chaos*. SPC Press.
This 130-page book does not give enough detail to be the only SPC book on your shelf, but it is a very interesting and useful addition. It is aimed more at understanding why we need to use SPC rather than the tools themselves, and includes some interesting examples.

Wheeler, D.J. (2003) *Making Sense of Data*. SPC Press.
This is another excellent book by Wheeler which I recommend for the more advanced user of SPC.

Experimental Resources

The Rods Experiment

The Rods Experiment kits may still be available from:
Technical Prototypes Ltd
2 New Park Street
Leicester
LE3 5NH, UK
Tel: 01162 548750
They also supply other tools for teaching statistics.

Quincunx

A quincunx is an extremely powerful physical model for teaching a variety of aspects of management, and in particular to teach the concepts of variation. A quincunx consists of a hopper of beads. Beads can be released one at a time from the hopper and drop into a funnel. Once through the opening at the bottom of the funnel, the bead bounces through a grid of pins eventually arriving at one of a series of slots at the bottom of the quincunx. As beads are dropped through the quincunx they randomly end up in different slots. Most beads end in the slot directly below the bottom of the funnel, whilst progressively less beads end in the slots further away.

The pins represent the normal operating of a stable process. The funnel can be moved to simulate process changes, and the results monitored by looking at the slots where the beads finish.

By moving the funnel, changing the position of pins and various other alterations, it is possible to simulate common and special causes of variation.

Quincunxes may be obtained from:

http://www.qualitytng.com/shop
http://www.4ulr.com/products/statisticalanalysis/trainingaids_1.html

Useful web sites

Tim Stapenhurst can be reached at tim@sigma-c.co.uk
Juran Institute: http://www.juran.com/

Juran Institute is a management consultancy founded by Dr Joseph Juran. They provide training and consultancy in a range of quality management issues including Process Performance Improvement, Six Sigma, Change Management and Benchmarking. They have offices in the US, Canada and Europe.

The Deming Learning Network (DLN) Focuses is the official site for those wanting to know about the *management guru* Dr Edwards Deming. The site includes a number of useful links and recommends appropriate books and other resources:

In the UK: http://www.dln.org.uk/dlinks.html and http://www.deming.org.uk/
In the US: http://deming.ces.clemson.edu/pub/den/

Appendix A

Constants used in the control charts

The constants used in the control charts in particular for average and range charts, and for average and standard deviation charts, are given in detail in the following table. The * in the table denotes values used in the case studies estimated by interpolation. Values not shown in the table can be interpolated. The American system is to use control limits only, and the British system uses action and warning limits.

Sample size (n)	d_2	For average charts A_2	For range charts LCL D_3	UCL D_4	UAL D_5	LAL D_6	UWL D_7	LWL D_8	For average charts A_3	For standard deviation charts LCL B_3	UCL B_4	UAL B_5	LAL B_6	UWL B_7	LWL B_8	For median charts A_4
1	1.128	–	0.000	3.270	–	–	–	–	–	0.000	–	–	–	–	–	–
2	1.128	1.880	0.000	3.270	4.12	0.00	2.81	0.04	2.659	0.000	3.267	4.12	0.02	2.80	0.04	1.880
3	1.693	1.023	0.000	2.574	2.98	0.04	2.17	0.18	1.954	0.000	2.568	2.96	0.04	2.17	0.18	1.187
4	2.059	0.729	0.000	2.282	2.57	0.10	1.93	0.29	1.628	0.000	2.266	2.52	0.10	1.91	0.29	0.796
5	2.326	0.577	0.000	2.114	2.34	0.16	1.81	0.37	1.427	0.000	2.089	2.28	0.16	1.78	0.37	0.691
6	2.534	0.483	0.000	2.004	2.21	0.21	1.72	0.42	1.287	0.030	1.970	2.13	0.22	1.69	0.43	0.548
7	2.704	0.419	0.076	1.924	2.11	0.26	1.66	0.46	1.182	0.118	1.882	2.01	0.26	1.61	0.47	0.508
8	2.847	0.373	0.136	1.864	2.04	0.29	1.62	0.50	1.099	0.185	1.815	1.93	0.30	1.57	0.51	0.433
9	2.970	0.337	0.184	1.816	1.99	0.32	1.58	0.52	1.032	0.235	1.761	1.87	0.34	1.53	0.54	0.412
10	3.078	0.308	0.223	1.777	1.95	0.35	1.56	0.54	0.975	0.284	1.716	1.81	0.37	1.49	0.56	0.362
11	3.173	0.285	0.256	1.744	1.91	0.38	1.53	0.56	0.927	0.321	1.679	1.78	0.39	1.46	0.58	
12	3.258	0.266	0.283	1.717	1.87	0.40	1.51	0.58	0.886	0.354	1.646	1.73	0.42	1.44	0.60	
13	3.336	0.249	0.307	1.693					0.850	0.382	1.618	1.69	0.44	1.42	0.62	
14	3.407	0.235	0.328	1.672					0.817	0.406	1.594	1.67	0.46	1.41	0.63	
15	3.472	0.223	0.347	1.653					0.789	0.428	1.572	1.64	0.47	1.40	0.65	
16	3.532	0.212	0.363	1.637					0.763	0.448	1.552	1.63	0.49	1.38	0.66	
17	3.588	0.203	0.378	1.622					0.739	0.466	1.534	1.61	0.50	1.36	0.67	
18	3.640	0.194	0.391	1.608					0.718	0.482	1.518	1.59	0.52	1.35	0.68	
19	3.689	0.187	0.403	1.597					0.698	0.497	1.503	1.57	0.53	1.34	0.69	
20	3.735	0.180	0.415	1.585					0.680	0.510	1.490	1.54	0.54	1.34	0.69	
21	3.778	0.173	0.425	1.575					0.663	0.523	1.477	1.52	0.55	1.33	0.70	
22	3.819	0.167	0.434	1.566					0.647	0.534	1.466	1.51	0.56	1.32	0.71	
23	3.858	0.162	0.443	1.557					0.633	0.545	1.455	1.50	0.57	1.31	0.72	
24	3.895	0.157	0.451	1.548					0.619	0.555	1.445	1.49	0.58	1.30	0.72	
25	3.931	0.153	0.459	1.541					0.606	0.565	1.435	1.48	0.59	1.30	0.73	
36	4.293*															
40	4.393*															

LAL: lower action limit; LCL: lower control limit; LWL: lower warning limit; UAL: upper action limit; UCL: upper control limit; UWL: upper warning limit.

Glossary of terms and symbols

For information of different chart types, see inside front cover.

Symbols and acronyms

ABX	(Six sigma case study) Antimicrobial prophylaxis.
A_2, A_3	Constants used in determining limits for average charts.
ARL	Average run length.
BB	Black Belt.
B_3, B_4	Constants used in determining limits for standard deviation charts.
COPQ	Cost of Poor Quality.
CTQ	Critical to Quality.
D_3, D_4	Constants used in determining limits for range charts.
d_2, d_n	Hartley's constant. A constant used to estimate a standard deviation from a range and vice versa.
DMAIC	A "Six Sigma" methodology for improving processes. The letters stand for the five phases of a "Six Sigma" improvement project: Define, Measure, Analyse, Improve, Control.
FMEA	Failure Modes and Effects Analysis.
I	Individuals, see X.
LAL	Lower action limit, sometimes suffixed to indicate the chart for which it is being used, for example LAL_X for the X chart.
LCL	Lower control limit, sometimes suffixed to indicate the chart for which it is being used, for example LAL_X for the X chart.
LSL	Lower specification limit.
LWL	Lower warning limit, sometimes suffixed to indicate the chart for which it is being used, for example LAL_X for the X chart.
MA	Moving average (not to be confused with the moving range).
$\overline{MR}$	The average moving range.
p	The proportion of items having the attribute being monitored. For example, the proportion of forms incorrectly completed = number of forms incorrectly completed/number inspected.
np	The number of items having the attribute being monitored. For example, the number of incorrectly completed forms.
c	The number of events occurring, plotted on a c chart.
$\bar{c}$	The average number of events occurring.
n	The number of observations.
$\bar{n}$	The average number of observations.
R	The range.
$\bar{R}$	The average range.
$\tilde{R}$	The median of a set of ranges.
s	An estimate of the true standard deviation, σ.
$\bar{s}$	The average of several standard deviations.
S_{MR}, S_{mr}	The standard deviation calculated from the moving ranges.

SPC   Statistical process control.
T   The target in a cusum chart.
u The number of events occurring per number inspected $= c/n$.
$\bar{u}$ The number of events occurring per number inspected $= \Sigma c/\Sigma n$.
UAL   Upper action limit, sometimes suffixed to indicate the chart for which it is being used, for example LAL_R for the R chart.
UCL   Upper control limit, sometimes suffixed to indicate the chart for which it is being used, for example LAL_X for the X chart.
USL   Upper specification limit.
UWL   Upper warning limit, sometimes suffixed to indicate the chart for which it is being used, for example LAL_X for the X chart.
x An individual observation of a variables data item, plotted on an X chart. Alternatively, some texts use I meaning individuals.
x_i The ith observation of a set of data.
$\bar{X}$ The average of X, plotted on an $\bar{X}$ (average) chart.
X_s (Six sigma) the causes of problems or variable effecting process outputs.
$\bar{\bar{X}}$ The average of a set of averages.
$\tilde{X}$ The median (middle value) of a set of data.
$\tilde{\bar{X}}$ The average of a group of medians.
Y_s (Six sigma) the process outputs and measures.
Z   The transformed value of $x = (x - \text{target})/s$. The target values used is often the mean, or taken from the specification.
σ The Greek letter used to denote the standard deviation of a distribution. Usually we do not know the true value of σ, so we collect a sample of data and estimate it. The estimate is denoted by s.

Definitions

Action limits (action lines)	See **control limits**.
Assignable causes	See **special cause**.
Attributes data	A term often applied to data that take integer values only. Also known as **counts data**, attributes data are data that count the number of times an event occurs. Another term often used synonymously for attributes data is **discrete** data. Attributes can be broadly split into two types: **defects** and **defective**. Data that are not attributes data are termed **variables** or **continuous data**.
Average	The sum of values divided by the number (**sample size**) of them. Also known as the **mean**.
Average run length	The average number of observations required to identify a change in the process.
Bar chart	A chart showing the relative frequency of occurrence of data that are categorised in non-numeric groups (e.g. the number of late flights for different airlines).

In a ranked bar chart the bars are grouped in descending (or less frequently in ascending) order.

Bias
Systematic errors that lead to under- or overestimating statistics such as the **mean** and **standard deviation**.

Binomial data
Data that can take only one of two values, frequently termed true or false. Examples: whether an item conforms (true) or does not conform (false) to a requirement, passes or fails an inspection, a delivery is on time or late, a budget was exceeded or not, a procedure was followed or not. These data follow the **binomial distribution** and are analysed with **np** (for analysing the number of occurrences) and **p** (for analysing the proportion of occurrences) charts. See **defectives data**.

Binomial distribution
The distribution that describes the probability of occurrences of conforming and non-conforming observations. This distribution underlies the np and p charts.

Characteristic
See **process characteristic**.

Check sheet
A simple sheet for recording the frequency with which different events occur.

Common cause variation
Variation that is always present and randomly affects all observations in a set of data. A process exhibiting only common cause variation is said to be "in control".

Continuous data
See **variables data**.

Control
See **statistical control**.

Control chart
A **run chart** (plot of a process characteristic usually through time which include a line representing the average) with statistically determined control limits.

Control limits (control lines)
The calculated values against which data are compared to determine if a process is in a state of **statistical control**. The values are drawn on control charts as control lines.

Historically the British system uses:

– Action lines (upper and lower) set at the **mean** $\pm$ 3.09 (**standard deviations**).
– Warning lines (upper and lower) set at the **mean** $\pm$ 1.96 (**standard deviations**).

The American system uses:

– Control lines (upper and lower) set at the **mean** $\pm$ 3 (**standard deviations**).

In practice there is little difference between the action and control limits, and the two terms are used interchangeably in this book.

Whilst control charts always use action (control) limits, the use of warning limits are far less consistent.

Some chartists also include lines at ± 1 standard deviation.

	The theoretical probability of obtaining a value outside the control (action) limits is about 1 in 1000. If a value is obtained beyond these limits, it is assumed that a change has occurred in the process and action is taken to investigate and correct (if appropriate) the causes.
Correlation	A statistic showing how closely two variables co-vary, such that systematic changes on one variable are reflected by systematic changes in the other. The strength of correlation or interdependencies indicated by the correlation coefficient, which varies from -1 (as one variable increases the other decreases) to 0 (no correlation) to $+1$ (both variables increase or decrease together).
Counts data	See **attributes** data.
Cross plot	See **scatter diagram**.
Cusum (cumulative sum) chart	An advanced type of chart which plots the cumulative difference between each observation and a target value. Cusum charts are extremely efficient at detecting small changes in the mean but not very efficient at detecting single out-of-control observations.

A weighted cusum chart takes into account the "area of opportunity". For example, the number of items rejected at inspection could be plotted on a cusum chart if the number of items inspected remains the same. However, if the number of items varies, a weighted cusum is used.

Non-weighted cusum charts are usually used with the "parent" control chart, for example np, c or variables chart. Weighted cusum charts are usually used with the "parent" control chart, for example p, u or variables chart.

Cusum mask	A sheet of paper marked and cut out to aid in interpreting a **cusum chart**.
Decision lines	Lines drawn on a **cusum chart** to help determine whether a process change has occurred.
Defectives data	Data that can take only one of two values, also known as **binomial data**. Defectives data follow a **binomial distribution**, and are charted using a p or np chart. Examples include:

– During an inspection, an item can either pass or fail.
– A train is either late or on time.
– A job application is either accepted or not accepted.

Binomial data contains the less information than both **attributes** and **variables** data. However, where there is a choice, they are usually the easiest and cheapest to collect.

| Defects data | Defects data count the number of times an event occurs. For example: |

– the number of accidents per million hours worked,
– the number of mistakes on a form,

- the number of flaws per unit of production,
- the number of errors made per week.

Defects data follow a **Poisson** distribution and are charted using a c or u chart. However, where the counts are high, for example, the number of passengers on a train, variables charts can be used.

Defects data contain more information than **defectives** data, but less than **variables** data.

Deviation	See **standard deviation**.
Discrete data	Data that can take only a countable set of values, often integers. The term discrete is often used synonymously with counts or **attributes data**.
Distribution	A way of describing a set of data, often shown as a histogram.

A distribution is characterised by its:

- location (usually measured by the **average**, **mode** or **median**);
- variability (usually measured by the **standard deviation** or **range**);
- shape (e.g. number of peaks; there are measures for describing some types of shape, but they are outside the scope of this book).

Data for different types of distribution include:

- **Variables** data frequently follow a **normal distribution**.
- **Defects** data frequently follow a **Poisson distribution**.
- **Defectives** data frequently follow a **binomial distribution**.

A change in any of these aspects of a distribution indicates a change in the process which produced the data being analysed. Control charts are designed to identify changes in the distribution, and hence show when the process has changed.

Frequency distribution	A table or graph which displays the frequency with which certain values occurred. Usually the data is displayed as a **histogram**.
Grand mean	Usually the **average** of a set of averages. Used with average charts.
Hawthorn	Improved process performance that results from operatives who know their performance is being monitored and exercise more care in the execution of the process than they would normally.
Histogram	A **bar chart** that represents the frequency **distribution** of a set of data. The bar widths represent the ranges of the values. The heights of the bars are proportional to the observed frequencies.

In control	See **statistical control**.
Individuals	Individuals data is a term applied to variables data that are charted one at a time, for example, the time taken to maintain a pump. This type of data is usually monitored using an X/MR chart. The individuals chart is another name for the X chart.
Limit	See **control limits**.
Lines	See **control lines**.
Lower action limit	See **control limits**.
Lower control limit	See **control limits**.
Lower specification limit	See **specification limits**.
Lower warning limit	See **control limits**.
Mask	See **cusum mask**.
Mean	See **average**.
Mean range	The average of a set of **ranges**, used in the range chart.
Median	The middle value of a set of data when the data are arranged in increasing (or decreasing) order. If there are even number of values, the median is found by averaging the middle two values.
Mode	The most frequently occurring value in a set of data.
Moving average (moving mean)	A statistic calculated by averaging the last n values. For example, in a 12-month moving average, the last $n = 12$ monthly values are averaged. The weighted moving average is similar to the moving average, but different values are given different weightings to reflect their perceived influence on the current process. Usually the most recent figure is given a higher weighting.
Moving range	Used with the X (individuals) chart, the moving ranges are calculated as difference between two consecutive data values. The sign (+ or −) is ignored. The moving ranges are used to monitor the variation of the process. The moving range should not be confused with the **moving average**.
Normal distribution	A frequency distribution that is followed by many data sets. It is characterised by a "bell"-shaped curve and is symmetrical about the average.
Out of (statistical) control Pareto chart/analysis	See **statistical control**. A simple chart used to separate the vital few (i.e. main) from the trivial many (i.e. minor) constituents of a total value. It is used in process improvement to identify, for example, the main causes of a problem.
Poisson distribution	The distribution in which **defects** data are assumed to follow that underlies the c and u charts.
Process	A process is everything (e.g. people, materials, equipment, procedures) required to turn an input into an output for a customer. Most processes have several customers, who may be internal or external (or both) to the company.

A payroll system, for example, includes both the staff (internal) and tax office (external) as customers.

Process characteristic	A feature of a process about which we could collect data.
Process control	See **statistical process control**.
Quartile	A set of data may be split into four parts each containing a quarter of the data observations. The top, or first, quartile is the value at which 25% of the data lie above, and 75% below. The second quartile is the values between which 50% of the data lie above, and 50% below, etc.
Random	If data are random, it is not possible to predict the value of one individual value from another, although they may come from a known definable distribution.
Random variation	See **common cause variation**.
Range	The difference between the maximum and minimum values of a set of data.
Ranked bar chart	See **bar chart**.
Regression analysis	A method of predicting the value of one variable from one (or more) other corresponding value(s).
Run	A set of data that appear to form an ordered series (e.g. a run of eight values above the average).
Run chart	A plot of a process characteristic usually over time. Usually the run chart includes a line that represents the **average**. If **control lines** are added the run chart becomes a **control chart**.
Sample	A set of observations taken from a process.
Sample size	The number of observations, n, in a sample.
Scatter diagram	A diagram which shows the relationship between two numeric **variables** usually termed X and Y. Also known as a cross plot or XY chart.
Shewhart charts	The group of control charts including the c, u, np, p and X developed by Walter Shewhart.
Sigma	The Greek letter, σ, used to denote the **standard deviation** of a population. This is estimated by taking samples of size n, and calculating the sample **standard deviation**, s.
Six Sigma	A process with "Six Sigma" capability means having 12 standard deviations of process output between the upper and lower specification limits. Essentially, process variation is reduced so that no more than 3.4 parts per million fall outside of the specification limits. The "Six Sigma" term also refers to a philosophy, goal and/or methodology utilized to drive out waste and improve the quality, cost and time performance of a business.
Skewed distribution	Any distribution that is not symmetrical about its **mean**. The **normal distribution** is not skewed, but the **binomial** and **Poisson distributions** are.

Special cause (of variation)	A special cause of variation is one that is not part of the normal process. It may be temporary (e.g. identified by a single point outside the control limits) or permanent (e.g. identified by a change in the process average). A process exhibiting special causes of variation is not in a state of **statistical control**.
Specification	The stated requirement of a process.
Specification limits	Sometimes simply referred to as the specification, these are the upper and lower limits within which the output of a process is required to lie. The specification limits, which tell us what we want to achieve, must not be confused with the **control limits**, which tell us what the in-control process will deliver.
Stable process	A process that is in a state of **statistical control**.
Standard deviation	Denoted by s. A measure of the variability of a distribution. In general, virtually all data values observed when measuring a stable process will lie between the **mean $\pm$ 3 (standard deviations)**. s is an estimate of the true standard deviation, **sigma (σ)**.
Statistical control	A process is said to be in a state of statistical control (often shortened to "in control") if it is subject only to **common cause variation** (i.e. all **special causes** have been removed). A **process** is deemed to be in control if the data on a **control chart are**: – **randomly** distributed around an **average**; – with no obvious trends, **runs** above/below the average or other non-random patterns; – with less data further away from the average; – within **control limits**. A process exhibiting both common and special causes of variation is said to be "out of control", "not in control" or, more correctly, "not in statistical control".
Statistical process control	Process control is the management of a process by observation and analysis to limit **variation** of the outputs. Statistical process control is the use of statistical techniques to help in process control.
Stratification	The process of separating data into classes or strata according to one or more defining variables.
Tally chart/sheet	See **check sheet**.
Top quartile	See **quartile**.
Type I error	The erroneous conclusion that a **special cause of variation** has occurred when it has not.
Type II error	The erroneous conclusion that a **special cause of variation** has not occurred when it has.
Upper action limit	See **control limits**.

Upper control limit	See **control limits**.
Upper specification limit	See **specification limits**.
Upper warning limit	See **control limits**.
Useful many	See **Pareto**.
Variables data	Variables, or continuous, data are measured values such as length, weight and time. In contrast, **attributes** data count numbers of occurrences of events.
	Variables data are monitored using the variables charts, such as the X/MR chart. Where there is a choice, variables data contain more information than attributes data.
Variance	The square of the **standard deviation**.
Variation	In inevitable difference between individual outputs of a **process**.
Vital few	See **Pareto**.
Weighted cusum chart	See **cusum chart**.
XY chart	See **scatter diagram**.

Index

(For an index of case studies, examples and charts see pages xi–xxi)

Advice desk, 128
Action:
  limits, 260, 277
    see also Control limits
  manager, 259
Admissions example, *see* Casualty department
    example of grouping
Airport check-in system example, 355–6
Analysts, *see* Comparing different groups
ARL, *see* Average run length
Attributes:
  charts, 263–70
  data, 263–8
    treated as variables, 265
Auto-correlation, *see* Correlation
Automatic adjustments, *see* Tampering
Average, *see* mean
Average of averages, method of calculation,
    100
Average run length (ARL), 321–6

Balanced scorecard, 355
Banding of data, 98, 107–9
  solutions to, 98, 108, 130
  see also Sample size
Bar charts, 344–7
  inappropriate use of, 142–3, 148, 151
  ranked, 347
Batch case study, 163–4
Beads experiment, 404
Belts, 423, 424
Best practice forums, 32
Black Belts, *see* Belts
Blending workshop, 401–2, 410, 412–14

c chart, 300–1
  see also Control chart selection
Capability, 195, 200, 204–5
Case Studies:
  how to use, 62
  index of studies, *see* xx–xxiii
  layout of, 62
  sources of, 59–60
Cast of customers, 355

Casualty department example of grouping,
    331–5
Category data, 265
Celto case study, 153–64
Central limit theorem, 271
Chart:
  formats, 61
  in case studies, *see* case study index
    xx–xxiii
  power of, *see* Power of chart types
  purpose of, 23, 149
  see also Control chart
Charting, how much to chart, 79
Check sheet, 348–9
  as a histogram, 122, 381
  rods experiment, 380–1
Chemical concentration case study, 63–8
Chunkiness, *see* Banding of data
Coin, spinning experiment, 191
Common cause variation, *see* Variation,
    common cause
Comparing:
  different groups with a control chart, 115,
    201–4
    case studies, 91–101, 103–17
  two or more numbers, 23–9
Control:
  importance of subprocess being in, 149
  in-control definition, 256
    see also Process in control
  out-of-control
    definition, 257
    examples, 257–8
    see also Process out of control
Control chart:
  attributes, 263–70
    vs. variables workshop, 404
  as history/memory of the process, 27, 205
  causes of out of control signals, 260–1
  comparison of results using different charts,
    case study, 153–64, 175–86, 193
  definition, 1
  difficulty in choosing, 153
  historic analysis of, 205

Control chart: (cont'd)
  implementing, see monitoring system
  interpreting, 15–16, 399, 407
    cusum, 312–17
    errors in, 259
    guidelines, 257–60
  limits, 20
  myths, 16–19
  non-time sequenced data, 143–9, 151
  purpose, 14, 259
  selection, 122, 125, 163, 397–8, 405–6
    c or u chart, 225–6, 269–70
    case study, 153–64
    p or np chart, 268–9
    powerful, the most, 275
    R or s chart, 274
    X or $\bar{X}$ chart, 270–1
  setting up, 278
  signals, see Control chart interpreting
  simplicity of keeping, 27
  two variables on one chart, 135–8
  use with different distributions, 5
  uses and applications, 17–18
  variables, 263–5
  with low variability, 108
Control limits:
  American system, 260
  British system, 260
  comparison of c and X/MR charts, 179
  establishing, 360
  moving, importance of, 191–2
  relation with sample size, 191
  relation with specification limits, 196, 198
  why 3 standard deviations, 256
Correlation, 328, 350–1
  auto, 278, 328–9, 369
  coefficient, 350–1
Cost of Poor Quality, 356, 424
Cost per foot, see Drilling case studies
Counts data, see Attributes data
Critical Success Factors, 355
Culture, 191–2
  fear, 30
  monitoring safety example, 230
  required, 31
Cumulative sum charts, 274, 307–20
  decision lines, 315–17
  interpretation of, 312–17
  masks, 313–15
  rods experiment, 389–91
  setting up, 309–12

  targets, 194
  weighted, 317–20
Customer:
  feedback, 356
  needs, 355–6

Data:
  checking, 234, 239
  constipation, 355
  defects, 266–8
  defectives, 266–8
  discrete, see Attributes
  tables, difficulty of interpreting, 25–7
  variables, see variables data
Decision lines for cusum charts, see
        cumulative sum charts
Difference chart, 273, 289
  rods experiment, 391–2
Distribution:
  attributes data, 326
  non-normal, 254–5
    see also Non-normal data
  shapes of, 5
  see also Normal distribution
Downtime:
  cusum example, 309–12
  workshop, 399, 407–8
Drilling case studies, 141–51, 165–73

Effect of choosing "incorrect" control chart,
        see Control chart comparison
Emergency hospital admissions example,
        344–9
Esquire Night Club, see Night Club
Established metrics, 354–5
Events:
  converting to rates, 86, 7
  grouping, see sample size
  need to adjust for opportunity, 76
  see also Rare events
Exam results case study, 69–74

False signals, 107, 259
Falsifying data, how to spot it, 29–30
Fear culture, see Culture
Framework for measurement, creating, 357–9
Frequency of measurement, see measurement
        frequency

General Electric, 423
Generating theories, 195, 201–4

Goal setting, 14, 30
  as top quartile, 32–5
  from benchmarking, 32
  see also target setting
Golf practice example of tampering, *see*
    Tampering
Graininess of data, *see* banding
Grand average, 148
Grouping, 330–5
  see also sample size

Hawthorn effect, 136, 138
Help desk, *see* Advice desk
Histograms, 337–43
  interpretation of, 339–40
  rods experiment, 381
Hospital admissions example, *see* Emergency
    hospital admissions example
Hospitality workshop, 403

Improvement, 364–6
  selecting areas for using Pareto chart, 218
Incidents, *see* Events
Independence, *see* Correlation
Information, amount of, 266
Ingredient in medication example, 349–50
Investigating out of control signals, *see* Out
    of control conditions
Invoices:
  paid late example, 299–300
  workshop, 401, 410–11

Jack Welch, 423
Job times example, 290–3
Journey to work (example), 9

Limits:
  action, *see* Action limits
  control, *see* Control limits
  specification, *see* Specification limits
  warning, *see* Warning limits
Loss Function, *see* Specification limits
Losses example, 301–4, 317–20
Lubricant blend workshop, *see* blending
    workshop

Management, interfering, 83
Mandatory requirements, 354
Masks, *see* Cumulative sum charts
Mean:
  definition, 6
  estimate of, 173, 247

Measurement:
  framework, *see* Framework for
    measurement
  frequency, 278, 369
  see also data
Measures:
  and recording process, 359–62
  lists of, 370–3
  outcome vs process, 369
  selecting, 354–7, 396–7
Median, 6
  control chart, 272, 287–8
  median moving range chart, 273
  median range chart, 287–8
Medical errors:
  case study, 103–17
  workshop, 403
Medication, ingredient of, *see* Ingredient in
    medication example
Metrics *see* Measures
Missing data, 188
Mode, 6
Monitoring, 256
  in real time – rods experiment, 384–6
  process monitoring administration process
    case study, 119–32
  routine, 361–2
  system, 353–66
Monthly report example, 25–7
Motorola, 423
Moving Average:
  case study:
    drilling, 165–73
    incidents, 75–83
    theoretical, 37–43
  problems with using, 46, 75, 83
  reasons for use, 37, 75
  response to:
    out of control point, 40–1
    process change, 38–9, 80, 82
    trends, 42
  seasonality, 43
  suppression of variation, 37, 44
  trends, 43, 168
  uses of, 44, 46
  vs. control charts, 37–43, 45, 79–82
  with varying time periods, 166–72
  workshop, 403
Moving mean – moving range charts,
    293–4
Moving Range chart, *see* MR chart

MR Chart, 271
    see also Non-sequential data
    see also X/MR chart
Multivariate chart, 305
Myths about SPC, 16–19

Night Club (Esquire), 17
No report, 131
Noise in data, 9
Non-normal data, 265, 326–7
Non-sequential data, 18
    case studies, 69–74, 233–9
    order of data can be important, 239, 244
    order of plotting, 283
    trends not applicable, 234
Normal distribution, 253
    source, 4
Normality of data, 245–6
Normalisation of data, 368
np chart, 299–300
    see also Control chart selection

OC curves, *see* Operating Characteristic
        curves
Off-centered process, *see* process off
        centered
Off-specification:
    batches workshop, 403
    blends workshop *see* Blending
        workshop
    material, reaction to, 200
    possible causes, 261
Operating Characteristic curves, 324–6
Order of plotting, *see* Charts Non sequential
        data
Organisational review activities, 395–6
Over control, *see* Tampering
Out of Control conditions:
    frequent, 138, 180–1
    investigating, 361–4
Outliers *see* outlying data
Outlying data, 146–9

p-chart, 294–8
    short cut for calculations, 191
    see also Control chart selection
Paper manufacturing case study, 195–212
Pareto charts, 347–8
Performance monitoring:
    indicators, *see* Measures
    paying for, 27

proposal case study, 119–32
    purpose, 21
Pharmaceutical data example, 265–6
Power of chart types, 267–8
Predicting with a control chart, 17–18, 123
Process:
    aim, 20
    analysis, 356–7
    average, *see* process mean
    change:
        difficulty of pin pointing, 136
        may be gradual, 136
        risk of confusing with seasonality, 226
        with seasonality case study, 213–32
    definition, 3
    improvement:
        how not to, 364–5
        process, 10, 364–6
        selecting what to improve, 230, 232
    in control:
        definition, 16
        explanation, 7
        management implications of, 10
    information, recording on chart, 124–5
    mean, 20
    model, 3
    monitoring, 353–64
    off-centered, 198
    out of control:
        explanation, 8
        management implications of, 10
    predictable, 7
    target, 20
    unpredictable, 8
Proportions, 294

Quartiles, 32–5
    definition, 32
    difficulties with, 34–5

R chart:
    case study, 141–51
    varying sample size calculations, 150–1
    see also $\bar{X}$/R chart
    see also Control chart selection
Railway supplies example, 340–3
Range:
    chart, *see* R chart
    definition, 6–7
Rare events, 329–30, 369
    case study, 85–90, 103–17

definition of, 329
examples, 85, 138
medical errors workshop, *see* Medical
    errors workshop
see also sample size
Red rods, *see* Rods experiment
Regression, 166–73
Rejected:
    pipes example, 295–7
    product example, 280–6
    tenders workshop, 402, 414–15
Related variables:
    as surrogates, 184–6
    investigation of, 156–8, 178, 224–5
    see also Correlation
Repairs workshop, 400, 407, 409
Review, organisational, *see* organisational
    review
Rods experiment, 377–94
Run chart, 343
    comparison with raw data and tables,
        24–31
    rods experiment, 382

s chart, 286–7
    see also Control chart selection
Sample size, 98, 108–9, 279, 369
    effect on charts, 110, 117
    estimating minimum, 116
    varying, 268
    see also $\bar{X}$/R chart variable sample size
Scaling of axes, 171
Scatter diagram, 349–51
    checking for auto-correlation, 328
Seasonality:
    looking for, 215–18
    taking account of, 221–4
    with process change case study, 213–32
Selecting the appropriate chart, 263–75
Shift systems, *see* Comparing different groups
Short runs, 19, 273
Signals in data, 9
    see also Control chart interpretation
Six Sigma:
    basis of, 424
    case study, 430–7
    design for, 425
    efficiency and effectiveness, 425
    improvement methodology, 425–9
        analyse, 428
        control, 429

        define, 426–7
        improve, 427–8
        measure, 427–8
    meaning of, 424–5
    origin of, 423
    relationship with SPC, 19–20
    selecting projects, 425
    what is, 423–4
Skewed distributions, identifying on a scatter
    diagram, 134
Specification limits, 16, 20
    case study, 195–8, 204
    railway supplies example, 340–3
    relation with control limits, *see* Control
        limits
    vs. Loss Function, 204–5, 245
Standard deviation:
    definition, 7
    derivation of formula, 254
    formula:
        standard, 7
        for averaging, 249
        for non-normal distributions, 254–5
    importance of, 255–6
State of (statistical) control, *see* Control
Statistical control, *see* Control
Statistical measures:
    of location, 5–6
    of spread, 6–7
Statistical Process Control (SPC) definition, 1
Sub-groups, *see* grouping
Sub-process:
    comparison, 99
    importance of being aware of
        (case studies), 64–8, 93–101
Surgical complications workshop, 403,
    415–20
Surrogate metrics, 357

Tables of data, *see* Data tables
Taguchi Loss Function, *see* Specification
    limits
Tally chart, *see* Check sheet
Tampering:
    automatic adjustments, 14
    examples, 11, 204
    explanation, 11, 259
    golf practice worked example, 12–14
    result of off-specification material, *see*
        Off-specification material
    types of, 14

Target setting, 14, 20
  Cusums, *see* cumulative sum charts
  for incapable processes, 204
  for training, 245
  three ways of meeting, 31
  see also Goal setting
  see also Falsifying data
Targets in cumulative sum charts, *see*
      Cumulative sum charts
Tenders, rejected, example, 304–5
Tick sheet, *see* check sheet
Training, 245
  administration process case study,
      119–32
Trends case study, 165–73
Twist-offs, case study, 133–9

u chart, 302–5
  see also Control chart selection

Variables:
  charts, 263–6
  data, 263–5
Variance, 7
Variation:
  changes in variation over time, 7–9
  causes of, 9
  common cause, definition, 8
  explanation, 4
  importance of understanding, 1, 7–9
  measures of, 6–7
  special cause definition, 8
Viscosity example:
  histogram, 337–9
  run chart, 343

Warning limits, 16, 257–8
Washouts case study, 133–9
Weight factors for seasonality, 221–4, 227–9
Weighted cusum charts, *see* Cumulative sum
      charts

$\bar{X}$/R chart, 279–83
  for investigating seasonality, 182–4, 215–18
  limits closer than X chart, 205
  rods experiment, 387–9
  variable sample size, 207–11, 284–6
    case study, 195–212, 237–41
  $\bar{X}$/R vs $\bar{X}$/s chart, 237–8
  see also Control chart selection
X/MR chart, 290–3
  as a "safe" option, 160–3, 186
  in place of attributes charts, 272
  limits further apart than $\bar{X}$ chart, 205
  rods experiment, 383–4
  see also Control chart selection

Year to date (YTD), 47–57
  case study (theoretical):
    YTD average, 53–4
    YTD vs. last years YTD, 50–1
    YTD vs. plan, 47–9
  definition, 47
  difficulties in using, 52
  vs. control charts, 50–2, 54–7
Yellow rods, *see* Rods experiment

Z chart, 274, 289
  rods experiment, 392–3
Zero, frequent occurrences of, *see* Rare
      events

Breinigsville, PA USA
20 November 2010
249617BV00005B/2/P